S0-BMG-584

TWO-DIMENSIONAL TURBULENCE IN PLASMAS AND FLUIDS

TWO-DIMENSIONAL TURBULENCE IN PLASMAS AND FLUIDS

Research Workshop

Canberra, Australia June-July 1997

EDITORS

Robert L. Dewar

Research School of Physical Sciences & Engineering,
The Australian National University

Ross W. Griffiths

Research School of Earth Sciences,
The Australian National University

AIP CONFERENCE PROCEEDINGS 414

American Institute of Physics   Woodbury, New York

L.C. Catalog Card No. 97-77177
ISBN 1-56396-764-2
ISSN 0094-243X
DOE CONF- 9706162

Printed in the United States of America

CONTENTS

EXPOSITORY ARTICLES

RESEARCH PAPERS

Preface

This volume contains expository overview articles and contributed papers from an interdisciplinary workshop held in the Institute of Advanced Studies (IAS) of the Australian National University (ANU). The IAS is a research and graduate teaching section of the University and contains the Research School of Earth Sciences, in which the Geophysical Fluid Dynamics group performs laboratory experiments for simulating geophysical flows, and the Research School of Physical Sciences and Engineering, containing both the Plasma Research Laboratory and the Department of Theoretical Physics.

The Plasma Research Laboratory administers the National Plasma Fusion Research Facility, which is charged with developing and exploiting a helical axis stellarator, the H-1NF Heliac, for advanced toroidal plasma confinement experiments. The Department of Theoretical Physics administers the National Centre for Theoretical Physics (NCTP), set up in response to an acute need identified early in 1993 by the National Committee for Physics of the Australian Academy of Science for a national Institute for Theoretical Physics along the lines, for instance, of that at Santa Barbara or the Isaac Newton Institute at Cambridge.

In 1994 the ANU took the initiative of creating a pilot version of such an institute using small initial grants from the University and the Australian Research Council. Its activities consist of the planning and organization of annual Physics Summer Schools and of topical Research Workshops. These run over an extended period (of the order of one month) in order that collaborative research that can be undertaken during the workshops, and are intended to involve experimentalists as well as theoreticians. For current activities of the NCTP (and an Erratum list for this proceedings should it prove necessary) see the World Wide Web site http://rsphysse.anu.edu.au/~grp105/NCTP/NCTP.html . A link can also be found there to the National Institute for Theoretical Physics at the University of Adelaide, which was selected by a national panel to be the central node for the national institute following a joint proposal with the ANU and the University of New South Wales. The Universities of Melbourne and Tasmania have also joined the consortium in a bid to generate continuing funding for this initiative.

The Workshop on Two-Dimensional Turbulence in Plasmas and Fluids, the proceedings of which this volume represents, was one such NCTP workshop. It provided a natural meeting ground for theoreticians and experimentalists from both geophysical fluid dynamics and plasma physics to establish a dialogue on the underlying physics common to the two areas, exploiting striking similarities in the mathematical description of two complex systems — the atmospheres (or oceans) of planets, such as the Earth or Jupiter; and plasmas in magnetically confined fusion devices, such as tokamaks or stellarators.

Coriolis forces and magnetic fields play a similar role in causing fluid motions to be quasi two dimensional and vortical. In both cases heat is deposited centrally (near the equator, or near the magnetic axis) and exhausted at the periphery (near the poles, or at the wall of the confinement vessel), driving turbulence that mediates an anomalous transport of energy across the system. Both waves (Rossby waves and drift waves) and self-organized structures (vortex solitons) are involved.

Due to the phenomenon of inverse cascade, energy is coupled from small scales into larger scales, and thus the details of large-scale coherent structures become of prime importance. This is particularly true in meteorology where, for example, the fate of a town can depend on the path taken by a tropical cyclone! On the other hand, for the design of fusion power plants, and for global climate studies, the time-average transport, and its scaling with changing parameters, is what is of interest. Thus a constant juxtaposition of particular realization against statistical ensemble formed a feature of the workshop, providing a wealth of examples and insights.

A great deal of expertise has been built up in the two communities through observation, experiment, and numerical simulation, but the different jargon in the two fields has formed a barrier to communication (e.g. the Charney–Obukhov equation in geophysical fluid dynamics is called the Hasegawa–Mima equation in plasma physics). A number of internationally known speakers with an overview of both fields acted as discussion leaders, and helped bridge the communication gap through a series of expository lectures. This volume collects these lectures and also contains a representative sample of the research papers contributed by the workshop participants.

The organizers would like to thank the participants in the workshop for helping make it happen, and the contributors to this volume for volunteering and providing their manuscripts in a timely fashion. Special thanks are due for the sterling efforts of Ms Heli Jackson and Ms Martina Landsmann in the organization of the workshop and the preparation of this Proceedings.

The financial support of the two IAS Research Schools involved and of the IAS itself is also gratefully acknowledged. It is to be hoped that eventually a truly national Institute for Theoretical Physics will be funded, to make more such cross-disciplinary workshops posssible.

Robert L. Dewar
Ross W. Griffiths

Canberra, 16 September 1997

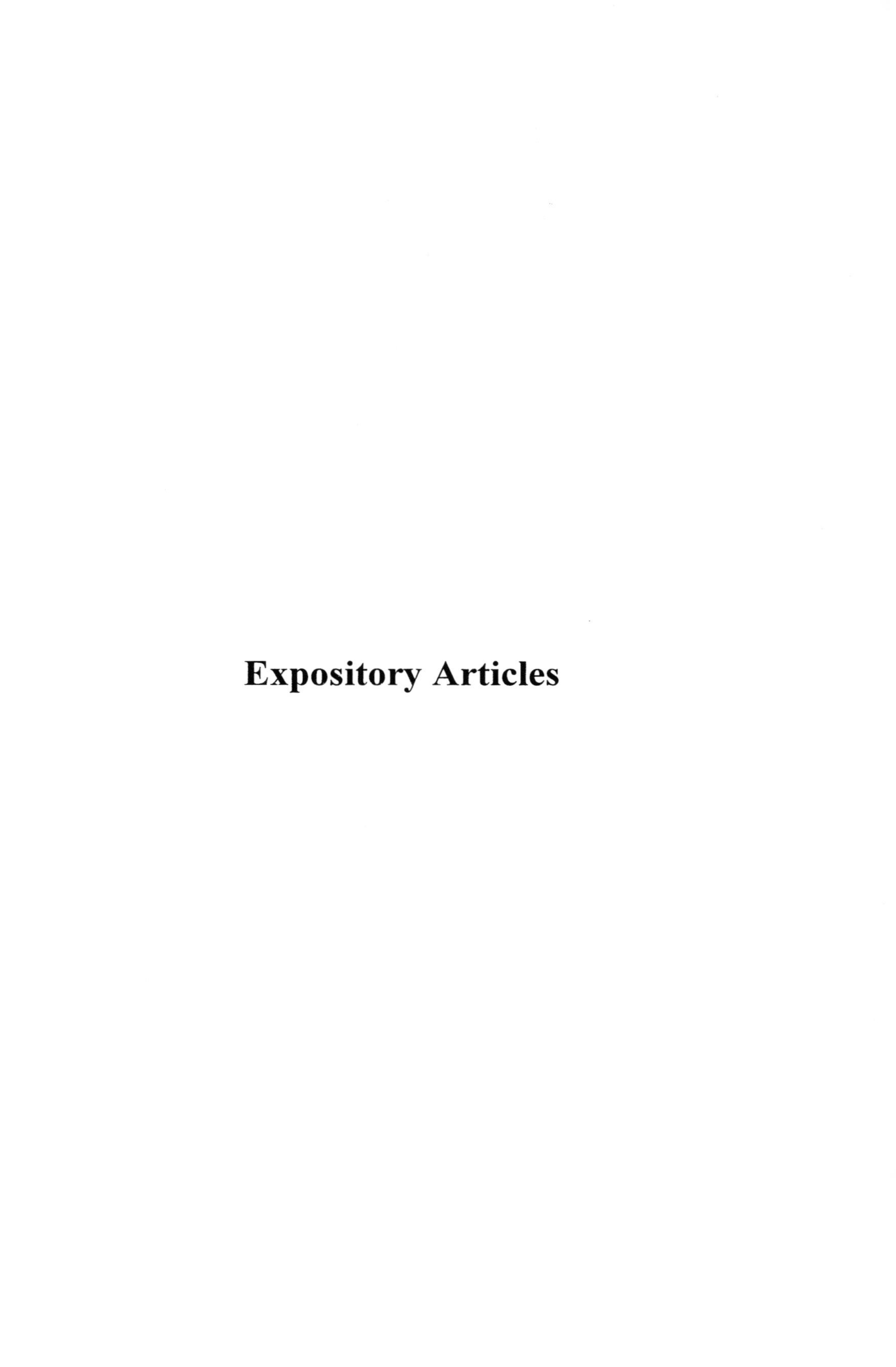

Expository Articles

Vortex-Wave Dynamics in the Drift Wave-Rossby Wave Problem with Temperature Gradients

W. Horton*

*Institute for Fusion Studies
The University of Texas at Austin
Austin, TX 78712

Abstract. The processes governing the propagation of low frequency vortex-wave convective disturbances in the two different physical systems of neutral fluids on rotating planets and plasmas confined by magnetic fields are explored with (i) physical descriptions of the convective transport, (ii) establishing the relevant conservation laws and (iii) computer simulations. The role of a global, ambient temperature gradient in driving the three-dimensional baroclinic instability is compared with the ion temperature gradient instability in magnetically confined plasma. Steady-state power balance and the turbulent viscosities and thermal diffusivities are analyzed using the same class of turbulent transport formulas.

DRIFT WAVES–ROSSBY WAVES

There are important similarities in the geophysical fluid dynamics and the low frequency drift wave dynamics in magnetized plasmas. In both systems the motions are quasi-two-dimensional meaning that there are large horizontal velocities $\boldsymbol{v}_\perp$ that can often be taken nearly independent of the vertical coordinate z or the direction parallel to the magnetic field in the plasma. The fundamental reason for the similar structure of the low-frequency dynamics in the two systems with completely different physical properties, neutral fluids versus the charged particle system, is the mathematical form of the horizontal acceleration. In a fluid system rotating with angular velocity $\boldsymbol{\Omega}$ the Coriolis force $2\rho\boldsymbol{v} \times \boldsymbol{\Omega}$ where ρ is the mass density, has exactly the same mathematical form as the Lorentz force $e_a n_a \boldsymbol{v}_a \times \boldsymbol{B}/c$ acting on the charged fluid with a cyclotron frequency $\boldsymbol{\Omega}_a = e_a\boldsymbol{B}/m_a c$ for charge species e_a, m_a with density n_a and pressure $p_a = n_a k_B T_a$.

The degree of the correspondence of the two systems is made precise by establishing the form of the potential vorticity $q(x, y, t)$ in the two systems

CP414, *Two-Dimensional Turbulence in Plasmas and Fluids:* Research Workshop
edited by R. L. Dewar and R. W. Griffiths

and the form of Ertel's theorem $dq/dt = 0$. The parallel between the drift wave dynamics and the geophysical fluid dynamics was first emphasized by Hasegawa and Mima (1978), Hasegawa *et al.* (1979). The homology was clarified through the use of Ertel's theorem in Meiss and Horton (1983). Meiss and Horton introduce the name Charney-Hasegawa-Mima (CHM) equation to describe the isomorphism between the two systems that occurs in the strictly 2D limit where the Taylor-Proudman theorem holds in the strong sense. Meiss and Horton (1983) also describe the generalization of the system to include the motion parallel to the rotation axis. When the coupling to the vertical velocity $w(v_{\|})$ is important the GFD and plasma dynamics begin to differ due to the stable stratification from the buoyancy effects in the atmosphere and oceans. To the extent that the stratification dynamics given by the Brunt-Väisälä frequency N is fast compared to the Coriolis frequency parameter $f = 2\Omega \sin\theta$ where θ is the latitude ($N \gg f$) the net effect of the stratification on the long-time scale $T > 1/f$ dynamics is maintained in the quasi-two-dimensionality of the large-scale ($L > \rho_R$) dynamics. Here ρ_R is the Rossby radius measuring the horizontal scale for dispersion of the Rossby waves and the exponential decay length for localized vortex structures. The effect of the vertical stratification is to decrease the effective Rossby radius from the shallow water value $\rho_R = (gH)^{1/2}/f$ to the internal deformation radius $\rho_I = NH/f$. The origin of this change in the Rossby radius due to the vertical stratification of the fluid is explained in the section on vertical motion in stratified medium.

In Sec. 2 we introduce the instabilities that occur from the horizontal temperature gradient. In the GFD system this instability is called the baroclinic instability and is perhaps the most important and universal form of instability in the atmosphere (Gill, 1982, ch. 13). The fastest growth rate occurs for the horizontal wavenumber $k_\perp$ such that the $k_\perp \rho_R \sim 1$ (Pedlosky, 1987, p. 521) which for $\rho_R = 1000\,\mathrm{km}$ gives the azimuthal mode number $m = 6$ for the Earth. Monin (1972) attributes the peak in the power spectrum of the kinetic energy fluctuations at the period of four days to this source of turbulence. These facts are similar to the situation in confined plasmas when appropriately scaled. The ion temperature gradient instability (Horton *et al.*, 1980, 1981, and 1992) has a peak in the wavenumber spectrum at $k_\perp \rho_s \sim 0.3$ with $\omega/\Omega_i \simeq 0.3\rho_s/L_T$. Here Ω_i is the ion cyclotron frequency and $\rho_s = c_s/\Omega_i$ is the ion inertial scale length which are the plasma analogs of f and ρ_R as shown in Table 1. The scale length of the radial temperature gradient is L_T. Examples of the large-scale quasi-two-dimensional convection cells obtained from direct numerical simulations are presented.

We conclude in Sec. 3 by comparing the turbulent thermal transport problem for the atmosphere and the plasma. We discuss the alternative scaling laws for the effective thermal diffusivity χ. Arguments are given to show that both for the atmosphere and the plasma the Prandtl number constructed from the ratio of the large-scale turbulent viscosity ($L > 1000\,\mathrm{km}$ using the Richardson four-thirds law) and the turbulent thermal diffusivity χ is comparable to

TABLE 1. Analogy Between Drift Wave and Rossby Wave

Drift Wave	Rossby Wave
H-M equation: $(1-\nabla^2)\frac{\partial\phi}{\partial t} + v_d\,\frac{\partial\phi}{\partial y} - [\phi, \nabla^2\phi] = 0$	Charney equation: $(1-\nabla^2)\frac{\partial h}{\partial t} - v_R\,\frac{\partial h}{\partial x} - [h, \nabla^2 h] = 0$
Electrostatic potential $\phi(x,y,t)$	Variable part of fluid depth: $h(x,y,t)$
$\phi(x,y,t) = \left(\frac{L_n}{\rho_s}\right) e\Phi\left(\frac{x}{\rho_s}, \frac{y}{\rho_s}, \frac{c_s}{r_n}t\right) \Big/ T_e$	$h(x,y,t) = \left(\frac{L_R}{\rho_g}\right)\delta h\left(\frac{x}{\rho_g}, \frac{y}{\rho_g}, \frac{c_g}{L_R}t\right) \Big/ H$
Lorentz force: $m_i\,\omega_{ci}\,\boldsymbol{v}_\perp \times \widehat{\boldsymbol{z}}$	Coriolis force: $\rho f \boldsymbol{v}_\perp \times \widehat{\boldsymbol{z}}$
$\boldsymbol{E} \times \boldsymbol{B}$ drift flow: $\boldsymbol{v}_\perp = \left(\frac{c}{B}\right)\widehat{\boldsymbol{z}} \times \boldsymbol{\nabla}\Phi$	Geostrophic flow: $\boldsymbol{v}_\perp = \left(\frac{g}{f}\right)\widehat{\boldsymbol{z}} \times \boldsymbol{\nabla}\delta h$
Cyclotron frequency: $\omega_{ci} = \frac{eB}{cm_i}$	Coriolis parameter: f
Drift coefficient: $L_n^{-1} = -\frac{\partial}{\partial x}\,\ell n\, n_0$	Rossby coefficient: $L_R^{-1} = \frac{\partial}{\partial y}\,\ell n\left(\frac{f}{H}\right)$
Larmor radius: $\rho_s = \frac{c_s}{\omega_{ci}}$	Rossby radius: $\rho_g = \frac{c_g}{f}$
Ion acoustic speed: $c_s = \left(\frac{T_e}{m_i}\right)^{1/2}$ where T_e is electron temperature	Gravity wave speed: $c_g = (gH)^{1/2}$ where H is depth of fluid layer.
Drift velocity: $v_d = c_s\,\rho_s\,\frac{\partial}{\partial x}\,\ell n\, n_0$	Rossby velocity: $v_R = c_g\,\rho_g\,\frac{\partial}{\partial y}\,\ell n\left(\frac{f}{H}\right)$
Dispersion relation: $\omega = \frac{k_y\,v_d}{1+k^2\,\rho_s^2}$	Dispersion relation: $\omega = -\frac{k_x\,v_R}{1+k^2\,\rho_g^2}$

unity.

Physical mechanism of the drift wave–Rossby wave

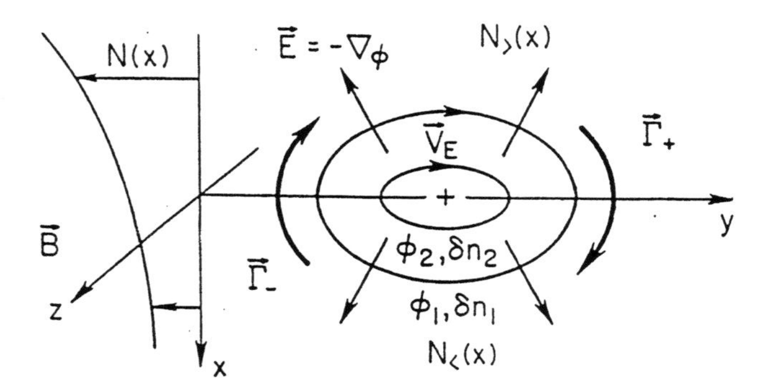

FIGURE 1. Drift wave mechanism showing $\boldsymbol{E} \times \boldsymbol{B}$ convection in a nonuniform, magnetized plasma. The local potential maximum Φ produces clockwise rotation that moves the structure to the right with the speed given in Eq. (4).

With Fig. 1 showing the positive potential associated with a local excess of positive charge, we can understand the drift wave propagation from the convection of the ion density. The convection of the plasma density is given by

$$\frac{\partial n}{\partial t} + \boldsymbol{v}_E \cdot \boldsymbol{\nabla} n = 0 \tag{1}$$

where

$$\boldsymbol{v}_E = \frac{c\boldsymbol{E} \times \boldsymbol{B}}{B^2} = \frac{c\hat{\boldsymbol{z}} \times \nabla\Phi}{B}. \tag{2}$$

Clearly the flow (2) is along the contours of constant electric potential Φ and from Eq. (1) the higher density at the top of the frame is brought to the right and the lower density beneath the structure is brought to the left. In time Δt the amount of excess density δn that accumulates to the right of the potential maximum Φ is $\delta n \Delta x \Delta y = (n_> \boldsymbol{v}_E - n_< \boldsymbol{v}_E) \cdot \hat{\boldsymbol{x}} \Delta y \Delta t$. The electron fluid is able to move freely and rapidly along the magnetic field $\boldsymbol{B}$ to neutralize the excess charge $n_i = n_e(x) \exp(e\Phi/T_e) \simeq n_0(1 + e\Phi/T_e)$ for small $\varphi = e\Phi/T_e$. Combining these results for δn with the flow through the midplane

$$\boldsymbol{v}_E \cdot \hat{\boldsymbol{x}} \Delta y = cE_y \Delta y/B = c\Phi/B \tag{3}$$

and the density excess

$$n_> - n_< = \Delta x \, \partial n / \partial x$$

gives that the potential maximum Φ moves to the position $y + \Delta y$ at the speed

$$\frac{\Delta y}{\Delta t} = \frac{-cT_e}{eBn} \frac{\partial n}{\partial x} = v_{de}. \tag{4}$$

Even with constant n the structure will translate when $v_E^> \neq v_E^<$ due to $\Delta x \partial B / \partial x$. This is the basic mechanism of the plasma drift wave.

For an incompressible fluid of depth $H(x, y, t)$ the hydrostatic pressure is $p = \rho g H(x, y, t)$ and the geostrophic flow velocity balances the pressure gradient $-\nabla p + \rho f \boldsymbol{v} \times \widehat{\boldsymbol{z}} = 0$ to give

$$\boldsymbol{v} = \frac{g}{f} \widehat{\boldsymbol{z}} \times \nabla H. \tag{5}$$

Now reconsider Fig. 1 with depth of the fluid $H_0(x)$ corresponding to the density $n(x)$. The local potential Φ now corresponds to the excess column height $g \delta H(x, y, t)$ of an anticyclone. Again there is clockwise rotation bringing the deeper fluid to the right and the shallow water to the left so that the bulge $\delta H > 0$ propagates to the right with the speed $v_R = (g/f)(\partial H / \partial x)$. Reversing the sign of the perturbation $\delta H < 0$ changes the direction of rotation for the low pressure cyclone. However, the depression again moves to the right with the same speed since the counter clockwise rotation brings shallower fluid to the right. For nonlinear finite amplitude disturbances $h = \delta H / H$ the deeper anticyclonic perturbation propagates faster than the cyclonic disturbance.

The dynamical equation for $H(x, y, t)$ follows from integrating the incompressibility condition

$$\nabla \cdot \boldsymbol{v}_\perp + \frac{\partial w}{\partial z} = 0 \tag{6}$$

from $z = H_B(x, y)$ the bottom to $z = H(x, y, t) + H_B$ the free surface and using that $w(z = H_B) \equiv 0$ and $w(z = H + H_B) = dH/dt$ while taking $\boldsymbol{v} = u \widehat{\boldsymbol{x}} + v \widehat{\boldsymbol{y}}$ independent of z (Pedlosky, 1987, p. 62). The shallow water equation is then

$$\frac{\partial H}{\partial t} + \boldsymbol{v}_\perp \cdot \nabla H + H \nabla \cdot \boldsymbol{v}_\perp = 0. \tag{7}$$

Using the geostrophic velocity with $f = f(y)$ and $H = H_0(y)$ as in the standard GFD coordinates, we obtain

$$\frac{\partial}{\partial t} \delta H + g \frac{\partial}{\partial y} \left(\frac{H_0(y)}{f(y)} \right) \frac{\partial \delta H}{\partial x} = 0 \tag{8}$$

for the long wavelength Rossby wave equation. Note that y points northward and x eastward with the Rossby wave propagating westward. The fusion coordinates in Fig. 1 are obtained by rotating the GFD coordinates anticlockwise about $\hat{z}$ by 90° so that the $\hat{x}$ points in the direction of the decreasing density $n(x)$ and $\hat{y}$ in the direction of the drift wave propagation. The simulations presented here are expressed in terms of the fusion coordinates while we leave the GFD equations in GFD coordinates.

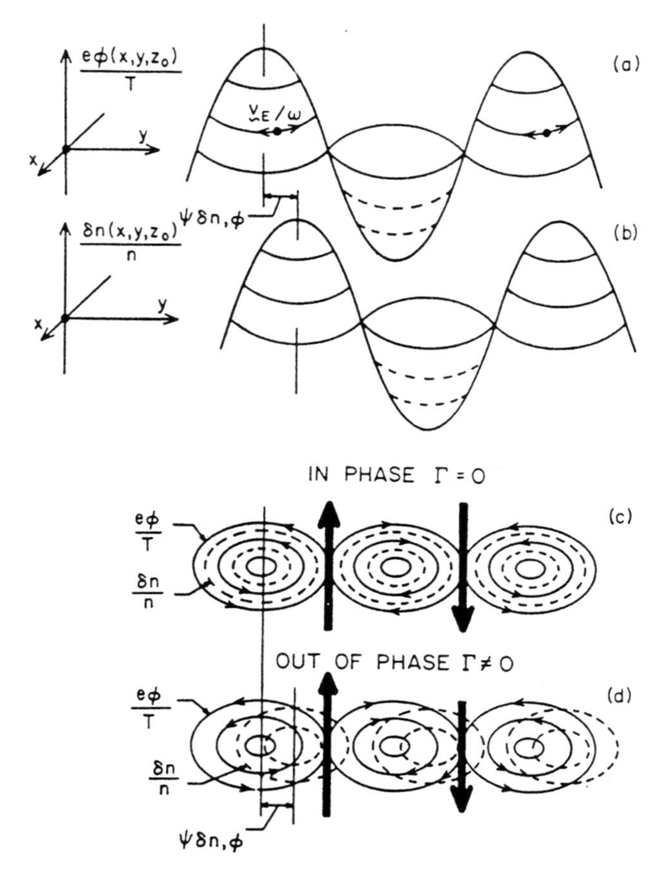

FIGURE 2. (a) A segment of a drift wave fluctuation showing the variation of the electrostatic potential perpendicular to the magnetic field at a given instant of time. The isopotential contours in the plane perpendicular to $B\hat{z}$ are the stream lines of the $\boldsymbol{E} \times \boldsymbol{B}$ particle motion. (b) A segment of the correlated but phase shifted density variation. (c) Top view of the potential and density contours in (a) and (b) in the case where the density and potential variation are in phase. (d) Top view in the case where the potential and density variation are out of phase by $\psi_{\delta n,\varphi}$.

In the presence of resistivity or the electron Landau resonance the density and potential fluctuations develop a phase difference as shown in Fig. 2. Repeating the above argument with the phase difference shows that there is now a net transport from $\langle n\boldsymbol{v}_E \cdot \widehat{\boldsymbol{x}}\rangle_y$ and that there is a growth or decay of the amplitude of the structure depending on the direction of the phase shift. The same situation will arise in Sec. 2 for the phase of the temperature fluctuation relative to that of the stream function for the $\boldsymbol{v}_\perp$ flow velocity. For unstable fluctuations the net transport is down the relevant density or temperature gradient which determines the direction of the phase shift between the stream function and the transported field. For large amplitude, coherent structures the phase shift tends to be small.

Ertel's theorem

The isomorphism between the drift wave and the Rossby wave is most clearly and usefully expressed by Ertel's theorem in the reduced form suitable for geostrophic and $\boldsymbol{E} \times \boldsymbol{B}$ flows. To see the action of the Coriolis force/Lorentz force in keeping the variation of $\boldsymbol{v}_\perp$ small with z, it is useful to examine the rotational part first for the full 3D momentum equation and then afterwards consider the rotational and the divergence parts of the horizontal components of the acceleration equation separated from the vertical component. The full 3D acceleration equation is

$$\rho \frac{d\boldsymbol{v}}{dt} = -\nabla p + \rho \boldsymbol{v} \times \boldsymbol{f} + \rho \boldsymbol{g}. \tag{9}$$

Upon using the vector identity

$$\boldsymbol{v} \cdot \nabla \boldsymbol{v} = \nabla \left(\frac{v^2}{2}\right) - \boldsymbol{v} \times (\nabla \times \boldsymbol{v}) \tag{10}$$

to rewrite the convective derivative we see that the effective pressure is the sum of the thermal pressure and the dynamic pressure $p_{\text{eff}} = p + \rho v^2$ and the horizontal rotational acceleration becomes $\boldsymbol{v} \times (f + \boldsymbol{\nabla} \times \boldsymbol{v})$ where $\boldsymbol{\omega} = \nabla \times \boldsymbol{v}$ the vorticity in the rotating frame combines with $\boldsymbol{f}$ to give the absolute vorticity $\boldsymbol{f} + \boldsymbol{\omega}$.

Dividing Eq. (9) by the mass density and taking the rotational part yields

$$\frac{d}{dt}(\boldsymbol{\omega} + \boldsymbol{f}) = (\boldsymbol{\omega} + \boldsymbol{f}) \cdot \nabla \boldsymbol{v} - (\boldsymbol{\omega} + \boldsymbol{f})(\nabla \cdot \boldsymbol{v}) + \frac{\nabla \rho \times \nabla p}{\rho^2} \tag{11}$$

for the convection of the absolute vorticity. Equation (11) shows that in the rotating frame the absolute vorticity $\boldsymbol{f} + \boldsymbol{\omega}$ plays the same role as $\boldsymbol{\omega}$ in an inertial frame.

For $p = p(\rho)$ the last term in Eq. (11) vanishes and the alternative form of Eq. (11) familiar to plasma physicists is

$$\partial_t(\boldsymbol{\omega} + \boldsymbol{f}) = \nabla \times [\boldsymbol{v} \times (\boldsymbol{\omega} + \boldsymbol{f})] \,.$$

This is the same as the "frozen in law" for the magnetic flux $\int \boldsymbol{B} \cdot d\boldsymbol{a}$ in the limit of infinite conductivity. Thus, the corresponding flux $\int_S (\boldsymbol{\omega} + \boldsymbol{f}) \cdot d\boldsymbol{a}$ through a surface S moving with the fluid velocity $\boldsymbol{v}(\boldsymbol{x}, t)$ is constant. This is the general form of the Taylor-Proudman theorem according to Chandrasekhar (1961, p. 84).

The Taylor-Proudman theorem states that for stationary, incompressible flow with $\nabla p \times \nabla \rho = 0$ the flow velocity $\boldsymbol{v}$ is independent of the z. The first term on the right-hand side of Eq. (11) gives the differential form of the Taylor-Proudman theorem: $\boldsymbol{f} \cdot \nabla \boldsymbol{v} = 0$. For slow $d/dt \ll f, \Omega_i$ motions and with the vanishing of $\nabla p \times \nabla \rho$, the convective structure in fluids and plasmas are quasi-two-dimensional due to this theorem. From the x-y components of Eq. (11) we see that to keep the fluid vorticity $\boldsymbol{\omega}$ pointing in the $\hat{\boldsymbol{z}}$ direction in the presence of large $f\hat{\boldsymbol{z}}$, the horizontal velocity must satisfy $|\partial_z \boldsymbol{v}_\perp| \ll f$. The $\hat{\boldsymbol{z}}$ component of the vorticity changes according to the vortex stretching given by $\partial_z w - \nabla \cdot \boldsymbol{v} = -\nabla_\perp \cdot \boldsymbol{v}_\perp$. The 2D compression occurs from both the change in f given by $\beta = \partial f / \partial y$ and the ageostrophic (polarization) drift velocity.

For $p = p(\rho)$ the vertical ($\hat{\boldsymbol{z}}$) component of Eq. (11) reduces to

$$\frac{d}{dt}(f + \omega) = (f + \omega)\frac{\partial w}{\partial z} \tag{12}$$

describing the exponential growth and decay of the absolute vorticity for a given positive or negative value of stretching $\partial w / \partial z$. In the full 3D system we will determine $\partial w / \partial z$ from the total geostrophic convective time derivative of $\partial p / \partial z$ to rewrite Eq. (12) as $dq/dt = 0$. In the 2D limit of shallow water the derivation of $q_G(x, y, t)$ is simple. Integrating Eq. (12) over z from the bottom to the upper free surface and using that at the free surface $w = dH/dt$ gives $H\, d_t(f + \omega) = (f + \omega)(dH/dt)$. Thus $q_G = (f + \omega)/H$ is conserved.

In both the GFD and plasma dynamics of vorticity the finite, or first order, Rossby number R_0 and finite Larmor radius (FLR) effects are essential to determine the horizontal compression $\nabla_\perp \cdot \boldsymbol{v}_\perp$. Iterating on $1/f$ in the momentum Eq. (9) we obtain the next term in the $1/f$ expansion by evaluating the convective derivative with the geostrophic velocity $\boldsymbol{v} = \hat{\boldsymbol{z}} \times \nabla p / \rho f$ to obtain the ageostrophic velocity

$$\boldsymbol{v}_{ag} = -\frac{1}{f}\frac{d}{dt}\left(\frac{1}{f\rho}\nabla_\perp p\right). \tag{13}$$

For a single layer of shallow fluid with a free upper surface the height integrated divergence free condition (6) evaluated with ageostrophic flow $\boldsymbol{v}_{ag} = -(g/f^2)(d\nabla\delta H/dt)$ and the upper boundary condition $w = dH/dt$ yields the reduced form of Ertel's theorem

$$\frac{d}{dt}\left(\frac{f + \frac{g}{f}\nabla^2 \delta H}{H_0 + \delta H}\right) = 0 \tag{14}$$

describing the conservation of the potential vorticity q_G.

For the magnetized plasma the vorticity equation (11) and continuity Eq. (1) yield

$$\frac{d}{dt}\left[\frac{\Omega_i + \frac{c}{B}\nabla^2\Phi}{n(\boldsymbol{x},t)}\right] = 0 \tag{15}$$

(Meiss and Horton, 1983). Using the Boltzmann relation $n(x,t) = n(x)\exp(e\Phi/T_e(x))$ gives the single PDE for $\Phi(x,y,t)$. The details of this analysis and further comparisons with GFD are given in Horton and Ichikawa (1996, ch. 6).

Comparing Eqs. (14) and (15) we see that for the two analog systems the role of deep water is played by the high plasma density and the gradient of the Coriolis parameter f is equivalent to the gradient of the magnetic field. Both systems have a conserved potential vorticity q with $dq/dt = 0$. The invariants derived from integrals of q, q^2 and the x, y moments of q are key ingredients of the Lyapunov stability analysis of the systems. The Lyapunov stability is treated in Laedke and Spatschek (1986, 1988), Swaters (1986), Sakuma and Ghil (1991), and Nycander (1992), but is too technical to develop here.

Equations (14) and (15) are highly nonlinear and do not have symmetry with respect to the interchange of anticyclonic ($\delta H > 0$ or $\delta\Phi > 0$) and cyclonic ($\delta H < 0$ or $\delta\Phi < 0$) disturbances. Clearly, the anticyclonic (AC) disturbance has $\omega = (g/f)\nabla^2\delta H < 0$ opposite to f so that a localized disturbance may freely drift to larger f or smaller H_0 while conserving q_H by increasing its strength. The cyclonic (C) disturbance has ω of the same sign as f and may propagate to smaller f while gaining strength.

Reduced dynamical equations showing the broken AC-C symmetry were given in a number of works including Tasso (1967), Petviashvili (1977), Horton and Petviashvili (1993), Nezlin and Snezhkin (1993), and Horton and Ichikawa (1996). Su *et al.* (1991, 1992) investigate the properties of these equations in some detail showing how the Larichev-Reznik (1976) dipoles are split in part with only the anticyclonic vortex forming along lined coherent structure. Nezlin (1994) emphasizes the importance of the anticyclone over the cyclone in Jupiter's atmosphere and other geophysical vortices when the vortex radius r_0 exceeds the Rossby deformation radius ρ_R.

In the local limit where the potential vorticity in Eq. (14) reduces to

$$q_H = \nabla^2 h - \frac{1}{\rho_R^2} h + \beta y \tag{16}$$

and in Eq. (15) to

$$q_p = \rho_s^2 \nabla^2 \varphi - \varphi + v_d x \tag{17}$$

the equation $dq/dt = 0$ gives the locally homogeneous pde called the Charney-Hasegawa-Mima equation (hereafter called CHM). The relevant dimensionless coordinates are ρ_R or ρ_s for x, y and the time unit ρ_R/v_R or ρ_s/v_d such that the Rossby/drift speed is unity. In these space-time coordinates the amplitude is scaled as $\rho_R\beta/f \sim \rho_R/R_p \ll 1$ and $\rho_s/L_n \ll 1$ to give the nonlinearity unit strength. Here R_p is the radius of the planet. The CHM equation is then

$$(1-\nabla^2)\frac{\partial\varphi}{\partial t} + \frac{\partial\varphi}{\partial y} - [\varphi, \nabla^2\varphi] = 0 \tag{18}$$

where we have taken x in the direction of the inhomogeneity (northward on the planet and radially in the plasma) and y in the symmetry direction.

The nonlinear drift wave equation (18) exhibits both dispersive wave and coherent vortex propagation. The vortex properties dominate, trapping the wake fields, when the amplitude is such that the rotation period in the structure of size $r_0 = \pi/k_\perp$ is shorter than the corressponding wave period $2\pi/\omega_{k_\perp}$. Here r_0 is the radius of the incipient vortex and the trapping condition requires $\max(v_x) = (1/\rho_0 f)(\partial\delta p/\partial y) > v_R$ or $\delta H/H > r_0/R_p$ for the shallow fluid. These properties are illustrated in Fig. 3 which compares the propagation properties of a small amplitude disturbance in the left column with a large amplitude disturbance well above the trapping condition in the right column for an initial gaussian AC distribution. The upward (westward) propagating wave contains wave fronts of speeds from zero to v_R whereas the large amplitude disturbance traps most of the wave energy and propagates coherently at the speed v_R. The monopolar structure does not have the reflection symmetry required of solution of Eq. (18) $[\varphi(-x,y,t) = -\varphi(x,y,t)]$ so that an underlying odd dipolar component is created in one rotation period. This reflection system is why the exact, localized stationary solutions must have odd mirror or reflection symmetry like the Larichev-Reznik dipolar structures. However, in practice, as Fig. 3b shows that the dominant monopolar structure, with a small dipolar component can form a long-lived, coherent structure. We will see this again in the next example where a monopolar structure is the product of a collision between two dipoles. The wakefield of the linear structure can be computed in detail. The tilted chain of trailing waves is understood from the $\max(\partial\omega/\partial k_x) \cong -2\max(k_x k_y \rho_s^2)v_R$ which occurs for $\theta = \tan^{-1}(k_x/k_y) = 45°$.

In the atmosphere the trapped structure form by either the sudden creation of hot or cold spots by nonadiabatic processes and/or the confluence of counterstreaming flows. Holland (1995) describes both processes operating in the summer Asian monsoons in the western Pacific in the tropics at $\simeq 10°$ N latitude. This region during the summer has a high birthrate of mesoscale vortices, some of which grow into large, long-lived cyclonic depressions. An example at an early state of development given in Fig. 4 from Holland (1995, fig. 15) shows a structure that is similar to the first three vortices shown in the head of the wake in Fig. 3a. Now we consider the interaction and structural stability of the dipolar vortices.

To further illustrate the properties of Eq. (18) let us launch two Larichev-Reznik dipole solutions with opposite polarizations of the electric field or $\nabla\delta H$

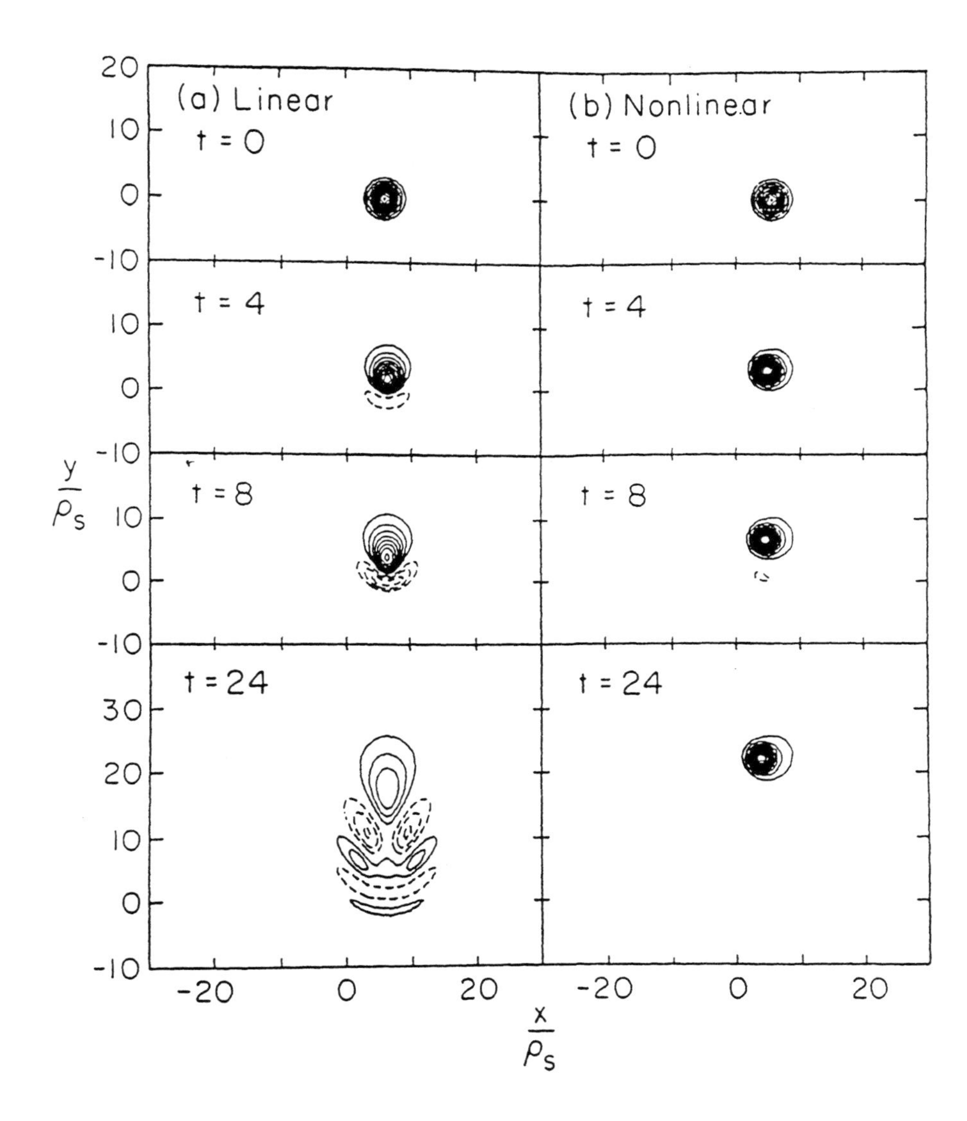

FIGURE 3. Comparison of the linear and nonlinear drift wave–Rossby wave propagation from an initial gaussian anticyclonic disturbance. The drift wave–Rossby wave speed is unity and in the $\hat{y}$ direction.

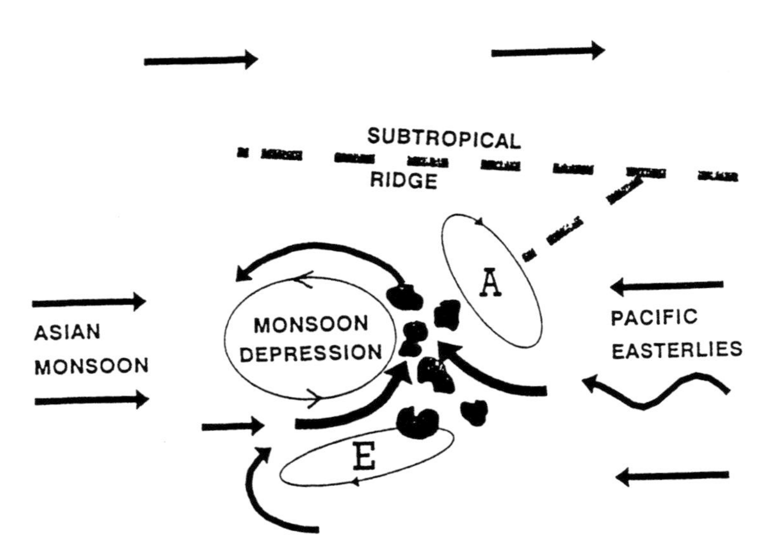

FIGURE 4. A strong leading cyclonic Rossby wave disturbance with its wakefield of two anticyclonic patches that appears to conform to the wakefield structure shown in Fig. 3a. Other conditions relevant to this southwast Pacific typhoon are given in Holland (1995) (courtesy of Holland). Reprinted with kind permission of Springer-Verlag ©1995.

— so that they collide. Rather than taking the symmetric co-axial collision, it is more informative to have the dipole centers offset by the impact parameter b comparable to the radius r_0 of the Larichev-Reznik dipoles. The situation is shown in Fig. 5 with $r_0 = 6\rho_R$ and $b = 5\rho_R$. In the coordinates of Eq. (18) the Rossby-drift wave propagates with unit speed in the $+y$ direction (westward). This is also the direction of propagation of the lower dipole in Fig. 5 with the anticyclone on the right and the cyclotron on the left. The upper dipole has the AC on the left and C on the right, and, by itself, will propagate downward (eastward). What we see happening is the merging or coalescence of the two cyclonic regions to form a new, stronger cyclone. The merger is of the type shown in Griffiths and Hopfinger (1986, 1987).

Now, which of the two AC's pair up with the large cyclone depends on the stability of the dipole system with out-of-balance monopolar components. This problem is investigated by Javonović and Horton (1993) using Lyapunov stability theory and simulations. The stability question can be answered, however, using Ertel's theorem with more clarity than using the heavy mathematical machinery of Lyapunov stability theory (Muzylev and Reznik, 1992). First consider the left side AC with the large, central cyclone. The anticlockwise velocity field of the large cyclone rotates the left AC down under itself which strengthens it. The strengthened left AC then rotates the central C to

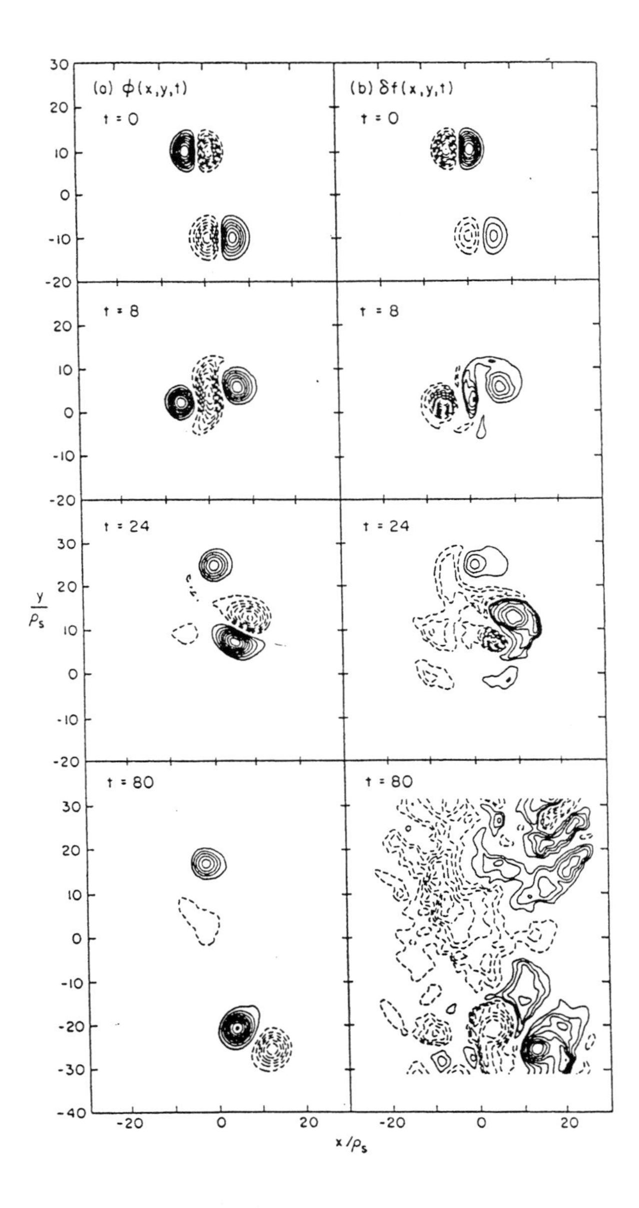

FIGURE 5. Collision of opposite polarity drift wave vortices with impact parameter $b = 5\rho_s$ and dipole radius $a = 6\rho_s$. (a) 15 contours of potential with the range ± 54.7 at $t = 0$ and -49.5 to 65.5 at $t = 80$; (b) 15 contours for perturbed distribution δf with the range ± 54.7 at $t = 0$ and -10.4 to 13.3 at $t = 80$. Positive values of the functions are solid lines and negative values are dashed.

the left, allowing it to strengthen from the reduced f. This result is a stable, oscillatory drift of this C-AC pair while propagating downward (eastward). Now, if one considers the alternative possibility for the central C and the right hand AC joining together, the system falls apart. The right side AC is weakened by its rotation from the large cyclone. Thus, they (the C and right-hand AC) do not bind. In fact, the same arguments are easy to use to show why the westward propagating dipole is unstable to a slight de-symmetrization of the dipole lobe strengthens. The motion is such that the initially stronger lobe continues to gain strength by drifting farther to the north (for the $AC > C$) or to the south (for the $C > AC$).

This is in accordance with the mathematical stability analysis (Laedke and Spatschek, 1988) indicating that the eastward (ion-diamagnetic direction) dipole is stable. Difficulties in giving a complete mathematical proof of stability are described by Nycander (1992). The simulations and the point vortex models (Kono and Horton, 1991) and Hobson (1991) give clear physical pictures for why the westward dipole is unstable. These stability properties may account for the rather uncommon occurrence of the dipole structures in the atmosphere. The effect of the temperature gradient is another reason for the lack of the dipoles (Jovanović and Horton, 1993).

The collision is also inelastic with a wakefield generated similar to that shown in Fig. 3. The right-hand side of Fig. 5 shows the contours of a passive field f convected by the flow. The initial value of the passive f is trapped inside the dipoles.

Vertical motion in the stably stratified medium

The stable vertical stratification of the fluid introduces the restoring force from the buoyancy of the fluid. This buoyancy frequency

$$N = (-g\partial_z \ell n\, \rho)^{1/2} \tag{19}$$

follows from the convection of the density perturbation $\delta\rho$ by the vertical velocity w

$$\frac{d\delta\rho}{dt} + w\,\frac{\partial\rho_0}{\partial z} = 0 \tag{20}$$

and the vertical acceleration

$$\rho_0\,\frac{dw}{dt} = -\frac{\partial}{\partial z}\,\delta p - g\delta\rho. \tag{21}$$

Here d/dt is the horizontal convective derivative with the geostrophic velocity. Taking d/dt of Eq. (21) and using Eq. (20) for $d\delta\rho/dt$ gives

$$\frac{d^2}{dt^2}w = -N^2 w - \frac{d}{dt}\left(\frac{1}{\rho}\frac{\partial}{\partial z}\,\delta p\right). \tag{22}$$

Thus, perturbed vertical pressure gradients drive vertical oscillations at the frequency N about the initial stratified equilibrium position. For motions on long-time scales $\Delta t \gg 1/N$ compared to the rapid (small amplitude) vertical motion, the time-averaged vertical velocity is

$$w = \frac{-1}{N^2}\frac{d}{dt}\left(\frac{1}{\rho}\frac{\partial \delta p}{\partial z}\right), \tag{23}$$

following from the time-average of Eq. (22).

This determination of w is the primary difference with the plasma confinement where the equilibrium is uniform along $\boldsymbol{\Omega} = \Omega\hat{\mathbf{z}}$ with neutral stability with respect to translations along $\hat{\mathbf{z}}$. There are stable oscillations along $\hat{\mathbf{z}}$ in the plasma, but they are associated with the ion-acoustic waves $k_{\|}^2 c_s^2$ with $c_s^2 = T_e/m_i$ not with stratification or an equilibrium restoring force. The acoustic waves are low in frequency compared with $\Omega_i = eB/m_i c$. The acoustic waves occur both above and below the drift wave frequency depending on the ratio of $k_{\|} c_s / k_y v_{de} = k_{\|} L_n / k_y \rho_s$.

Equation (23) describes the vertical outflow from a growing, localized high pressure (AC) region. There is a balancing inward horizontal flow from the ageostrophic velocity. Balancing the divergence of the inward horizontal flow with the outward vertical flows gives the basic quasigeostrophic baroclinic equation for the eigenmodes of the system. Substituting Eq. (23) into Eq. (12) gives

$$\frac{d}{dt}\left\{f(y) + \frac{1}{f\rho_0}\nabla_\perp^2\delta p + \frac{\partial}{\partial z}\frac{f}{N^2\rho_0}\frac{\partial \delta p}{\partial z}\right\} = 0 \tag{24}$$

where $df/dt = (\beta/f\rho_0)\partial_x \delta p$. The vertical eigenmodes with $w = 0$ at $z = 0$ and $z = D$ are approximately $\delta p_n = \delta p_n(x, y, t)\cos(n\pi z/D)$ giving the horizontal dynamics

$$\frac{d}{dt}\left[\frac{1}{\rho_0}\nabla_\perp^2\delta p_n - \frac{f^2}{N^2 D_n^2}\delta p_n\right] + \frac{\beta}{\rho_0}\frac{\partial}{\partial x}\delta p_n(x, y, t) = 0. \tag{25}$$

For a sufficiently smooth $N(z)$-profile the eigenvalues are $D_n = D/n\pi$. For a strongly localized $N(z)$ of height Δz_m there are trapped internal modes when $k_\perp N_{\max}\Delta z_m > \pi f$ where $k_\perp = (k_x^2 + k_y^2)^{1/2}$ is the horizontal wavenumber.

Thus, the internal Rossby deformation radius ρ_R is

$$\rho_R(n) = ND_n/f \tag{26}$$

where $\rho_R(n = 0) \to \infty$ for the barotropic (flute) mode and $\rho_R(n = 1) \simeq ND/\pi f$ for the first baroclinic mode. The relation with the shallow water

Rossby radius is found by noting that $N = [(\Delta\rho/\rho)(g/D)]^{1/2}$ so that $\rho_R(n = 1) \to [Dg(\Delta\rho/\rho)]^{1/2}/f < (gH)^{1/2}/f$. For the oceans the effect of $\Delta\rho/\rho$ and the eigenmode calculation is to lower $\rho_R(n = 1)$ to about 80 km. Thus, the oceans have a much smaller value of $\rho_R/R_p \sim 10^{-2}$ more analogous to the magnetic confinement experiments. Another, unfortunate, analogy of the oceans with the plasmas is that detecting the Rossby wave structures in the oceans is much more difficult than in the atmosphere.

HORIZONTAL TEMPERATURE GRADIENTS

Both magnetized plasmas and geophysical fluids are subjected to localized heating and cooling resulting in substantial temperature gradients. The resulting temperature gradients can drive large-scale convective motions that serve to transport thermal energy and momentum. The stability analysis of the various equilibrium models is a classical problem in both fields. Here we briefly compare the systems before showing the results for the plasma simulations.

The linear stability analysis gives an expression for the growth rate $\gamma^\ell(\boldsymbol{k}, \{\mu\})$ for the mode as a function of wave vector $\boldsymbol{k}$ and the system parameters $\{\mu\}$. Generally, the fluctuations that maximize $\gamma_{\boldsymbol{k}}^\ell$ provide the dominant source of energy into the convection system. The temperature gradient modes have a well-defined maximum growth rate at the wavenumber $k_\perp \sim \rho^{-1}$ where $\rho = \rho_I = ND/f$ in the baroclinic instability and $\rho = \rho_s = (m_i T_e)^{1/2}/eB$ in the magnetized plasma.

Baroclinic instability

The baroclinic instability mechanism is a modification of the Rayleigh-Benard instability taking into account the role of the Coriolis force in the momentum equation and the constraints of the vertical stratification. The thermal energy release is driven by the gravitational potential energy obtained by interchanging lighter and heavier fluid parcels taking into account their change in density $\rho = \rho(p, T)$ with temperature and pressure. For liquids the equation of state is simpler with $\rho = \rho(T)$ for the usual range of pressures.

For the oceans and the rotating water tanks (liquids) the fractional change in density is very small with $\rho = \rho_0(1 - \alpha(T - T_0))$ with $\alpha = 2 \times 10^{-3}/°C$ for water. Thus the Boussinesq approximation applies where $\rho = \rho_0 =$ const except in the buoyancy term where $g\delta\rho = -g\rho_0\alpha\delta T$. Gravitational potential energy is released to drive convection when the lower level fluid is heated becoming lighter or the upper level is cooled becoming heavier.

For the atmosphere and plasma (gases) the compressibility of the gas determined by the ratio of the specific heats $\Gamma = C_p/C_v$ must be taken into account. For the diatomic atmosphere gases $\Gamma = 1.4$ while for the collisionless plasma the value of Γ ranges from 5/3 to 3 as the effective number of degrees of freedom involved in the dynamics changes from three to one. For the ITG instability the effective Γ is given in Kim and Horton (1991) from kinetic theory considerations. For the ideal gas equation of state $\rho = P/nRT$ so that the volume expansion coefficient $\alpha = -\partial \ell n\, \rho/\partial T)_p = 1/T \sim 3 \times 10^{-3}/°C$ at $T \sim 300°$ K.

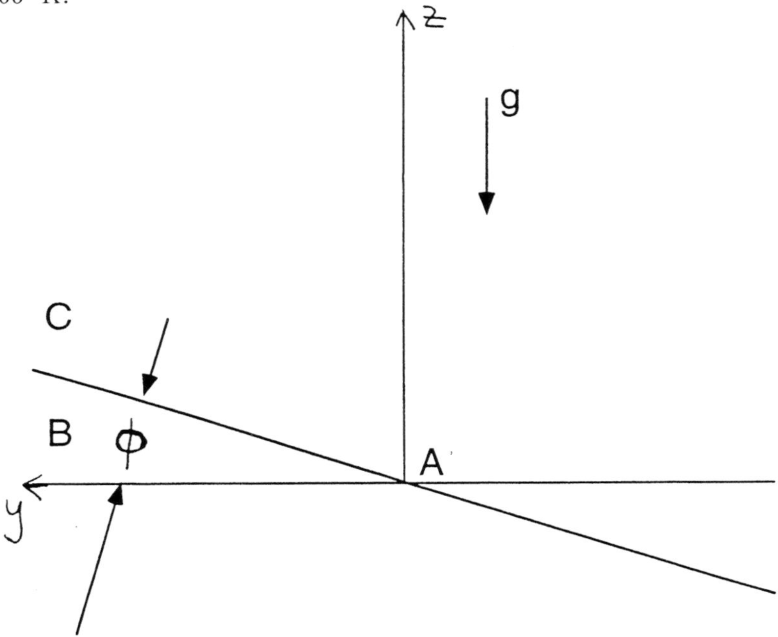

FIGURE 6. For the sloping constant density surfaces at angle ϕ from the horizontal, gravitational potential energy is released by the interchange of fluid elements between A and B.

In the geophysical problem the effect of the horizontal temperature instability is complicated by the vertical stratification of the fluid described in Sec. 1.3. For instability the buoyancy must be overcome by suitable restrictions on $\boldsymbol{k}, \omega$ in geostrophic motions to give rise to growth of the convection. To understand the origin of these restrictions consider the sloping contours of the potential temperature and density shown in Fig. 6. Both Pedlosky (1987, p. 519) and Gill (1982, p. 555) begin their discussion of the baroclinic instability by calculating the gravitational potential energy released resulting from interchange motions for the sloping contours shown in Fig. 6. The angle ϕ (in radians) of the slope of the isotherms and isopycnals (constant density surfaces) required for fast-growing modes is very small. It is evident (see Pedlosky and Gill for the calculation) that only interchanges of fluid parcels A and B, within the small wedge of angle ϕ, will release potential energy. Interchange of A and C

requires work be down against g due to the stable stratification.

In the toroidal plasma confinement device the equilibrium pressure, density and temperature are independent of z having gradients perpendicular to $\boldsymbol{B}$. The role of gravity g is played by the curvature and gradient-B drifts of the charged particles producing through charge separation an effective gravity term from the electric acceleration acting across the magnetic field $\boldsymbol{\Omega}_i$. The electric field produces an interchange of plasma that is mathematically of the same form as that with the gravitational acceleration component $\boldsymbol{k}\cdot\boldsymbol{g}\times\nabla\rho = k_x g\,\partial\rho/\partial y$ of the neutral fluid. The interchange is effective in lowering the plasma thermal energy. The effective plasma g is given by $g = c_s^2/R_c$ where R_c is the radius of curvature of the field lines (Hazeltine and Meiss, 1992).

To understand the conditions on the $\boldsymbol{k}, \omega$ for the release of potential energy we analyze the temperature equation $dT/dt = 0$ for the incompressible liquid or the adiabatic equation $d\theta/dt = 0$ for the compressible gas or plasma. Here $\theta = \theta_0(p/p_0)^{1/\Gamma}(n_0/n)$ is the potential temperature and $\Gamma = C_p/C_v$ the ratio of specific heats.

For liquids with $\Delta\rho/\rho = \alpha\Delta T < 10^{-2}$ the Boussinesq approximation is valid. In the Boussinesq equations $\rho = \rho_0 =$ constant everywhere except in the buoyancy term where

$$\delta\rho g = -\alpha\rho_0\delta T = -\partial(\delta p)/\partial z \tag{27}$$

for slow motions that avoid the stabilizing vertical oscillations.

For the atmosphere the compression due to Γ is retained by working with $\theta = \theta_0 + \delta\theta$ where $\delta\theta/\theta_0 = -\delta\rho/\rho_0 + \delta p/\Gamma p_0$ for adiabatic motion. Pedlosky (1987, p. 365) introduces the horizontal velocity streamfunction $\psi = \delta p/f\rho_0$ and uses $g\delta\rho = -\partial_z\delta p$ to show that

$$\left(\frac{g}{f}\right)\frac{\delta\theta_0}{\theta_0} = \partial_z\left(\frac{\delta p}{f\rho_0}\right) \tag{28}$$

when $\partial z\ell n\,\rho_0 = -g/c^2$ with $c^2 = \Gamma p_0/\rho_0$.

The equilibrium constant-slope surfaces of Fig. 6 are given by $\theta = \theta_0(z-\phi y)$ and $\rho = \rho_0(z-\phi y)$. The perturbation in the temperature δT, or perturbed potential temperature $\delta\theta$, in the absence of thermal diffusion $\omega \gg k^2\chi$ is governed by

$$\left(\frac{\partial}{\partial t} + u\,\frac{\partial}{\partial x}\right)\delta\theta + (-v\phi + w)\,\frac{\partial\theta_0}{\partial z} = 0 \tag{29}$$

where we use $\partial\theta_0/\partial y = -\phi\partial\theta_0/\partial z$ due to the slope ϕ. Multiplying Eq. (29) by g/θ_0 and recognizing that the Brunt-Väisälä buoyancy frequency is

$$N^2 = \frac{g}{\theta_0}\,\frac{\partial\theta_0}{\partial z} \tag{30}$$

and using Eq. (28) we obtain

$$-i(\omega - k_x u)\frac{\partial}{\partial z}\left(\frac{\delta p}{\rho_0}\right) + N^2(-\phi v + w) = 0. \tag{31}$$

¿From Eq. (31) it is clear that the stable vertical oscillations from the buoyancy restoring effect $N^2 w$ in Eq. (31) are lost when

$$w(-\phi v + w) < 0. \tag{32}$$

A simple estimate of the conditions on $\boldsymbol{k}, \omega$ implied by condition Eq. (32) is obtained from the shallow water equations $w = d\delta H/\delta t = -i\omega\delta H$ and $v = (g/f)\partial_x \delta H = i(k_x g/f)\delta H$ giving the necessary condition

$$\omega(\omega - \omega_*) < 0. \tag{33}$$

for instability where $\omega_* = -k_x g\phi/f$. A sufficient condition for instability requires finding the actual dynamics required to release the energy by solving for the wave function and eigenvalues as in the Eady problem, for example. The shallow water approximation shows that for $\phi \sim 10^{-4}$ and $k_x \sim 10^{-4} m^{-1}$ waves with $\omega \lesssim 10^{-4}\, s^{-1}$ Hz satisfy $\omega < \omega_*$. The phase velocity ω/k_x must be parallel to $\nabla\rho \times \boldsymbol{g}$ and low enough to release the potential energy ($\phi v > w$). Without the $\beta = \partial f/\partial y$ effect the $\widehat{\boldsymbol{y}}$-direction is the direction of the local temperature gradient without regard to the north-south direction. An estimate for ω_* taking into account the 3D nature of the baroclinic motion is obtained by using w from Eq. (23) and $v = ik_x \delta p/\rho_0 f$ in Eq. (31). The result is $\omega_*(3D) = k_x\, DN^2\phi/f$ whereupon using $\rho_I = ND/f$ gives

$$\omega_* \cong k_x \rho_R N\phi. \tag{34}$$

Thus, for $k_x\rho_R \sim 1$ the upper limit of ω_* and the growth rate γ of the unstable mode is $N\phi$. For large slopes $\phi \sim f/N \sim 10^{-2}$ the instability breaks the quasigeostrophic condition $|\omega| \ll f$. This regime may occur in sharp fronts.

The classical baroclinic instability analysis (the Eady and Charney problems, Gill, 1982, pp. 556–563) is briefly given for completeness. The linearization of Eq. (24) gives vertical eigenvalue problem

$$\left[\frac{d^2}{dz^2} - \frac{N^2 k_\perp^2}{f^2}\left(1 + \frac{k_x\beta}{\omega k_\perp^2}\right)\right]\frac{\delta p}{\rho_0} = 0 \tag{35}$$

with solutions

$$\frac{\delta p}{\rho_0} = A\sinh(qz) + B\cosh(qz) \tag{36}$$

where $q = (k_\perp N/f)(1 + k_x\beta/\omega k_\perp^2)^{1/2}$. Equation (31) determines $w_k(z)$ since all other terms are known from Eq. (36) with $v = (ik_x/f)(\delta p/\rho_0)$. Now the vertical equilibrium $g\rho(z - \phi y) = -\partial p/\partial z$ demands that there is shear in the horizontal velocity $u(z) = -(1/\rho_0 f)\partial p/\partial y$ with $\partial u/\partial z = \phi N^2/f$.

The simplest case is the Eady problem which has a rigid lid so that $w = 0$ at both $z = \pm D/2$ and has $\beta = 0$. The eigenvalues follow from substituting Eq. (36) into Eq. (31) and evaluating the equation at $z = \pm D/2$. The determinant of the 2×2 system in the A, B coefficients yields

$$\omega = \pm i \frac{|k_x|\phi N^2 D}{f} \left[\left(\coth\left(\frac{qD}{2}\right) - \frac{qD}{2} \right) \left(\frac{qD}{2} - \tanh\left(\frac{qD}{2}\right) \right) \right]^{1/2} \tag{37}$$

for the unstable complex conjugate modes for $qD < 2.4$. The modes (37) coalesce and go to stable modes for $qD = k_\perp ND/f > 2.4$. The maximum growth rate $\gamma_{\max} = 0.31 fu'/N = 0.31 N\phi$ occurs at $k_y \to 0$ and $k_x(ND/f) = 1.6$ corresponding to a wavelength $2\pi/k_x \simeq 4\rho_I = 4(ND/f)$. Pedlosky states that for $\rho_I \cong 10^3$ km this wavelength is in excellent agreement with the most energetic synoptic scale atmospheric disturbances. Monin (1972) shows the energy spectrum in Fig. 7 computed from wind fluctuations with a pronounced peak at a period corresponding to approximately 100 hr $\sim$ 4 days that agrees with the baroclinic instability as the source of energy into the spectrum. The

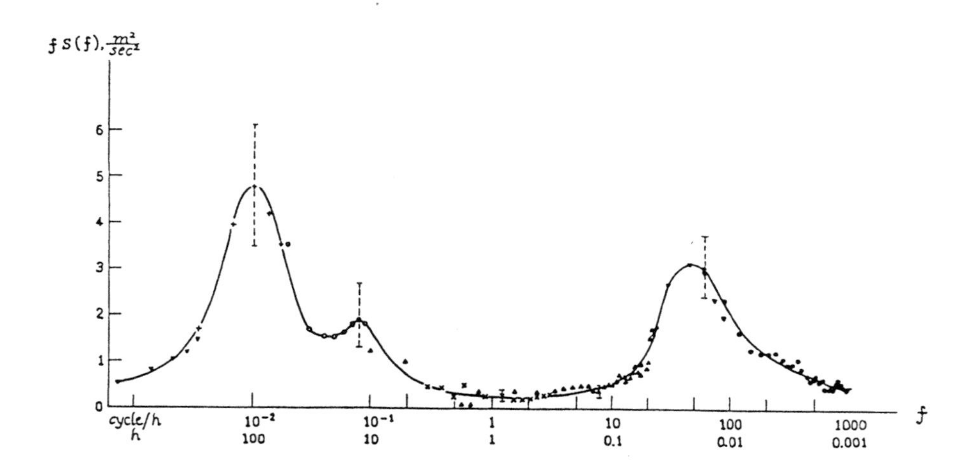

FIGURE 7. Energy-like spectrum $fS(f)$ constructed from the power spectrum $S(f)$ of the horizontal wind velocity time series taken from Monin (1972). The dominant peak in the spectrum is at a period of $100\,h \approx 4$ days. Monin interprets the peak as the position of the energy injection from the large baroclinic instability. For $\rho_R = 10^3$ km the peak corresponds to $k_x\rho_R = 1$ for a longitudinal mode number $m = 6$ (from Monin, 1972). Reprinted from A. Monin, *Weather Prediction as a Problem in Physics*, Cambridge Mass: The MIT Press, ©1972, page 9.

wavelength corresponding to the peak is approximately such that $k_x \simeq m/R_0$ with $m \cong 6$ consistent with $\lambda \sim 4000\,\mathrm{km}$ and $R_0 = R_E \cos\theta \approx 4000\,\mathrm{km}$ at $\theta = 30°$.

The situation for tokamaks is that with substantial ion heating a peak in the fluctuation spectrum appears at $k_\perp \rho_s \sim 0.5$ that is identified (Brower *et al.*, 1987, 1989) with the ion temperature gradient by the direction of propagation being in the direction parallel to $\boldsymbol{B} \times \nabla p_i$ (corresponding to eastward in the Rossby Wave–Drift Wave analog). The electron drift wave spectrum (westward Rossby wave) is universally present: whereas the ion feature in the spectrum occurs only when the ion temperature gradient exceeds the adiabatic gas threshold condition.

Nonlinear dynamical equations for the Baroclinic instability

For the case of a liquid the equation of state is simpler than for a gas or plasma with the standard model taking the linear relation $\rho = \rho(T) = \rho_0\,[1 - \alpha(T - T_0)]$. In this case $d\rho/dt = \rho' dT/dt = \rho'\chi\nabla^2 T$ where χ is the thermal diffusivity. In this model the fluctuation dynamics for fixed $\partial\rho/\partial y \propto \partial T/\partial y$ suitable for numerical integration are given by

$$\frac{d}{dt}\left(\frac{\nabla_\perp^2 \delta p}{f\rho_0}\right) + \frac{\beta}{f\rho_0}\frac{\partial}{\partial x}\delta p = f\,\frac{\partial w}{\partial z} + \frac{\mu}{f\rho_0}\nabla_\perp^4 \delta p \tag{38}$$

$$\rho_0\,\frac{dw}{dt} = -g\delta\rho - \frac{\partial}{\partial z}\delta p \tag{39}$$

$$\frac{d}{dt}\,\delta\rho + \frac{1}{f\rho_0}\frac{\partial\rho_0}{\partial y}\frac{\partial}{\partial x}\,\delta p + \frac{\partial\rho_0}{\partial z}\,w = \chi\nabla^2\delta\rho. \tag{40}$$

For the liquid we may take $\rho_0, \partial\rho_0/\partial y$ and $\partial\rho_0/\partial z$ constant throughout the volume Ω of the system which simplifies the analysis and simulations. Equation (38) is the same vorticity equation as in the Rossby waves, but now the vortex stretching by the vertical velocity w is controlled by the change in $\delta\rho$ both through horizontal convection and vertical convection.

The stable vertical stratification is overcome and the total fluctuation energy grows exponentially due to the horizontal temperature gradient $\partial T/\partial y = (-1/\alpha\rho_0)(\partial\rho_0/\partial y)$. In the steady state the energy production by the meridional thermal flux balances the dissipation in Eqs. (38)–(40). To demonstrate this balance we need to analyze the flow of fluctuation energy through the system (38)–(40). Multiply Eq. (38) by $\delta p/f_0$, Eq. (39) by w and Eq. (40) by $\delta\rho$ and integrating over the volume Ω yields

$$\frac{\partial}{\partial t}\left[\frac{1}{2}\int_\Omega \frac{(\nabla_\perp \delta p)^2}{f^2\rho_0}\,d^3x\right] = -\int_\Omega \delta p\,\frac{\partial w}{\partial z}\,d^3x - \frac{\mu}{f^2\rho_0}\int_\Omega (\nabla_\perp^2 \delta p)^2 d^3x \tag{41}$$

$$\frac{\partial}{\partial t}\left[\frac{1}{2}\int\limits_{\Omega} \rho_0 w^2\, d^3x\right] = -g\int\limits_{\Omega} w\delta\rho\, d^3x - \int\limits_{\Omega} w\frac{\partial\delta p}{dz}d^3x \tag{42}$$

$$\frac{\partial}{\partial t}\left[\frac{1}{2}\int\limits_{\Omega} \delta\rho^2 d^3x\right] = -\frac{\partial\rho_0}{\partial z}\int\limits_{\Omega} \delta\rho w d^3x - \frac{\partial\rho_0}{\partial y}\int\limits_{\Omega} \delta\rho v d^3x - \chi\int\limits_{\Omega}(\nabla\delta\rho)^2 d^3x. \tag{43}$$

For simplicity the boundary surface integrals over $\partial\Omega$ have been dropped. Thus the three fluxes $\langle\delta p\, \partial w/\partial z\rangle, \langle w\delta p\rangle$ and $\langle\delta\rho\, v\rangle$ transfer energy between the kinetic and potential energy components

$$K_\perp = \frac{1}{2}\int\limits_{\Omega} \rho_0(u^2+v^2)d^3x, \tag{44}$$

$$K_z = \frac{1}{2}\int\limits_{\Omega} \rho_0 w^2 d^3x. \tag{45}$$

and

$$U = g\int\limits_{\Omega} z\rho d^3x = \left[\frac{-g}{(\partial\rho_0/\partial z)}\right]\frac{1}{2}\int\limits_{\Omega} \delta\rho^2 d^3x \tag{46}$$

is the potential energy. To show the last relationship for the potential energy one uses that for $d\rho/dt = 0$ the Casimir $\int_\Omega \rho^2 d^3x = \int_\Omega (\rho_0 + z\rho_0' + \delta\rho)^2 d^3x$ is a constant of the motion.

Combining Eqs. (40)–(43) yields

$$\frac{d}{dt}\int\limits_{\Omega}\left[\frac{\rho_0}{2}(\nabla_\perp\psi)^2 + \frac{\rho_0}{2}w^2 - \frac{g}{\partial\rho_0/\partial z}\frac{(\delta\rho)^2}{2}\right]d^3x \tag{47}$$

$$= g\phi\int\limits_{\Omega} v\delta\rho d^3x - \mu\rho_0\int\limits_{\Omega}(\nabla^2\psi)^2 d^3x + \frac{g\chi}{\partial\rho_0/\partial z}\int\limits_{\Omega}(\nabla\delta\rho)^2 d^3x. \tag{48}$$

¿From the first energy integral we see that the requirement of stable vertical stratification leads to the positive definiteness of the energy integral W. For stable stratification the system has a linear instability driven by the slope $\phi = -(\partial\rho_0/\partial y)/(\partial\rho_0/\partial z)$. In the nonlinear steady saturated state of this instability Eq. (48) states that the product of the thermal flux proportional to $\phi\int v\delta\rho d^3x$ which multiplied by g balances the dissipation due to small scale viscosity μ and thermal diffusivity χ to produce the saturated state.

In the form given here with *three* pdes in Eqs. (38)–(40) the baroclinic instability dynamics is similar in form to the ITG plasma equations. This is not the case when the further reduction to the quasi-geostrophic form is made as follows. For the stably stratified medium the vertical velocity w adjusts adiabatically to $\partial_z\partial p$ such that Eq. (23) for w applies. Then $\delta\rho = -g^{-1}\partial_z\delta p$ and the energy integral w now reduces to $W \rightarrow W_g$ with

$$W_g = \int_\Omega \left[\frac{\rho_0}{2} (\nabla_\perp \psi)^2 + \frac{\rho_0 f^2}{2N^2} (\partial_z \psi)^2 \right] d^3x \tag{49}$$

which is the energy associated with gyrostrophic motions. The minimization of W with respect to ψ gives the elliptic baroclinic ($\beta = 0$) equation for $\psi(x, y, z)$.

Temperature gradient driven convection in magnetized plasma

The plasma equations for temperature gradient driven convection have $f \to \Omega_i(\sim 10^8/\text{s})$, $\rho_R \to \rho_s(\sim 0.1\,\text{cm})$ and $\varepsilon = \rho_s/L_n \sim 0.1\,\text{cm}/10\,\text{cm} = 10^{-2}$. The growing linear modes occur for $k_\| L_n < k_y \rho_s < 1$ with $k_y^{\max} \sim 1/2\rho_s \sim 5\,\text{cm}^{-1}$ and $\gamma^{\max} \sim 0.1 c_s/L_T \sim 10^5/\text{s}$ being typical. The signature of the ITG mode in the laboratory plasma is propagation of part of the drift wave fluctuation spectrum in the ion diamagnetic direction ($\sim$ eastward). This is opposed to the resistivity and electron-wave Landau resonance instability that occurs for the usual universal drift wave fluctuations propagating in the electron diamagnetic direction ($\sim$ westward). For example, in the Ohmically heated TEXT tokamaks ($R/a = 1\,\text{m}/0.27\,\text{m}, B = 2\,\text{T}, I = 0.3\,\text{MA}$ the electron drift wave fluctuations are always present, but the fluctuations characteristic of the ion temperature gradient only appear when the conditions are such as to have $\eta_i = d\ell n\, T_i/d\ell n = L_n/L_{T_i} > 1$ and $T_i \simeq T_e$.

With powerful auxiliary heating in the large fusion devices ($I > 1\,\text{MA}$), conditions with $T_i/T_e > 3$ and $\eta_i \gg 1$ are achieved and the signatures of the ion temperature gradient driven turbulence are widely reported. In these devices the power balance analysis reveals that the effective ion thermal diffusivity $\chi_i > \chi_e$ and χ_i values consistent with the ∇T_i-driven drift waves. The value of χ_i inferred from power balance is typically consistent with the mixing length estimate $\chi_i = \gamma^{\max}/k_\perp^2 = (10^5/\text{s})/(5\,\text{cm}^{-1})^2 = 5 \times 10^3\,\text{cm}^2/\text{s} = 0.5\text{m}^2/\text{s}$. The ratio of this turbulent χ_i to the collisional (neoclassical) χ_i^{neo} varies widely from machine to machine and over the radius of a given discharge but is substantially greater than unity (Scott *et al.*, 1994). A detailed stability-transport analysis of a key ITG experiment in the TFTR tokamak may be found in the team project Horton *et al.* (1992). A critique of ITG theory is found in Ottaviani *et al.* (1997).

Here we briefly describe the ITG dynamics indicating some similarities and differences with the baroclinic instability. The vorticity equation arises from the condition $\nabla \cdot \boldsymbol{j} = 0$ stating that current loops must be closed in the quasineutral plasma. The large $\boldsymbol{v}_E = c\boldsymbol{E} \times \boldsymbol{B}/B^2 = c\hat{\boldsymbol{z}} \times \nabla\Phi/B$ velocities cancel in the current $\boldsymbol{j}$ so that the ion current from the finite inertia (polarization) drift $\boldsymbol{v}_p$, corresponding to the ageostrophic drift,

$$\boldsymbol{v}_p = \frac{-c^2 m_i}{eB^2}\left(\frac{\partial}{\partial t} + (\boldsymbol{v}_E + \boldsymbol{v}_D)\cdot\boldsymbol{\nabla}\right)\boldsymbol{\nabla}_\perp\Phi, \tag{50}$$

balances the divergence of the parallel current $\nabla_{\|} j_{\|}$. Since the ion fluid is hot $T_i/T_e \gtrsim 1$ the ion gyroradius $\rho_i = c(m_i T_i)^{1/2}/eB \gtrsim \rho_s$ and there are important collisionless stresses from $\nabla\cdot\boldsymbol{\pi}_i(\boldsymbol{v}_E)$ — the divergence of the off-diagonal terms in the momentum stress tensor $m_i n_i \langle v_\alpha v_\beta\rangle = p_i\delta_{\alpha\beta} + \pi_{i,\alpha\beta}$ where $\pi_{i,\alpha\beta}$ is a linear function of $\partial v_E/\partial x$. The effect of the stress tensor shows up as the new term $K\nabla_\perp^2\partial_y\phi$ in the vorticity equation giving the change of the drift wave propagation at short wavelengths to the ion diamagnetic direction. In Fig. 8 we show the result of repeating the initial value experiment given in Fig. 3 for the ITG equations. Now we see that the long wavelength waves propagate in the electron direction (upward) while the short wavelength modes propagate in the ion diamagnetic direction (downward). The right panel shows that the nonlinear binding effect still applies to the ITG mode fluctuations.

The appropriate dimensionless space-time variables are

$$\widetilde{x} = \frac{x - x_0}{\rho_s}, \quad \widetilde{y} = \frac{y}{\rho_s}, \quad \widetilde{z} = \frac{z}{L_n}, \quad \tau = \frac{tc_s}{L_n}. \tag{51}$$

and the amplitudes of the fluctuations scale as

$$\phi = \left(\frac{e\widetilde{\Phi}}{T_e}\right)\left(\frac{L_n}{\rho_s}\right),$$

$$v = \left(\frac{\widetilde{v}_{\|}}{c_s}\right)\left(\frac{L_n}{\rho_s}\right), \tag{52}$$

$$p = \left(\frac{\widetilde{p}_i}{p_{i0}}\right)\left(\frac{L_n}{\rho_s}\right)\left(\frac{T_i}{T_e}\right). \tag{53}$$

The $\boldsymbol{E}\times\boldsymbol{B}$ convective derivative is

$$\{f,g\} = \widehat{\boldsymbol{z}}\cdot\nabla_\perp f\times\boldsymbol{\nabla}_\perp g = \frac{\partial f}{\partial\widetilde{x}}\frac{\partial g}{\partial\widetilde{y}} - \frac{\partial f}{\partial\widetilde{y}}\frac{\partial g}{\partial\widetilde{x}}, \tag{54}$$

and due to the shearing of the helical magnetic field

$$\boldsymbol{B} = B\left\{\widehat{\boldsymbol{z}} + \left[\frac{(x - x_0)}{L_s}\right]\widehat{\boldsymbol{y}}\right\}, \tag{55}$$

the parallel derivative $\boldsymbol{B}\cdot\boldsymbol{\nabla} = B\nabla_{\|}$ is given by

$$\boldsymbol{\nabla}_{\|} = \frac{\partial}{\partial\widetilde{z}} + S\widetilde{x}\frac{\partial}{\partial\widetilde{y}} \tag{56}$$

where $S = L_n/L_s$ measures the strengths of the magnetic shear. (Note that $S = 0$ is a well-defined limit applicable to cylindrical devices.)

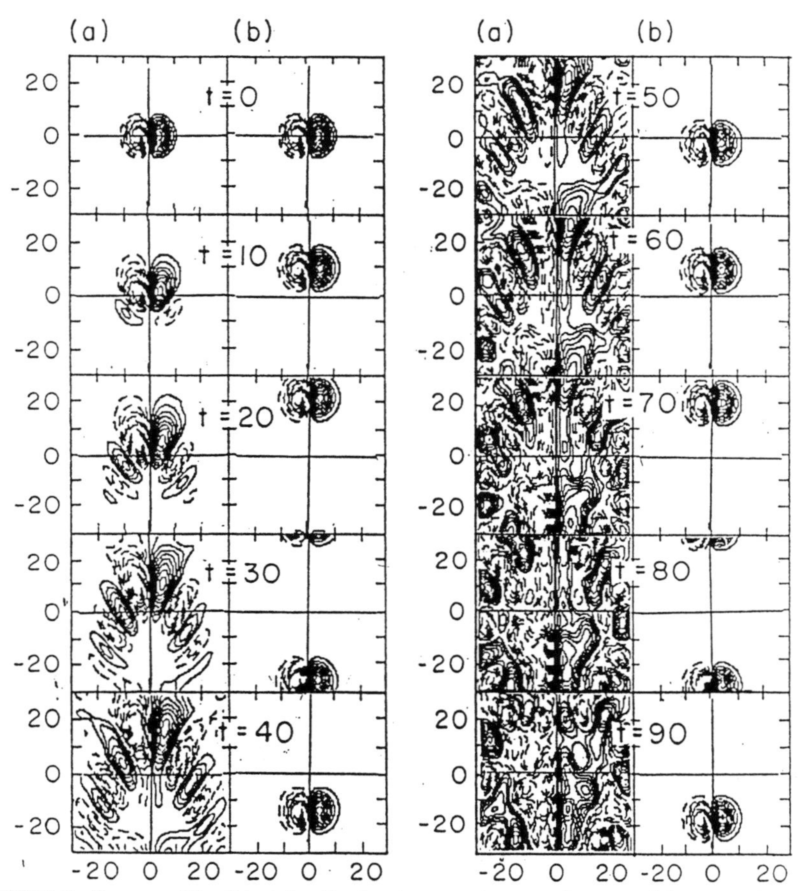

FIGURE 8. Same as Fig. 3 but for the ion temperature gradient vorticity equation. Now the long wavelength components propagate upward (westward) and the short wavelength components propagate downward (eastward). The strong nonlinear self-binding is still a dominant effect.

The ITG convective turbulence is then described by

$$(1 - \nabla_\perp^2)\frac{\partial \phi}{\partial \tau} = -(1 + K\nabla_\perp^2)\frac{\partial \phi}{\partial \widehat{y}} - \boldsymbol{\nabla}_\| v + \{\phi, \nabla_\perp^2 \phi\} - \mu_\perp \nabla^4 \phi, \tag{57}$$

$$\frac{\partial v}{\partial \tau} = -\boldsymbol{\nabla}_\| (\phi + p) - \{\phi, v\} + \mu_\perp \nabla_\perp^2 v + \mu_\| \nabla_\|^2 v, \tag{58}$$

$$\frac{\partial p}{\partial \tau} = -K\frac{\partial \phi}{\partial \tilde{y}} - \Gamma \boldsymbol{\nabla}_{\|} v - \{\phi, p\} + \chi_{\perp}\nabla_{\perp}^2 p + \chi_{\|}\nabla_{\|}^2 p. \tag{59}$$

An example of the turbulent fields created by this system is shown in Fig. 9. In this case $K = 3, \Gamma = 2, S = 0.3$ and $\mu_{\|} = \chi_{\|} = 1$, $\mu_{\perp} = \chi_{\perp} = 0.01$. The variation of the turbulence with the system parameters and the shooting code solutions of the linear eigenvalue problem are thoroughly developed in Hamaguchi and Horton (1990, 1992). The first 3D simulation is found in Horton-Estes-Biskamp (1980).

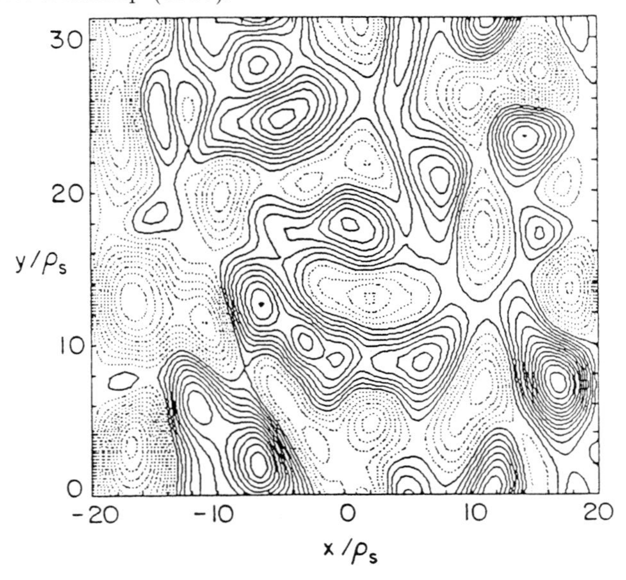

FIGURE 9. The isopotential contours in the saturated state of a 3D simulation of the ITG equations. Size of the convective cells is $\simeq 6\rho_s$ only slightly larger than $\pi/k_y^{\max}$ for maximizing the linear growth rate. The $\boldsymbol{E} \times \boldsymbol{B}$ rotation rate $\Omega_E \simeq 0.9\ [c_s/L_n]$ larger than the maximum growth rate and wave frequency by a factor of 1.5.

Examining the convection in Fig. 9 shows that the saturation level is such that the convection around the vortical structure is completed within the correlation time τ_c of the field. This is a $\tilde{\phi}$-level such that the $\boldsymbol{E} \times \boldsymbol{B}$-rotation number $R_E = \Omega_E/\Delta\omega \gtrsim 1$ where $\Omega_E = \boldsymbol{k} \cdot \boldsymbol{v}_E \sim k_x k_y c\tilde{\phi}/B$. The time $2\pi/\Omega_E$ can also be understood by computing the time to convect around the rectangular cell given by $\lambda_x/|v_x| + \lambda_y/|v_y| = 2\pi/k_x k_y(c\tilde{\phi}/B)$. As shown earlier, the nonlinear self-binding becomes effective for $R_E > R_{\text{crit}} \gtrsim 1$. Thus, this particular saturated state is just into the nonlinear self-binding regime.

The saturation level $R_E = 1$ may also be understood as the level where the mean-square fluctuating pressure gradient just balances the ambient background gradient driven by the auxiliary heating. This level is called the mixing length level and is the standard methods of calculating the turbulence level driven by ambient gradients.

For a given amplitude level of the fluctuations the convective transport depends on the phase relation between δn or δp_i and $\delta\phi$ as discussed in Sec. 1.1. As the rotation number R_E becomes appreciably larger than unity the pressure fluctuation becomes aligned with ϕ: $\delta p_i \approx f(\phi)$ and there is only transport across the separatrices between the trapped regions (Ottaviani, 1997). For $R_E < 1$, however, the phase relation remains not too far from the linear relation between δp_i and ϕ. Thus, quasilinear calculations for the convective flux are typical with the linear $\delta p_i/\delta\phi$-formula used for calculating the phase relations.

In the simulations the effective diffusivity χ_i is defined by

$$\chi_i = \frac{\overline{\langle \tilde{p}_i \tilde{v}_{ir} \rangle}}{-p'_{i0}} = \frac{\rho_s}{L_n}\left(\frac{cT_e}{eB}\right)\overline{\left\langle p\frac{\partial\phi}{\partial y}\right\rangle}K^{-1}. \tag{60}$$

Here the

$$\overline{g(t)} = \lim_{T\to\infty}\frac{1}{T}\int_0^T g(t)dt$$

and

$$\langle f \rangle = \frac{1}{\Delta L_y L_z}\int_{-L_x}^{L_x} d\tilde{x}\int_0^{L_y} d\tilde{y}\int_0^{L_z} d\tilde{z},\ f(\boldsymbol{x} \tag{61}$$

where Δ denotes the width of the region of appreciable turbulence, i.e. the "support" of a function $f(x)$ for homogeneous turbulence $\Delta \to 2L_x$. Figure 10 shows a typical result for the temperature gradient dependence of χ_i in the units of the gyro-Bohm diffusivity $\chi_{gB} = (\rho_s/L_n)(cT_e/eB)$ for the model in Eqs. (57)–(59) and for the parameters given above.

The energy integral E_T for system (57)–(59) is

$$E_T = \frac{1}{2}\int_\Omega d^3x\left[\phi^2 + (\nabla\phi)^2 + v^2 + \frac{1}{\Gamma}p^2\right].$$

Similar to Eq. (47) for the barotropic instability before the hydrostatic equilibrium approximation is used. The power transfer fluxes between the three energy components are $\left\langle v\nabla_{\|}\phi\right\rangle, \left\langle p\nabla_{\|}v\right\rangle$ and $Q = -\langle p\partial_y\phi\rangle$ with rate of change of the total energy given by

$$\frac{dE_T}{dt} = \frac{KQ}{\Gamma} - \sum_{\alpha=1}^{5} P_\alpha$$

where P_α are the positive definite dissipation integrals $\mu_\perp\langle(\nabla^2\phi)^2\rangle$, $\mu_\perp\langle(\nabla_\perp v)^2\rangle, \mu_{\|}\left\langle(\nabla_{\|}v)^2\right\rangle$, $\chi_\perp\langle(\nabla_\perp p)^2\rangle$ and $\chi_{\|}\left\langle(\nabla_{\|}p)^2\right\rangle$. In the turbulent

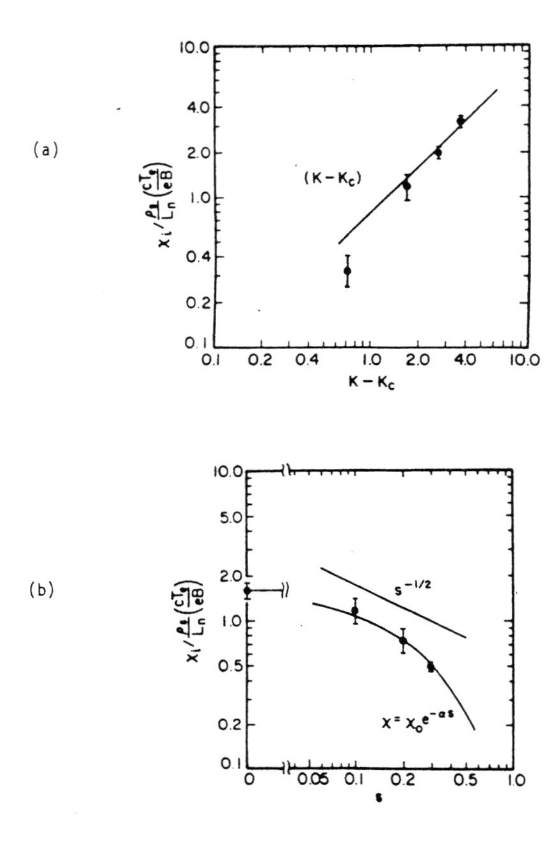

FIGURE 10. The turbulent ion thermal conductivity in units of the gyroBohm conductivity as a function of the deviation from the critical adiabatic gradient.

steady state the thermal flux Q times the temperature gradient K balances the viscous-thermal diffusive dissipation.

Since the time of this work (Hamaguchi and Horton, 1990) the ITG modeling has received much attention due to its almost unique ability to explain the power balance in the large fusion confinement devices. Now much more sophisticated fluid descriptions called gyrofluids using up to 13 pdes are used to describe the turbulence. The turbulence was also chosen as the topic for the fusion Grand Challenge project in super computing. In this project large particle simulation codes in the full 3D torus are used for the simulations of drift wave turbulence. A typical result is shown in Fig. 11 from Parker *et al.* (1996). Waltz *et al.* (1994) gives a comprehensive analysis with the gyro-Landau equations and compares the results with three discharges from TFTR.

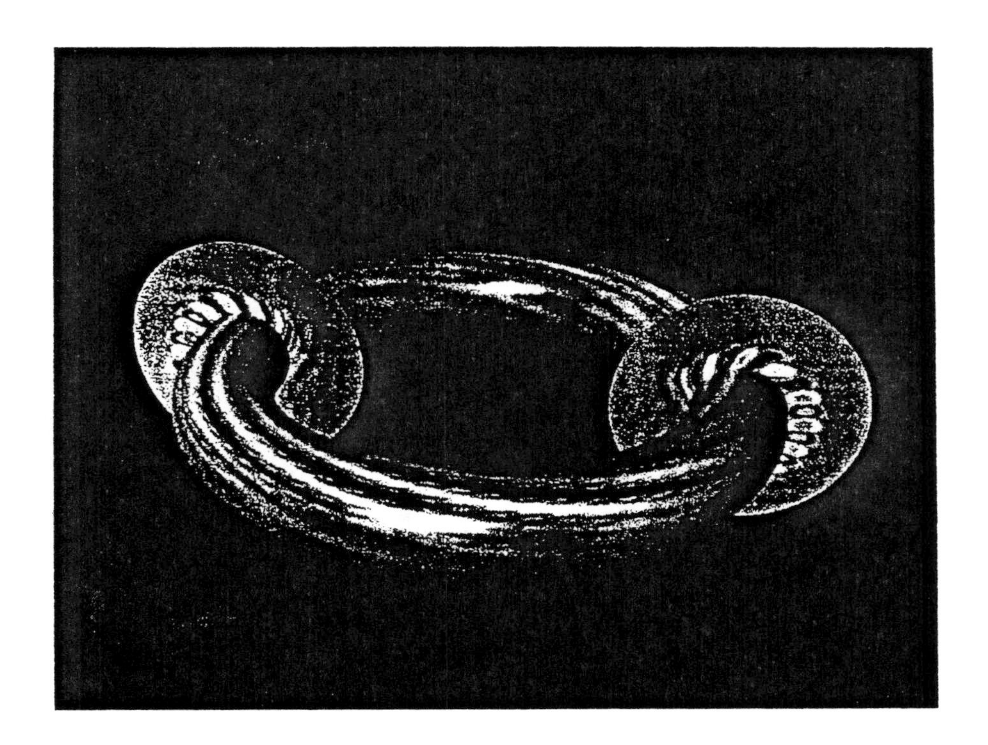

FIGURE 11. The linear phase of a global ITG mode in a tokamak obtained by the Parker group as part of the fusion Grand Challenge Supercomputer project. The method used is an extension to guiding centers of the classical particle-in-cell simulation technique.

THERMAL BALANCE IN THE ATMOSPHERE

Solar energy is the primary source of atmospheric heating. Other thermal sources from the upper mantle and dissipation of ionospheric currents are negligible in comparison. Of the total solar power intercepted by the Earth, about 40% is estimated to be reflected back into space and the remaining power $P_0 \cong 10^{17}\,\mathrm{W}$ is preferentially absorbed in the equatorial zone where angle of incidence is normal. The distribution of the absorbed solar power is about 27% into the atmosphere and 73% into the ocean and land (Gill, 1982, p. 10). The atmosphere and the oceans each transport northward from the equitorial region a power of about $3 \times 10^{15}\,\mathrm{W}$. For comparison the power dissipated by ionospheric currents reaches $10^{12}\,\mathrm{W}$ during magnetic substorms that occur on average every few hours (Horton and Doxas, 1997). The result is a substantial temperature gradient between the equatorial zone and the polar regions which derives atmospheric zonal flows and large three-dimensional convection cells. The atmospheric turbulence is divided into large space scale-long-time scale synoptic motions ($L \sim \rho_R > 1000\,\text{km}, \Delta t > 1/f \sim 1\,\text{day}$) and smaller scale-faster motions for which the gradient of the Ciriolos parameter $\beta = \partial f/\partial y$ is not important. The large scale motions are directly analogous to the low frequency drift instabilities of the magnetized plasma. The role of the baroclinic instability driven by the horizontal temperature gradient $\partial T/\partial y$ is the analog of the ion temperature gradient instability in plasma. Thus, it is important to compare the role of turbulent thermal convection in the two systems.

We now estimate the thermal diffusivity χ implied by the solar input power to the equatorial atmosphere and the meridional temperature gradient showing that the result is consistent with corresponding plasma thermal diffusivity formulas used in magnetic fusion research. In plasma confinement studies there are two forms of the thermal diffusivity scaling (gH/f and $(gH)^{3/2}/f^2 R_p$) in plasma confinement studies called Bohm and gyro-Bohm, respectively, that are currently used. The gyro-Bohm is the formula shown in Fig. 10. In the Bohm scaling the system is near the critical gradient for convection so that there is a small coefficient in the Bohm formula. For the drift wave or gyro-Bohm, thermal diffusivity formula the coefficient is order unity.

In terms of the atmospheric parameters these two thermal diffusivity formulas are $\chi_B = 5 \times 10^{-3}(gH/f)$ and $\chi_{gB} = 0.3(\rho_R/R_p)(gH/f) = 0.3(gH)^{3/2}/R_p f^2$ respectively. Let us take $(gH)^{1/2} = 300\,\text{m/s}$ and $f = 7.3 \times 10^{-5}/\text{s}$ and the planetary radius $R_p = 6 \times 10^6\,\text{m}$. These formulas give $\chi_B = 10^6\,\text{m}^2/\text{s}$ and $\chi_{gB} = 1.5 \times 10^7\,\text{m}^2/\text{s}$. The small coefficient in the Bohm formula is interpreted to express the fact that the large scale convective cells $L = (\rho_R R_p)^{1/2}$ observed in the global simulations force the temperature profile to relax, keeping the system close to marginal stability.

The empirical case for the Bohm transport model is presented in Erba *et al.* (1995) for discharges up to 7 MA of plasma current. The stored plasma thermal energy is approximately 10^6 J with power flux through the plasma of order $20\ \text{MW}/80\,\text{m}^2 \cong 2 \times 10^5\ \text{W/m}^2$. The thermal energy confinement (storage) times is $\tau_E = E/P \cong 0.5\,\text{s}$. The ion thermal diffusivity χ_i is approximately three times the electron χ_e.

In the drift wave (gyro-Bohm) formula the system is in a state of well-developed turbulence (well away from marginal stability) with the horizontal scales $L \simeq 4\rho_R$ and the correlation time $\tau_c = 1/\max(\omega_k) \simeq 5R_p/\sqrt{gH}$ corresponding to the maximum wave frequency at $k_\perp \rho_R = 1$. Correlation times of this order correspond to a few days.

Power balance in the atmosphere

The energy content of the atmosphere is divided into thermal energy $E_p = (5/2)nk_B T$ and mean kinetic energy E_k. The kinetic energy is divided into mean zonal flows $\overline{E}_k$ and turbulent flows $\widetilde{E}_k$. We calculate for the energies $E_p = 10^{24}$ J, $\overline{E}_k = 10^{21}$J and $\widetilde{E}_k = 10^{20}$J. Monin (1972) considers the confinement time for the energy E_k arguing that 2% of the absorbed solar power $P_s = 10^{17}$ W is converted into kinetic energy $E_k = 10^{21}$ J giving $\tau = E_k/2 \times 10^{15}$ W $= 5 \times 10^5$ s$\cong$ 1 week. Monin notes that this decay time is consistent with the Richardson's four-thirds law $\nu_L = \varepsilon^{1/2} L^{4/3}$ for the $\widetilde{E}_k = 5 \times 10^{20}$ J. The turbulent viscosity ν_L is compared with the turbulent thermal diffusivity in the following subsection.

The thermal flux from the equatorial zones to the polar regions is

$$F = -n\chi \frac{dT}{dy} \simeq \frac{n\chi \Delta T}{R_E} = 3 \times 10^{15}\,\mathrm{W} \Big/ \left[1.3 \times 10^{11}\,\mathrm{m}^2\right] = 2 \times 10^4\,\mathrm{W/m}^2 \quad (62)$$

where we require for steady state that the flux F times the cross-sectional area $A = 2\pi R_E \cos\theta_L\, H$ at the latitude $\theta_L = 50°$ equals the fraction of the solar input power 3×10^{15} W that is transported northward through the atmosphere (Gill, 1982, p. 15) to the arctic region. Equation (62) introduces the definition of the effective thermal diffusivity χ due to the turbulent convection. Taking the equator-polar temperature gradient as $\Delta T/R_E = 10°\mathrm{C}/6 \times 10^6$ m, we infer the mean turbulent diffusivity from Eq. (62) is

$$\chi_{\mathrm{PB}} = 3 \times 10^7\,\mathrm{m}^2/\mathrm{s} \quad (63)$$

for steady-state power balance (PB). This power balance thermal conductivity is approximately two times the standard plasma physics analogs for temperature gradient driven thermal diffusivities estimated above for χ_{gB}.

Anomalous viscosity and the turbulent Prandtl number

The thermal diffusivity (63) may be compared to the large scale turbulent viscosity ν_L. Monin (1972) gives that $\nu_L = 10^7\,\mathrm{m}^2/\mathrm{s}$ at the scale $L = 2200$ km. Using this value for the turbulent viscosity and Eq. (63) for the turbulent thermal diffusivity gives the effective Prandtl number of $\nu_L/\chi = 1/3$. To extrapolate ν_L at $L = 2200$ km to $L = R_E = 6340$ km using the Richardson four-thirds law would increase the turbulent viscosity to $\nu_L = 10^7\,\mathrm{m}^2/\mathrm{s}(6340/2200)^{4/3} = 4 \times 10^7\,\mathrm{m}^2/\mathrm{s}$ making the effective Prandtl number close to unity.

The effective viscosity also follows from the Kolmogorov eddy lifetime. Monin, citing Palmén (1961), estimates the Kolmogorov energy injection constant $\varepsilon = 4\,\mathrm{cm}^2/s^3$ using 2% of the solar input power as driving the turbulent convection. This value of ε is close to that of Brunt (1926) $\varepsilon = 5\,\mathrm{cm}^2/s^3$ based on independent arguments.

In conclusion, the turbulent diffusivity of thermal energy from the equator to the poles is of the same order as the turbulent eddy viscosity. The same situation is found in toroidal plasmas where the toroidal flow velocities of order a $100 - 300\,\mathrm{km/s}$ are known to decay with a turbulent viscosity ν_ϕ that is approximately $2/3\,\chi$ where χ is the thermal diffusivity from the temperature gradient driven turbulence (Sugama and Horton, 1997). The reason that the Prandtl number for the large scale fluctuations is order unity for both the plasma and the geophysical fluid is that the mechanism for the transport of both the momentum and the thermal energy is the large scale geostrophic $\boldsymbol{E} \times \boldsymbol{B}$ horizontal turbulent flows.

Acknowledgments

The author wishes to express his thanks to R.L. Dewar and R.W. Griffiths for timely gathering of researchers in the field of geophysical fluid dynamics and plasma turbulence. Useful discussions with many workshop participants including B. Blackwell, G. Holland, M. Nezlin, V. Naulin, R. Kinney, and M. Wakatani are acknowledged.

This work was supported by the U.S. Dept. of Energy Contract No. DE–FG03–96ER–54346.

REFERENCES

1. Brower, D.L., Peebles, W.A., Kim, S.K., Luhmann, Jr., N.C., Tang, W.M., and Phillips, P.E., *Phys. Rev. Lett.* **54**, 689 (1987).
2. Brower, D.L., Redi, M.H., Tang, W.M., Bravenec, R.V. *et al.*, *Nucl. Fusion* **29**, 1247 (1989).
3. Brunt, D., *Physical and Dynamical Meteorology*, 2nd ed., Cambridge Univ. Press, New York, 1939.
4. Chandrasekhar, S., *Hydrodynamic and Hydromagnetic Stability*, Dover Publications, Inc., New York, 1961.
5. Charney, J.G., *Geofys. Publikasjoner, Norske Videnskaps-Akad. Oslo* **17**, 17 (1948).
6. Erba, M., Parail, V., Springman, E., and Taroni, A., *Plasma Phys. Control. Fusion* **37**, 1249 (1995).
7. Gill, A.E., *Atmosphere–Ocean Dynamics*, Academic Press, London, 1982.
8. Griffiths, R.W., and Hopfinger, E.J., *J. Fluid Mech.* **173**, 501 (1986).
9. Griffiths, R.W., and Hopfinger, E.J., *J. Fluid Mech.* **178**, 73 (1987).
10. Hamaguchi, S., and Horton, W., *Phys. Fluids B* **2**, 1833 (1990).
11. Hamaguchi, S., and Horton, W., *Phys. Fluids B* **4**, 319 (1992).
12. Hasegawa, A., and Mima, K., *Phys. Rev. Lett.* **39**, 205 (1977) and *Phys. Fluids* **87** (1978).
13. Hasegawa, A., Maclennan, C.G., and Kodama, Y., *Phys. Fluids* **22**, 2122 (1979).

14. Hazeltine, R.D., and Meiss, J.D., *Plasma Confinement*, Addison-Wesley, Redwood City, California, 1992.
15. Hobson, D.D., *Phys. Fluids A* **3**, 3027–3033 (1991).
16. Holland, G.J., *Meteorol. Atmos. Phys.* **56**, 57–79 (1995).
17. Horton, W., Estes, R., and Biskamp, D., *Plasma Phys.* **22**, 663 (1980).
18. Horton, W., Choi, D.-I., and Tang, W.M., *Phys. Fluids* **24**, 1077 (1981).
19. Horton, W., Lindberg, D., Kim, J-Y., Dong, J.-Q., Hammett, G.W., Scott, S.D., Zarnstorff, M.C., and Hamaguchi, S., *Phys. Fluids B* **4**, 953 (1992).
20. Horton, W., and Petviashvili, V., eds. Horton, W., Ichikawa, Y., Prigogine, I. (Ed. in-Chief), Zaslavsky, G., "On the trapping condition for planetary vortex structures," *Research Trends in Physics: Chaotic Dynamics and Transport in Fluids and Plasmas*, American Institute of Physics, New York, 1993.
21. Horton, W., and Ichikawa, Y.-H., *Chaos and Structures in Nonlinear Plasmas*, World Scientific, 1996, chap. 6, pp. 221–273.
22. Horton, W., and Doxas, I., *J. Geophys. Res.* **102**, (Dec. (1997), and **101**, 27223 (1996).
23. Jovanović, D., and Horton, W., *Phys. Fluids B* **5**, 9 (1993).
24. Kim, J-Y., and Horton, W., *Phys. Fluids B* **3**, 1167 (1991).
25. Kono, M., and Horton, W., *Phys. Fluids B* **3**, 3255–3262 (1991).
26. Laedke, E.W., and Spatschek, K.H., *Phys. Fluids* **29**, 133–142 (1986).
27. Laedke, E.W., and Spatschek, K.H., *Phys. Fluids* **31**, 1492 (1988).
28. Larichev, V.D., and Reznik, G.K., *Fiz. Plasmy* **3**, 270 (1976).
29. Meiss, J.D., and Horton, W., Jr., *Phys. Fluids* **26**, 990 (1983).
30. Monin, A.S., *Weather Forecasting as a Problem in Physics*, MIT Press, 1972.
31. Muzylev, S.V., and Reznik, G.M. *Phys. Fluids B* **4**, 2841 (1992).
32. Nezlin, M.V., and Snezhkin, E.N., in *Rossby Vortices, Spiral Structures, Solitons*, Springer-Verlag, 1993, pp. 196–203.
33. Nycander, J., *Phys. Fluids A* **4**, 467–476 (1992).
34. Ottaviani, M., Beer, M.A., Cowley, S.C., Horton, W., and Krommes, J.A., *Phys. Rep.* **283**, 121–146 (1997).
35. Palmén, E., *Geofis. Pura Appl.* **49**, 167 (1961).
36. Parker, S.E., Mynick, H.E., Artun, M., Cummings, J.C., *et al.*, *Phys. Plasmas* **3**, 1959 (1996).
37. Pedlosky, J., 1987, *Geophysical Fluid Dynamics*, Springer-Verlag, New York, pp. 518–532.
38. Petviashvili, V.I., *Fiz. Plazmy* **3**, 270 [Sov. *J. Plasma Phys.* **3**, 150 (1977).
39. Rossby, C.G., in *The Atmospheres of the Earth and Planets*, Univ. of Chicago Press, Chicago, 1948, p. 25.
40. Sakuma, H., and Ghil, M., *Phys. Fluids A* **3**, 408–414 (1991).
41. Su, X.N., Horton, W., and Morrison, P.J., *Phys. Fluids B* **3**, 921 (1991).
42. Su, X.N., Horton, W., and Morrison, P.J., *Phys. Fluids B* **4**, 1238 (1992).
43. Scott, S.D., Barnes, Cris W., Ernst, D., Schivell, J., Synakowski, E.J., Bell, M.G., Bell, R.E., Bush, C.E., Fredrickson, E.D., Grek, B., Hill, K.W., Janos, A., Jassby, D.L., Johnson, D., Mansfield, D.K., Owens, D.K., Park, H., Ramsey, A.T., Stratton, B.C., Thompson, M., and Zarnstorff, M.C., "Parametric

Variations of Ion Transport in TFTR," in *U.S.-Japan Workshop on Ion Temperature Gradient-Driven Turbulent Transport*, AIP Conference Proceedings, Austin, TX, 1993.
44. Sugama, H., and Horton, W., *Phys. Plasmas* **4**, 405 (1997).
45. Swaters, G.E., *Phys. Fluids* **29**, 1419–1422 (1986).
46. Tasso, H., *Phys. Lett. A* **24**, 618 (1967).
47. Waltz, R.E., Kerbel, G.D., and Milovich, J., *Phys. Plasmas* **1**, 2229 (1994).

Numerical Simulation of Geophysical Turbulence and Eddies

Olivier MÉTAIS

Laboratoire des Écoulements Géophysiques et Industriels de Grenoble, UMR CNRS 5519
Institut de Mécanique de Grenoble
Institut National Polytechnique et Université Joseph Fourier

Abstract. Most atmospheric and oceanic turbulence and eddies originate from the development of instabilities resulting from the combined effects of density gradients and rotation. We present here numerical Direct Numerical Simulations (DNS) and Large-Eddy Simulations (LES) of turbulent flows of atmospheric and oceanic interest. We first investigate the effects of solid-body rotation on free-shear flows, wall-bounded flows, and homogeneous turbulence, and show the drastic modification of the flow topology. Stably-stratified rotating turbulence is then numerically investigated with energy injection at small scales. We observe inverse cascades corresponding to a well defined $k^{-5/3}$ spectral range for the geostrophic part of both the kinetic and available potential energy spectra. The applications to the observed mesoscale atmospheric spectrum are discussed. We finally show how DNS and LES techniques can be applied to the computation of synoptic and frontal scale cyclogenesis.

INTRODUCTION

Most of the flows encountered in the atmosphere or the ocean are composed of interacting waves and turbulence. Geophysical eddies and turbulence often originate from the development of instabilities resulting from the combined effects of density gradients and rotation, and these strongly affect the dynamics over a large range of scales. We summarize here the results of direct numerical simulations (DNS) and large-eddy numerical simulations (LES) aimed at investigating the effects of stable density stratification and/or solid-body rotation on turbulence and coherent vortices, and we particularly focus our study on three-dimensional processes. First, we recall the formalism of DNS and LES. We then consider the dynamics of coherent vortices present in free and wall-bounded shear flows submitted to solid-body rotation. Next, we recall some results concerning rotating homogeneous turbulence with and without stable density stratification. Finally, we apply LES techniques to the computa-

CP414, *Two-Dimensional Turbulence in Plasmas and Fluids:* Research Workshop
edited by R. L. Dewar and R. W. Griffiths

tion of atmospheric and oceanic mesoscale eddies resulting from the baroclinic instability.

DIRECT NUMERICAL SIMULATIONS (DNS) AND LARGE-EDDY SIMULATIONS (LES)

Direct-numerical simulations of turbulence (DNS) consist in solving explicitly all the scales of motion, from the largest scales l_I to the Kolmogorov dissipative scale l_D. There are physical limitations attached to DNS: indeed, it is a classical result of the statistical theory of three-dimensional turbulence that l_I/l_D varies like $R_l^{3/4}$, where R_l is the large-scale Reynolds number $u'l_I/\nu$ based upon the rms velocity fluctuation u'. Therefore, the total number of degrees of freedom necessary to represent a turbulent flow through this whole span of scales is of the order of $R_l^{9/4}$ in three dimensions. Because of the limitations imposed by the presently available computers, the DNS are restricted to Reynolds numbers $R_l \approx 3000$. These are several orders of magnitude smaller than those encountered in the atmosphere, the ocean or most of the industrial facilities. For a weakly viscous fluid, it is, therefore, not possible in the near future to simulate explicitly all the scales of motion from the smallest to the largest. People are usually more interested in the large scales of the flow: those control turbulent diffusion of momentum or heat. This is no longer a direct-numerical simulation of the turbulence, but a large-eddy simulation (LES). These LES need the representation of the energy exchanges with the small scales which are not explicitly simulated, called the subgrid-scales. Thus, subgrid-scale models have to be developed. The first of these models was proposed by Smagorinsky [39] for numerical studies related to atmospheric flows. Since then, modern methods of LES have been proposed, which are presented in the review article by Lesieur and Métais [24] (see also [33]).

In the large-eddy simulation (LES) approach, one gets rid of the scales of wavelength smaller than the grid mesh Δx by applying an appropriately chosen low-pass filter: we designate as $\overline{a}(\vec{x})$ the large-scale (grid-scale) component of any quantity $a(\vec{x})$, and $a' = a - \overline{a}$ the subgrid-scale part. We will here consider the incompressible Navier-Stokes equations within the Boussinesq approximation, in a frame rotating with constant angular velocity Ω about the x_3 axis. The large-scale velocity field $\overline{u}(\vec{x},t) = (\overline{u}_1, \overline{u}_2, \overline{u}_3)$ then satisfies the filtered equations:

$$\frac{\partial \bar{u}_i}{\partial t} + \frac{\partial}{\partial x_j}(\bar{u}_i \bar{u}_j) = -\frac{1}{\rho_0}\frac{\partial \bar{p}}{\partial x_i} + 2\epsilon_{ij3}\Omega \bar{u}_j - \beta g_i \delta_{i3}(\bar{T} - T_0) \tag{1}$$

$$+ \frac{\partial}{\partial x_j}\left\{\nu \left(\frac{\partial \bar{u}_i}{\partial x_j} + \frac{\partial \bar{u}_j}{\partial x_i}\right) + T_{ij}\right\} \tag{2}$$

$$\frac{\partial \bar{T}}{\partial t} + \frac{\partial}{\partial x_j}(\bar{T}\bar{u}_j) = \frac{\partial}{\partial x_j}\left\{\kappa \frac{\partial \bar{T}}{\partial x_j} + R_{ij}\right\} \tag{3}$$

T is the temperature, and $\bar{p}$ a filtered modified pressure; β is the fluid expansivity. T_{ij} and R_{ij} are the subgrid-scale transfers which have to be modelled. An eddy-viscosity and eddy-diffusivity assumption is currently performed, namely:

$$T_{ij} = \nu_t \left(\frac{\partial \bar{u}_i}{\partial x_j} + \frac{\partial \bar{u}_j}{\partial x_i} \right) + \frac{1}{3} T_{ll} \, \delta_{ij} \quad , \tag{4}$$

$$R_{ij} = \kappa_t \, \frac{\partial \bar{T}}{\partial x_j} \quad . \tag{5}$$

Spectral eddy-viscosity models

We here present the various subgrid-scale models used in the next sections. We first work in Fourier space, considering three-dimensional isotropic turbulence and we call $k_C = \pi / \Delta x$ the cutoff wavenumber separating the grid-scales from the subgrid-scales. The concept of k-dependent eddy-viscosity was first introduced by Kraichnan [15] for three-dimensional isotropic turbulence, in the following way: if $T_{>k_C}(k,t)$ is the kinetic energy transfer across the cutoff k_C, corresponding to triadic interactions such that $k < k_C, p$ and (or) $q > k_C$ (see [21], for details), one poses

$$\nu_t(k|k_c) = - \frac{T_{>k_C}(k,t)}{2k^2 E(k,t)} \quad , \tag{6}$$

in such a way that the kinetic energy spectrum $E(k,t)$ in the resolved scales $(k \leq k_C)$ satisfies:

$$[\frac{\partial}{\partial t} + 2(\nu + \nu_t(k|k_c))k^2] \; E(k,t) = T_{<k_C}(k,t) \quad , \tag{7}$$

where $T_{<k_C}(k,t)$ is the kinetic energy transfer corresponding to resolved triads such that $k, p, q \leq k_C$.

Using the Eddy-Damped Quasi-Normal Markovian (E.D.Q.N.M.) approximation (see [21]), and assuming that k_C lies within a $k^{-5/3}$ Kolmogorov cascade, it can be shown that the eddy-viscosity is given by

$$\nu_t(k|k_c) = 0.441 \; C_K{}^{-3/2} \left[\frac{E(k_C,t)}{k_C} \right]^{1/2} \nu_t^*(\frac{k}{k_C}) \tag{8}$$

where $E(k_C,t)$ is the kinetic-energy spectrum at the cutoff k_C, and at time t. $\nu_t^*(k/k_C)$ is a non-dimensional eddy-viscosity, constant and equal to 1 for $k/k_C <\approx 0.3$, and rising for higher k up to $k/k_C = 1$ (cusp behaviour, [15]).

C_K is the Kolmogorov constant. This cusp is due to the predominance of "local" transfers across k_C. The constants and the form of ν_t^* can be determined using E.D.Q.N.M. theory [5]. It can approximately be expresssed as:

$$\nu_t^* \left(\frac{k}{k_C} \right) = 1. + 34.5 e^{-3.03 (k_C/k)} \quad . \tag{9}$$

Métais and Lesieur [34] proposed an improvement of this plateau-cusp model, in order to take into account spectra decreasing differently than $-5/3$ at the cutoff. They shown that the scaling with $\sqrt{E(k_C,t)/k_C}$ of the eddy-viscosity still holds when k_C lies within a k^{-m} spectral range as long as $m \leq 3$. Métais & Lesieur [34] proposed, on the basis of non-local expansions within the framework of the E.D.Q.N.M. theory, that the eddy-viscosity should be expressed as

$$\nu_t(k|k_c) = \nu_t^{+\infty} \left[\frac{E(k_C,t)}{k_C} \right]^{1/2} \nu_t^* \left(\frac{k}{k_C} \right) \tag{10}$$

with

$$\nu_t^{+\infty} = 0.31 \, \frac{5-m}{m+1} \sqrt{3-m} \, C_K{}^{-3/2} \tag{11}$$

Note that the value 0.441 is recovered for $m = 5/3$. This model will be applied to the LES of the rotating channel flow.

Structure-function model

This model, due to Métais & Lesieur [34], consists in using the spectral eddy-viscosity in physical space, while taking into account the intermittency of turbulence. If the cusp behaviour near k_C is discarded, the eddy-viscosity given by eq. (9) can be determined locally in physical space, through the local second-order velocity structure function. It is then obtained:

$$\nu_t(\vec{x}, \Delta, t) = 0.105 \, C_K^{-3/2} \Delta x \, [\bar{F}_2(\vec{x}, \Delta x, t)]^{1/2} \quad , \tag{12}$$

where $\bar{F}_2$ is the structure function of the filtered signal

$$\bar{F}_2(\vec{x}, \Delta x) = \left\langle ||\bar{\vec{u}}(\vec{x},t) - \bar{\vec{u}}(\vec{x}+\vec{r},t)||^2 \right\rangle_{||\vec{r}|| = \Delta x} \quad . \tag{13}$$

One of the drawback of this model is the absence of a cusp near k_C. However, E.D.Q.N.M. data show that the exponential form given in eq. (9) can be correctly approximated by a power law of the type:

$$\nu_t^* \left(\frac{k}{k_C}\right) = \left(1. + \nu_{tn}^* \left(\frac{k}{k_C}\right)^{2n}\right) \quad . \tag{14}$$

with $2n \approx 4$ (3.7). Lesieur and Métais [24] have shown that ν_{tn}^* can be determined on the basis of energy conservation arguments, which yields:

$$\nu_{tn}^* = 0.512 \left(\frac{3n}{2} + 1\right) \quad . \tag{15}$$

Lesieur and Métais [24] showed that from (14), a physical-space dissipative operator based upon the structure-function model and taking into account the "cusp" behaviour can be derived. It has the form:

$$2\frac{\partial}{\partial x_j}\left[\nu_t^{SF} \bar{S}_{ij}\right] + \nu_t^{(1)} \left(\frac{\partial^2}{\partial x_j^2}\right)^3 \bar{u}_i \quad , \tag{16}$$

where $\bar{S}_{ij}$ is the deformation tensor of the field $\bar{u}_i$. $\left(\partial^2/\partial x_j^2\right)^3$ designates the Laplacian operator iterated three-times. ν_t^{SF} is given by the r.h.s. of eq. (12) multiplied by $0.441/(2/3)$, and $\nu_t^{(1)} = \nu_t^{SF} \times \nu_{t2}^*$ with ν_{t2}^* given by eq. (15) (with $n = 2$). The expression (16) is interesting in the sense that it provides an eddy dissipation combining the structure function model with a hyperviscosity $(\nabla^2)^3\overline{u}_i$. The latter represents in physical space the action of the cusp in Kraichnan's spectral eddy viscosity. The importance of the presence of a cusp will be demonstrated in the section devoted to the large eddy simulation of baroclinic eddies.

Hyper-viscosity

The model given by eq. (16) bears some resemblance to hyper-viscosity models which are widely used in the study of geophysical flows because of their simplicity. Indeed, the hyperviscosity consists in replacing the molecular dissipative operator $\nu\nabla^2$ by $(-1)^{\alpha-1}\nu_\alpha(\nabla^2)^\alpha$, where α is a positive integer. As opposed to eq. (16), ν_α is here a constant (positive) coefficient which has to be adjusted. This has been widely used in two-dimensional isotropic turbulence [3], with $\alpha = 2$ or $\alpha = 8$, as a way to shift the dissipation to the neighbourhood of k_C. This allows for a reduction of the number of scales strongly affected by viscous effects, and has allowed, in the case of two-dimensional turbulence, to demonstrate the existence of coherent vortices. In three-dimensional turbulence, it was used by Bartello *et al.* [2] to study the influence of a solid-body rotation, with surprisingly good results.

ROTATING FREE-SHEAR FLOWS

We first consider free-shear flows of basic velocity, $\vec{u} = (\bar{u}(y), 0, 0)$ (x, y and z are respectively the longitudinal, shear and spanwise directions). One works in a frame rotating with a rotation vector $\vec{\Omega} = (0, 0, \Omega)$ oriented along the span (positive or negative). The vorticity vector associated with the basic velocity profile $\vec{\omega} = (0, 0, -d\bar{u}/dy)$ can be parallel or anti-parallel to $\vec{\Omega}$. We will refer to the first case as the cyclonic case, while the second will be called anticyclonic.

Both the rotating mixing layer and the rotating plane wake have been considered and we have studied, through DNS and LES the modification of the three-dimensional flow topology due to rotation. The reader is referred to [25,36,30,27,28] for complete details. A solid-body rotation does not influence a two-dimensional flow in a plane perpendicular to the rotation axis. Therefore, the phenomena observed in the laboratory experiments can only be explained by considering the influence of rotation on the growth of *three-dimensional* perturbations.

Lets call $R_o^{(i)}$, the Rossby number based upon the maximum ambient vorticity of the mean profile (hyperbolic-tangent for the mixing layer; gaussian for the wake). In the mixing layer, $R_o^{(i)}$ is positive for cyclonic rotation and negative for anticyclonic rotation. For the wake, one side is cyclonic, while the other is anticyclonic. In that case, one considers the modulus $|R_o^{(i)}|$ of the Rossby number.

Linear Stability Analysis

In order to describe the early stage of the flow development, a three-dimensional linear-stability analysis of planar free-shear flows (on the basis of the generalized Orr-Sommerfeld equations) has been carried out in [42]. For cyclonic rotation and for strong anticyclonic rotation, it was found that the flow was two-dimensionalized and the instability diagram in the k_x, k_z plane (k_x, k_y and k_z are the components of the wave-vector $\vec{k}$) is concentrated around the Kelvin-Helmholtz mode, which is not affected by the rotation. For moderate anticyclonic rotation $R_o^{(i)} < -1$, in addition of the Kelvin-Helmholtz instability, a new instability was discovered consisting of the strong amplification of a purely longitudinal mode (along the k_z-axis; $k_x = 0$), corresponding to a purely streamwise instability: the shear/Coriolis instability. For both the mixing layer and the wake, this shear/Coriolis instability has a larger amplification rate than the co-existing Kelvin-Helmholtz instability for roughly the range $-8 < R_o^{(i)} < -1.5$, and its effect is maximum for $R_o^{(i)} \simeq -2.5$ (*critical Rossby number*).

Non-Linear Regime

The analysis performed in [42] has allowed to describe the early linear stage of the perturbations growth. Further insight in the nonlinear regime, has been obtained, firstly by examining the vorticity stretching mechanisms, and secondly by performing three-dimensional simulations of the full Navier-Stokes equation. Lesieur *et al.* [25] (see also [21]) have emphasized the importance of considering the absolute vorticity, and not just the relative vorticity, since Kelvin's circulation theorem directly applies to the absolute vorticity. Indeed, if the relative vorticity is written as the sum of the ambient $-(d\bar{u}/dy)\vec{z}$ and fluctuating $\vec{\omega}'$ components. The absolute vorticity then writes $(2\Omega - d\bar{u}/dy)\,\vec{z} + \vec{\omega}'$. If the flow is locally cyclonic (i.e., Ω and $-(d\bar{u}/dy)$ have the same sign), then the absolute vortex lines are closer to the spanwise direction than the corresponding relative ones. Therefore, as compared to the non-rotating case, the effectiveness of vortex turning and stretching is reduced. Conversely, if the flow is locally anticyclonic, especially for the regions where 2Ω has a value close to $d\bar{u}/dy$ (weak absolute spanwise vorticity), absolute vortex lines are very convoluted, and will be very rapidly stretched out all over the flow, as a dye would do. It was thus predicted that in rotating shear flows, the vortex filaments of Rossby number -1 (hence anticyclonic) would be stretched into longitudinal alternate vortices (weak-absolute vorticity stretching principle).

Coherent Vortices Topology

To illustrate the drastic effect of rotation on the flow topology, we here present the three-dimensional coherent vortices obtained in a LES of a plane wake. More detailed computations of both the wake and the mixing-layer can be found in Métais *et al.* [30]. The subgrid-scale model is the *structure function model* previously described. Temporal shear flows are here considered with periodicity in the streamwise direction. Initially, a low-amplitude three-dimensional random noise is superposed upon the ambient velocity profile.

The LES confirm the global trends observed in the experiments and predicted by the linear-stability analysis: the Karman vortices are two-dimensionalized by the rotation when these are cyclonic. We have checked that this is also true for the anti-cyclonic vortices when rapid rotation is applied. Conversely, the primary anti-cyclonic vortices are disrupted when a moderate rotation is applied. Figure 1 shows, for $|R_o^{(i)}| = 2.5$ (corresponding to a maximum anticyclonic destabilization), an isosurface of low "ageostrophic" pressure for which the contribution of the Coriolis term has been discarded. Here, the initial perturbation is quasi-twodimensional and favours the appearance of the Karman rollers (forced transition; see [30], for details). The wake exhibits a strong topological asymmetry. On the anticyclonic side of the

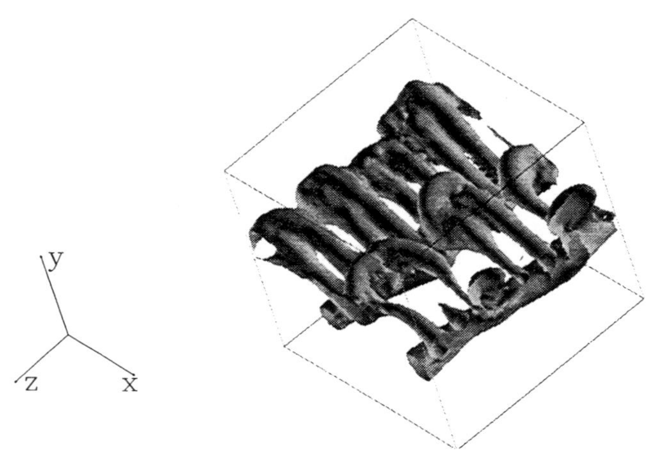

FIGURE 1. Plane wake at $|R_o^{(i)}| = 2.5$. Isosurfaces of low "ageostrophic" pressure.

wake, the Karman vortices are superposed with hairpin-shaped longitudinal vortices. The spanwise wavelength of the latter is approximately one-fourth of the computational domain width, in good agreement with the prediction of the linear stability analysis [42].

Vorticity Stretching Mechanisms

A close examination of the time evolution of the *absolute* vortex lines has been performed [30] in the destabilized anticyclonic region. It was shown that the flow undergoes very distinct stages. In the first stage, the vorticity dynamics are dominated by quasi-linear mechanisms yielding absolute vortex lines inclined at 45° with respect to the horizontal plane. These are in phase in the longitudinal direction. It has been furthermore checked that maximum longitudinal vorticity stretching is achieved in the flow regions with a local Rossby number of approximately -2.8. In a second stage, nonlinear stretching mechanisms yield quasi-horizontal longitudinal hairpins of absolute vorticity. These absolute vortex tubes correspond to a local Rossby number ≈ -1 and result from a strong longitudinal stretching of *absolute* vorticity. One of the striking features of the flow is the appearance, both for the mixing layer and for the plane wake, of a well-defined range of nearly constant mean shear whose vorticity exactly compensates the solid-body rotation vorticity. This is clearly demonstrated on Figure 2, which shows the profile of the local Rossby number $R_o(y,t) = -\frac{d\bar{u}(y,t)/dy}{2\Omega}$ obtained in a plane wake DNS at Reynolds number of 200 and for a rotation rate corresponding to $|R_o^{(i)}| = 2.5$.

In this $Ro(y) \approx -1$ range, the flow exhibits a very peculiar organization as we now illustrate through the DNS of an anticyclonic mixing layer at $R_o^{(i)} =$

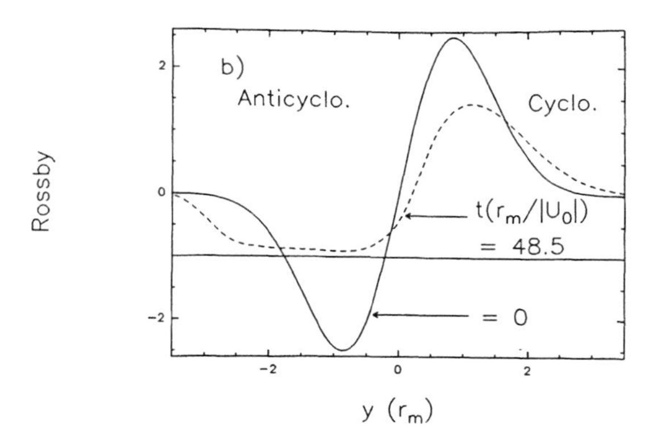

FIGURE 2. Plane wake $|R_o^{(i)}| = 2.5$. Rossby number profile $R_o(y,t)$ at $t = 0$ and at late time.

-5 (see Métais *et al.* [30], for details on the computation). Similarly to the previous wake computation, figure 3 clearly shows that, by the end of the computation, the mean velocity profile exhibits a long range of nearly constant shear of slope 2Ω.

In this range, the flow anisotropy is pronounced: the r.m.s. velocity component $\sqrt{\langle v'^2 \rangle}$ orientated along the shear direction y dominates the other two-components. The longitudinal component $\sqrt{\langle u'^2 \rangle}$ is very weak. This indicates that, in the $Ro(y) \approx -1$ range, the flow is mainly composed of alternatively upward and downward motions orientated in the y direction. This flow configuration yields very strong $\partial v'/\partial z$ gradients at the interface of the ascending and descending streams, implying a dominance of the longitudinal fluctuating vorticity component (see figure 4).

Thus the numerical simulations have clearly shown that a very efficient mechanism to create intense longitudinal vortices in rotating anticyclonic shear layers is provided, thanks to a linear longitudinal instability followed by a vigorous stretching of absolute vorticity taking place in regions of weak spanwise absolute vorticity in agreement with the phenomenological theory proposed in [25]. These numerical simulations results are presently compared with laboratory measurements of velocity field, vorticity field, and local Rossby numbers using *Particle Image Velocimetry* techniques. Indeed, laboratory experiments of high-Reynolds number turbulent wakes (Reynolds number $\approx 10^4$) are presently performed the Coriolis rotating platform in Grenoble. The first results have clearly confirmed some of the trends observed in our simulations: on the anticyclonic side, the velocity profile exhibits a linear range corresponding to a local Rossby number of -1, and the fluid motion is found highly

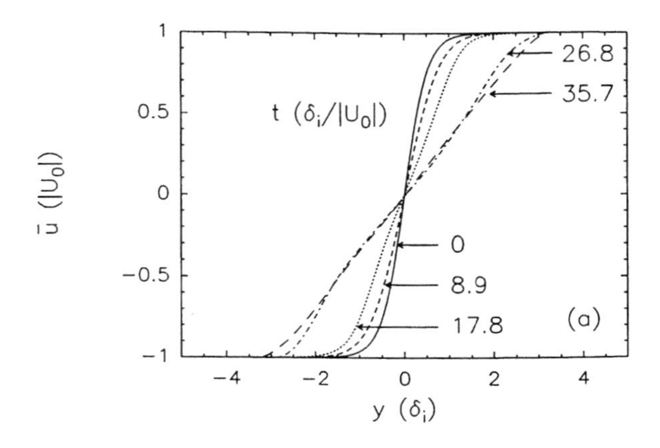

FIGURE 3. Anticyclonic mixing layer at $R_o^{(i)} = -5$, $Re = 100$, natural transition. Time evolution of the mean velocity profile $\bar{u}(y,t)$, normalized by $|U_0|$. y is non-dimensionalized by δ_i and the time unit is $\delta_i/|U_0|$.

three-dimensional in this region (see [41] for details).

TURBULENT CHANNEL FLOW

The previously described spectral eddy-viscosity varying with the exponent m of the spectrum near the cut-off wavenumber k_C is designated as the *spectral dynamic* model. Lamballais *et al.* [17,20] (see also [23]) have used this model for the computation of the non-rotating incompressible channel flow. The numerical code was based upon compact finite-difference scheme [22] in the transverse direction, while pseudo-spectral methods were used in the periodic longitudinal and spanwise direction. It was then possible to evaluate $E(k_C)$ by averaging on the planes parallel to the wall, which permits the determination of m ($E(k_C) \propto k_C^{-m}$). Note that for $m > 3$, the eddy-viscosity was taken equal to zero. Very good comparisons were obtained with the LES of Piomelli and Liu [37] using the dynamic model of Germano [12].

Lamballais *et al.* [17,20] have also used the spectral-dynamic eddy viscosity model (without any modification) to perform LES of the turbulent rotating channel flow. We here briefly recall the main results. The channel width is $2h$, and the macroscopic Reynolds number $R_e = 2hU_m/\nu$ is 14,000, which is supercritical with respect to the linear-stability analysis of the Poiseuille profile. The flow configuration is presented on Figure 5.

In previous works, the rotating channel flow problem has been studied by means of direct numerical simulation (DNS) at low Reynolds number

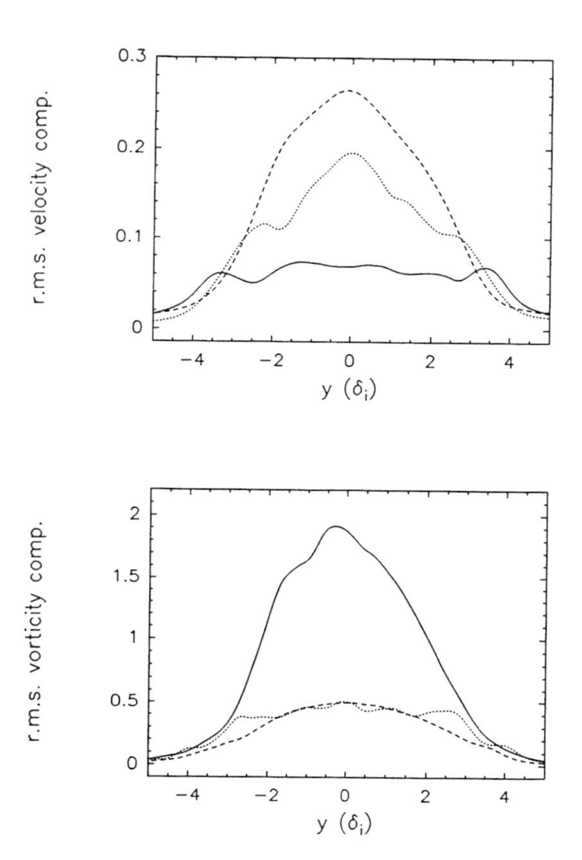

FIGURE 4. Anticyclonic mixing layer at $R_o^{(i)} = -5$, $Re = 100$, natural transition, $t = 32\delta_i/U_0$. a) profiles of the r.m.s. fluctuating velocity components $\sqrt{\langle u'^2 \rangle}$ (continuous line), $\sqrt{\langle v'^2 \rangle}$ (dashed line), $\sqrt{\langle w'^2 \rangle}$ (dotted line); b) profiles of the r.m.s. fluctuating vorticity components $\sqrt{\langle {\omega'_x}^2 \rangle}$ (continuous line), $\sqrt{\langle {\omega'_y}^2 \rangle}$ (dashed line), $\sqrt{\langle {\omega'_z}^2 \rangle}$ (dotted line)

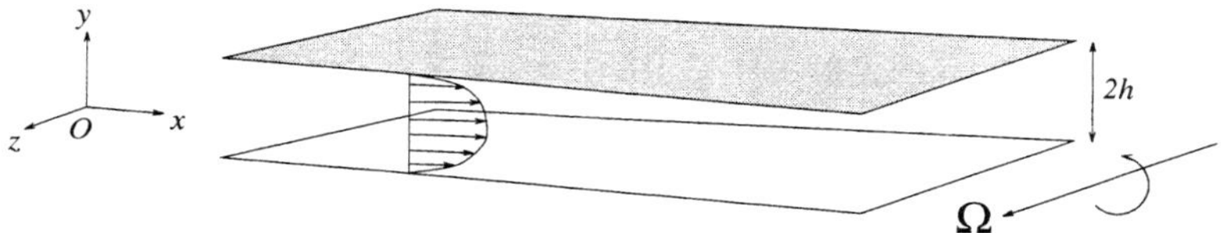

FIGURE 5. Schematic view of the rotating channel. h : half-channel height ; U_m : bulk velocity ; $Re = U_m 2h/\nu$: Reynolds number; $Ro_g = 3\ U_m/\Omega 2h$: global Rossby number. x, y, z are respectively the streamwise, transverse and spanwise directions.

[16,18,19]. It has been shown that rotation can considerably modify the vortex topology of the channel flow. Previous authors [7,40,37] have already successfully used LES to predict the principal modifications brought by a weak rotation upon turbulent statistics. Of the three studies, reference [37], where a variant of the dynamic procedure [12] applied to Smagorinsky's model was used, gave the best agreement with experimental data [14]. Here, we summarize results completing these previous works by a study which focuses upon a detailed study of vortex organization, and upon higher rotation rates and Reynolds numbers regimes. The global Rossby number $Ro_g = 3\ U_m/\Omega 2h$ is taken equal to 6. Detailed results are presented in [20].

Statistical Results

Figure 6 shows profiles of the r.m.s velocity fluctuations for the two simulations considered here (non-rotating and rotating cases). A tendency similar to what was observed previously in DNS [18,19,16,30] is recovered : when rotation is applied, the cyclonic region (corresponding to $y > 0$) of the flow is stabilized while turbulence remains very active in the anticyclonic region ($y < 0$). Similarly to the previous section on free-shear flows, the term cyclonic (respectively anticyclonic) designates a situation where the mean vorticity vector is paralell (respectively antiparallel) to the rotation vector $\vec{\Omega} = (0, 0, \Omega)$.

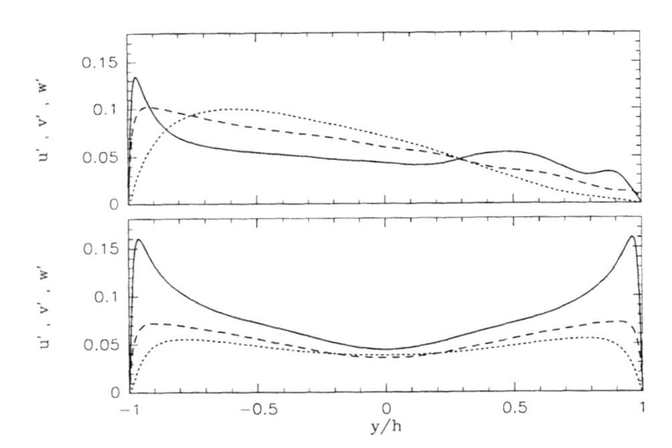

FIGURE 6. R.m.s of the velocity fluctuations in global units ($Re = 14000$) : —, $\sqrt{\langle \overline{u}'^2 \rangle}$; ..., $\sqrt{\langle \overline{v}'^2 \rangle}$; - - -, $\sqrt{\langle \overline{w}'^2 \rangle}$; (a) $Ro_g = 6$; (b) $Ro_g = \infty$.

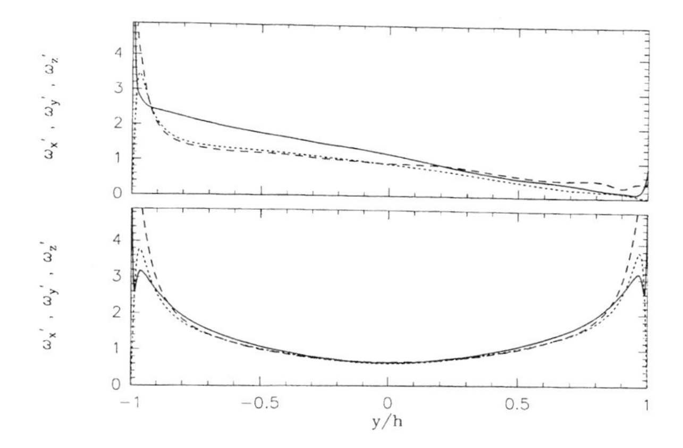

FIGURE 7. R.m.s of the vorticity fluctuations in global units ($Re = 14000$) : : —, $\sqrt{\langle \overline{\omega}_x'^2 \rangle}$; ..., $\sqrt{\langle \overline{\omega}_y'^2 \rangle}$; - - -, $\sqrt{\langle \overline{\omega}_z'^2 \rangle}$; (a) $Ro_g = 6$; (b) $Ro_g = \infty$.

One sees that, in the anticyclonic region (left part of figure 6), rotation reduces the longitudinal velocity fluctuations and increases the transverse ones, which indicates a tendency for the low-and high-speed streaks of the non-rotating case to transform into transverse and spanwise oscillations.

Similarly to the mixing layer and the plane wake, a striking characteristic of the rotating turbulent channel flow is the existence of a linear zone in the mean velocity profile, where the (constant) mean velocity gradient is equal to 2Ω (see [16,18]). As low Reynolds number DNS [18,19] previously shown, such a range is dynamically very important since it is associated with longitudinal vorticity production by absolute-vortex stretching mechanisms. The effect of this stretching can be observed in figure 7, where the x-component of the r.m.s vorticity fluctuations clearly dominates the other two components in the region where the mean velocity profile is linear. The ability for LES technique to account for such phenomena already observed via DNS is an important result.

Coherent-Structure Dynamics

In figure 8, we compare isosurfaces of the vorticity modulus $\omega = 4.5\ U_m/h$ for the non-rotating and rotating channel flow. Note that the coherence of motion is preserved in both LES, and well organized vortex structures can clearly be observed. The vortex topology modification in the anticyclonic region is similar to lower Reynolds number DNS [18,19]: as compared with the non-rotating case, the vortical structures are more inclined with respect

$Ro_g = \infty$ (Non-rotating case)

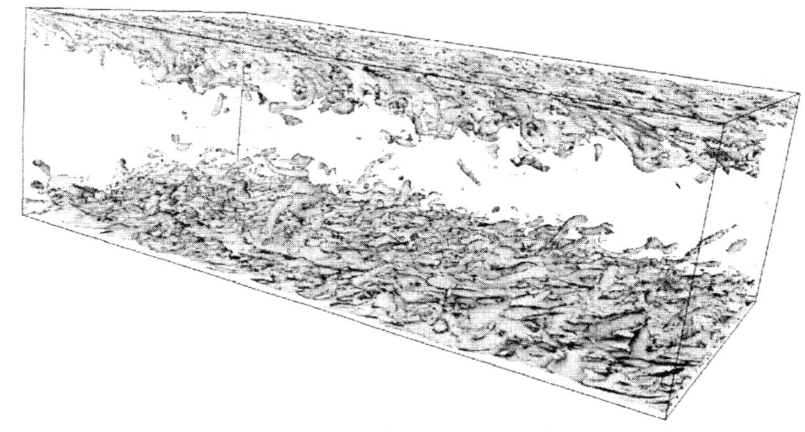

$Ro_g = 6$ (Rotating case)

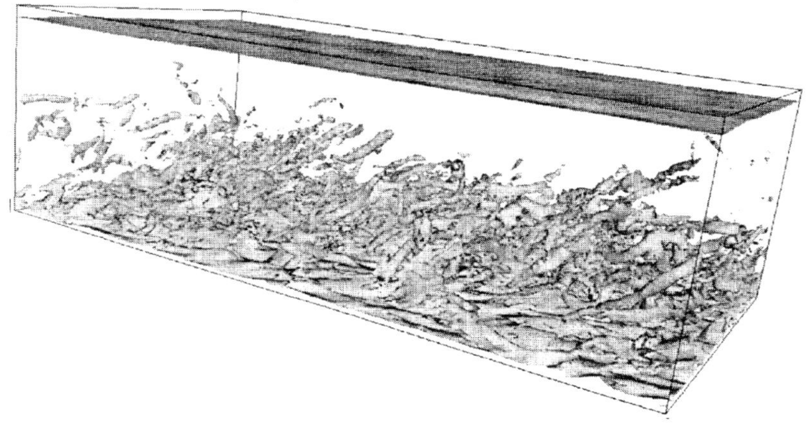

FIGURE 8. Isosurfaces of vorticity modulus $\omega = 4.5\ U_m/h$ ($Re = 14000$). Size of the computational domain : $(L_x, L_y, L_z) = (2\pi h, 2h, \pi h)$. Grid : $128 \times 97 \times 64$.

to the wall and more organized. Let θ be the angle between the wall and the projection of the vorticity vector on the (x, y)-plane. We have checked that the most probable values for θ obtained in DNS (at low Reynolds number) and in LES are identical (45° without rotation and nearly 25° for $Ro_g = 6$). This confirms the fact than the LES correctly reproduce the detailed vortex topology.

HOMOGENEOUS ROTATING TURBULENCE

Bartello *et al.* [2] have performed numerical simulations investigating the formation and stability of quasi-twodimensional coherent vortices in rotating homogeneous three-dimensional flow. Isolated coherent two-dimensional vortices obtained from a purely two-dimensional decay simulation, were superposed with a low-amplitude three-dimensional perurbation, and used to initialize a first set of simulations. In the non-rotating case, a three-dimensionalization of all vortices was observed. Conversely, when $2\Omega \approx [\vec{\omega} \cdot \vec{\Omega}]_{\mathrm{rms}}$, a rapid destablization of the anticyclones was observed to occur, whereas the initial two-dimensional cyclonic vortices persisted throughout the simulation. At larger Ω, both cyclones and anticyclones remained two-dimensional, consistent with the Taylor-Proudman theorem. A second set of simulations starting from isotropic three-dimensional fields was initialized by allowing a random field to evolve with $\Omega = 0$ to a fully-developed state. When the simulation were continued with $2\Omega \approx [\vec{\omega} \cdot \vec{\Omega}]_{\mathrm{rms}}$, the three-dimensional flow was observed to organize into two-dimensional cyclonic vortices. At large Ω, two-dimensional anticyclones also emerged from the initially-isotropic flow.

STABLY-STRATIFIED ROTATING TURBULENCE

Rotation and stable density stratification conjointly modify the turbulence dynamics in many geophysical situations and on a large range of scales. Riley *et al.* [38] and Lilly [26] have suggested that, in the limit of small Froude numbers, stably-stratified turbulence could obey a two-dimensional turbulence dynamics. However, the numerical studies by Herring and Métais [13] and Métais and Herring [31] have shown that the horizontal motion dominates in a strongly stably-stratified environment, but the flow develops a strong vertical variability and reorganizes itself into decoupled horizontal layers. The shear of the horizontal velocity at the interface between the layers leads to energy dissipation, and prevents the turbulence from exhibiting the characteristics of two-dimensional turbulence. Therefore, stable-stratification and rotation have antagonistic effects on turbulent flows: horizontal layering of the flow in one case and emergence of vertical quasi-two-dimensional rolls in the other one.

Métais *et al.* [30,35] and Bartello [1] have numerically investigated the effects of solid-body rotation on stably-stratified turbulence: both with energy injection at small scales and in a freely-decaying situation. Turbulence and non-linearily interacting inertial-gravity waves are simultaneously present in stably-stratified rotating flows. Métais *et al.* [29,35] have proposed a simple way of discriminating between turbulent and wave components. Indeed, they used the fact that inertial gravity waves have no potential vorticity, and thus simply extracted the geostrophic turbulent component of the velocity field as the part associated with the potential vorticity. This can be done in the limit of small Froude and Rossby numbers corresponding to Charney's [4] geostrophic turbulence. Note that, in the non-rotating case, one recovers the classical decomposition of the horizontal velocity field into rotational and divergent components. It is equivalent to Craya's [7] decomposition, which has been used to discriminate between stratified turbulence and internal gravity waves [38,31]. Subsequently, we call vortical mode the rotational component of the horizontal velocity field.

We here briefly recal some results described in details by Métais *et al.* [29,35]. The flow is forced at small scale with a forcing which is three-dimensional and which acts equally on the three velocity components. $\vec{\Omega}$ and the mean stratification are both oriented along the vertical direction. The attention is focused on the small Froude number régime (strong stratification) when the Rossby number ranges from large (slow rotation) to small (rapid rotation) values. Figure 9 displays, for various cases, the three-dimensional wavenumber spec-

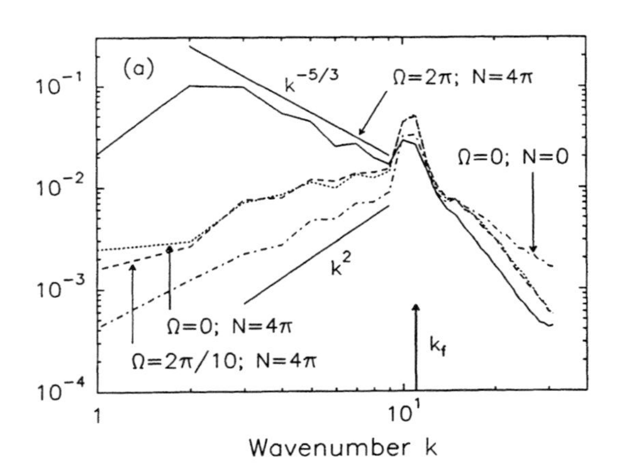

FIGURE 9. Forced simulation: three-dimensional wavenumber spectrum of the vortical kinetic energy for various values of Ω and of the Brunt-Vaissala frequency N.

trum of the vortical kinetic energy when the system has reached an equilibrium. In the non-rotating strongly-stratified case ($\Omega = 0; N = 4\pi; F_r \approx 0.2$), the vortical energy transfer towards the large scales is more efficient than in the non-stratified non-rotating case, but it remains weak. The buoyancy effects are still dominant when a weak rotation ($\Omega = 2\pi/10$; $N/f = 10$, $R_o \approx 1$) is imposed: the spectral behaviour remains almost unchanged. Conversely, a complete change is observed for strong rotation ($\Omega = 2\pi$; $N/f = 1$, $R_o \approx 0.1$): the spectrum now follows a $k^{-5/3}$ law for $k < k_f$ and the spectral slope is increased for $k > k_f$. Métais *et al.* [29] have checked that the observed $k^{-5/3}$ behaviour for small k is a manifestation of geostrophic turbulence dynamics (see [4]). This is demonstrated on figure 10 which compares, in the strongly-stratified, rapidly-rotating regime the three-dimensional wave-number spectrum of the total kinetic energy (KE_{total}) to the analogous spectra for the geostrophic kinetic energy (KE_{geo}) and the inertio-gravity wave kinetic energy (KE_{wave}). Due to the combined effects of rotation and stratification, the two kinds of motions are clearly segregated: the geostrophic energy dominates the $k^{-5/3}$ inverse cascade and reaches larger and larger scales. By contrast, the wave energy cascades towards the scales smaller than the injection scales and is therefore submitted to a strong dissipation. A similar picture can be drawn for the geostrophic and wave part of the available potential energy.

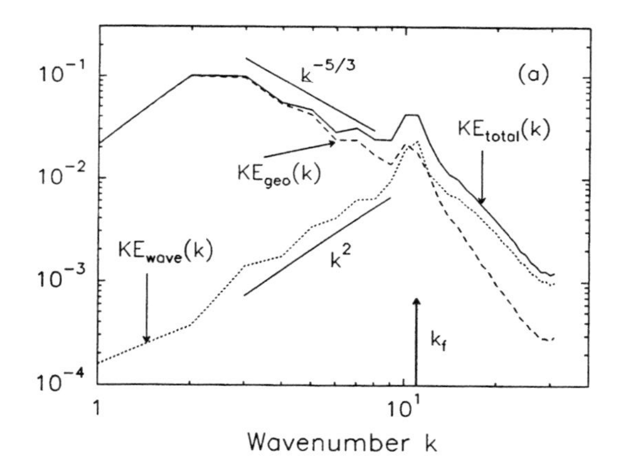

FIGURE 10. Forced simulation for $\Omega = 2\pi$; $N = 4\pi$: three-dimensional wave-number spectrum of the total kinetic energy (KE_{total}) to the analogous spectra for the geostrophic kinetic energy (KE_{geo}) and the inertio-gravity wave kinetic energy (KE_{wave}).

The atmospheric mesoscale spectra observed in the upper troposphere (see Gage and Nastrom [8], for a review) exhibit several features in common with the present numerical results such as $k^{-5/3}$ inverse cascades for both velocity and temperature spectra. These similarities might lead to believe that the mesoscale spectra do correspond to geostrophic turbulence propagating towards the large scales. There are however significant differences between

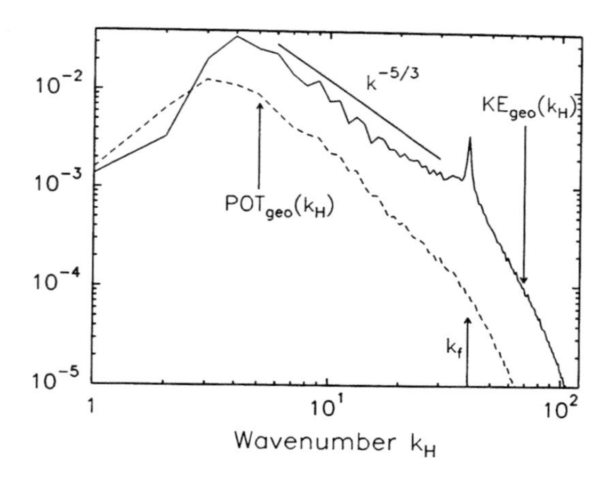

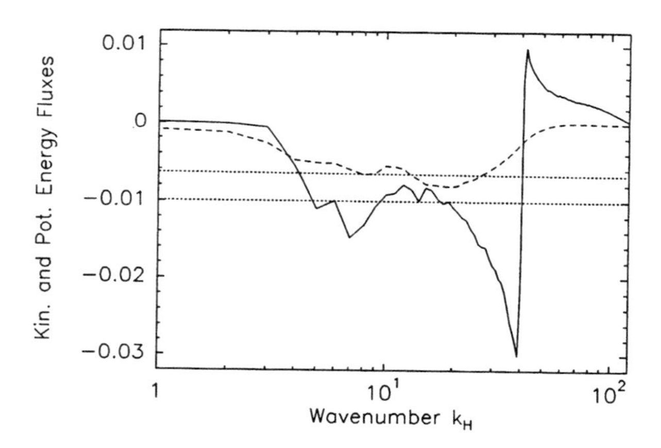

FIGURE 11. Forced simulation for $\Omega = 2\pi$; $N = 9\pi$: a) two-dimensional wave-number spectrum of the geostrophic kinetic energy (KE_{geo}) and the geopstrophic potential energy (POT_{geo}); b) Geostrophic kinetic energy flux (continuous line) and geostrophic potential energy flux (dashed line).

the present numerical simulations and the real atmospheric situation. Atmospheric flows are characterized by a large ratio of the Brunt-Vaissala frequency and the Coriolis parameter with $N/f \approx 100$, by a large aspect ratio between characteristic horizontal and vertical scales and by specific lower and upper boundary conditions. Moreover, the turbulent Rossby numbers extracted from the atmospheric spectra presented by Gage and Nastrom [8] are of order 1 at the supposed injection horizontal scale $10km$. This is at variance with the computations of Métais *et al.* [29] where $N/f \approx 1$, the aspect ratio is equal to one and periodic boundary conditions are used in all three directions. Furthermore, the injection scales are strongly influenced by the rotation effects with a corresponding Rossby number $Ro \approx 0.2$. We are presently undergoing high resolution computations based upon conditions closer to the atmospheric situation. The figure 11 shows the results of a recent simulation highly-resolved in the two horizontal directions with $256 \times 256 \times 16$ computational points. The upper and lower boundary conditions are free-slip conditions and the aspect ratio between the horizontal and vertical scales is 16. The forcing is purely two-dimensional but the flow is three-dimensional due to the presence of the stable density stratification. The simulation parameters are $\Omega = 2\pi$; $N = 9\pi$ which corresponds to a Rossby number of the order of 0.2 at the injection scale. Figure 11 a) shows that both geostrophic kinetic and potential energy spectra exhibit well defined $k_H^{-5/3}$ ranges at large scales (k_H horizontal wavenumber). This well-resolved computations have allowed us to verify that this characteristic spectral behaviour does indeed correspond with effective inverse cascades of geostrophic energy. Figure 11 b) displays both the kinetic and potential energy flux of the geostrophic component. Both curves exhibit a plateau at large-scale corresponding to a well-defined region of constant negative flux.

BAROCLINIC EDDIES

The baroclinic instability results from the combined effects of horizontal temperature gradients and fast rotation on a stably-stratified fluid. It corresponds to a very efficient mechanism of conversion of potential energy into horizontal kinetic energy. When one considers horizontal scales of the order of the internal Rossby radius of deformation ($\approx 1000km$ in the atmosphere and $\approx 50km$ in the ocean, at mid-latitude), this instability becomes very active and gives rise to "baroclinic" eddies. Garnier [9], Garnier *et al.* [10,11] have performed direct and large-eddy simulations of baroclinic jet flows instabilities with the goal to study the nature of the coherent vortices and in particular the asymmetry between cyclonic and anticyclonic eddies. Garnier *et al.* [10,11] have considered a stably-stratified medium associated with a constant vertical mean density gradient characterized by a constant Brunt-Vaissala frequency N:

$$N = \sqrt{-\left(\frac{g}{\rho_0}\right)\frac{d\bar{\rho}}{dz}} \quad . \tag{17}$$

The initial basic state is an horizontal density front oriented in the meridional direction $\vec{x}$:

$$\rho(x,z) = \Delta\rho_H \; tanh\left(\frac{2x}{\delta}\right) + \bar{\rho}(z) \quad , \tag{18}$$

where $2\Delta\rho_H$ is the density difference imposed between the two meridional boundaries of the computational domain and δ the front steepness.

The rotation vector $\vec{\Omega}$ is oriented in the z direction. Let $f = 2\Omega$ be the Coriolis parameter. In the limit of fast rotation and strong stratification, it can be shown that the density front has to be associated with a basic velocity profile. Indeed, the geostrophic equilibrium corresponding to a balance between the Coriolis force and the pressure gradient and the hydrostatic balance imply that this basic state has to satisfy the thermal wind equation:

$$\frac{\partial \vec{u}_H}{\partial z} = -\frac{g}{\rho_0 f}\vec{z} \times \vec{\nabla}_H \rho \quad , \tag{19}$$

where $\vec{a}_H$ stands for the horizontal projection of the vector $\vec{a}$ on the horizontal plane. This gives the following velocity profile:

$$\bar{V}(x,z)\vec{y} = -\frac{g}{\rho_0 f}\,\frac{2\Delta\rho_H}{\delta}\,\frac{1}{ch^2(2x/\delta)}\,(z-z_0)\,\vec{y} \;=\; V(z)\,\frac{1}{ch^2(2x/\delta)}\vec{y} \quad , \tag{20}$$

corresponding to a Bickley jet directed along $\vec{y}$ of vorticity thickness δ, sheared (with a constant shear) in the z direction.

When a small random perturbation is superposed to this basic state the flow becomes unstable. The nature of the instability is however very different depending upon the characteristic parameters. The two characteristic length scales for instabilities are the barotropic length scale δ and the baroclinic length scale $r_I = NH/f$ corresponding with the internal Rossby radius of deformation. z varies between z_B and z_T with $z_T - z_B = H$ the height of the computational domain. The two non-dimensionalized parameters are the Rossby (Ro) and Froude (Fr) numbers defined as:

$$Ro = \frac{\omega_{2D}^i}{f} = \frac{V(z_T)}{\delta f}; \quad Fr = \frac{V(z_T) - V(z_B)}{2NH} \quad . \tag{21}$$

where ω_{2D}^i stands for the vorticity maximum associated with the mean velocity profile. We then have :

$$\frac{Ro}{Fr} = \frac{r_I}{\delta} \quad . \tag{22}$$

First, Garnier *et al.* [11] have used direct numerical simulations to carry out linear stability studies. They have shown the existence of a critical value of the ratio $Ro/Fr = 1.5$ constituting the threshold between to distinct regimes:

1) $Ro/Fr > 1.5$ (or $1.5\delta < r_I$): the instability is weak and mainly barotropic: it is of Kelvin-Helmholtz type and is associated with the inflexional nature of the mean velocity profile.

2) $Ro/Fr \leq 1.5$ (or $1.5\delta \geq r_I$): the baroclinic instability corresponding to a conversion of potential energy associated with the horizontal density gradient into horizontal kinetic energy can develop. The amplification of the perturbations is much stronger than in the barotropic case.

We now concentrate on the second regime $Ro/Fr \leq 1.5$: here $Ro/Fr = 0.5$. The Rossby number is fixed to 0.1. The numerical code is similar to the channel flow study previously described except that compact differences schemes are here used into two spatial directions. The resolution is $193 \times 64 \times 20$ points both for the DNS and the LES. The Reynolds number $Re = V(z_T)\delta/\nu$ is 400 in the DNS 10,000 in the LES.

Synoptic-scale instability

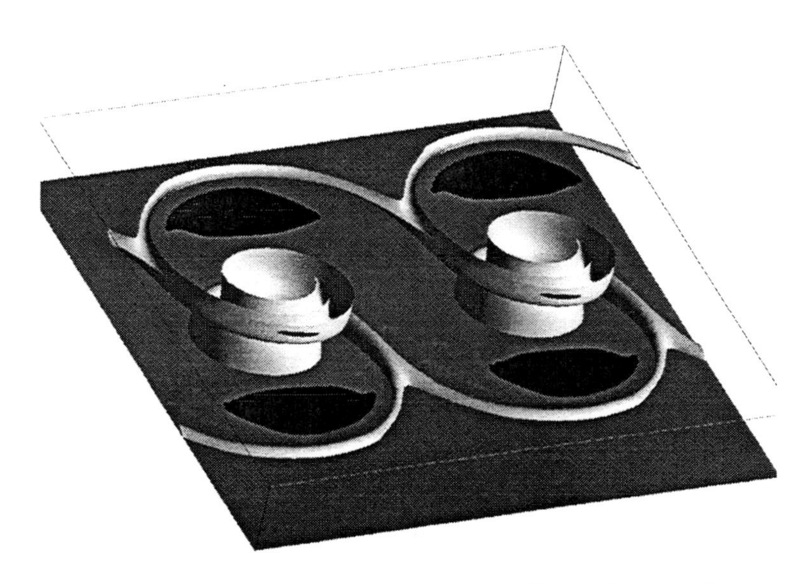

FIGURE 12. Iso-surfaces of vertical vorticity corresponding to $|\omega_z| = 2\omega_{2D}^i$; light-gray: cyclonic vorticity, black: anticyclonic vorticity. $t = 47 r_I / V(z_T)$.

Figure 12 shows the vorticity structure obtained by DNS once the instability has fully developed. We observe the formation of cyclonic eddies of strong intensity, composed of nearly two-dimensional cores between which braids of very high cyclonic vorticity are formed. The vorticity maxima are observed within these braids and correspond here to ≈ 8 times the vorticity maximum of the initial mean velocity profile. The vorticity intensification in the anticyclonic eddies is weaker (3 times the initial vorticity): we have checked that those are far more three-dimensional than the cyclonic eddies and strongy stretched by them. The asymmetry cyclones/anticyclones is well illustrated

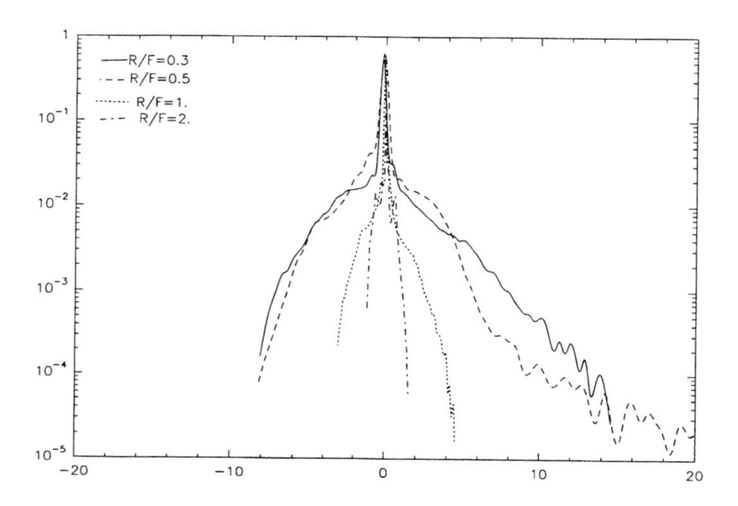

FIGURE 13. Probability density functions of the vertical vorticity component ω_z for several values of R_o/Fr.

when considering the probability density functions of the vertical vorticity for several value of the ratio Ro/Fr (figure 13). This clearly shows that for $Ro/Fr \leq 1$ the vorticity maxima are cyclonic and are localized in very concentrated regions of the space (low probability). Contrarily, the anticyclonic vorticity exhibit a quasi-gaussian behaviour indicating that, for moderate vorticity intensity, the most probable values are negative.

The preferential amplification of the cyclones can be easily explained by considering the various production terms in the vorticity dynamics. Neglecting dissipation effects, the vertical vorticity equation within the Boussinesq approximation may be written as follows:

$$\frac{d\omega_z}{dt} = \vec{\omega}_h \cdot \vec{\nabla}_h w + (\omega_z + f)\frac{\partial w}{\partial z} \tag{23}$$

where h denotes the horizontal direction. Vorticity might only be produced by the horizontal vorticity tilting or by the absolute vertical vorticity stretching. However, one may show that, for small Ro, the vorticity is mainly produced by the absolute vorticity stretching. It is responsible for the asymmetry between cyclonic and anticyclonic vorticity amplification. Indeed, positive feed-back effects accelerate the cyclonic production, as $(\omega_z + f)$ increases while ω_z grows. Conversely, when anticyclonic vorticity increases (for $-f < \omega_z < 0$), its production term decreases.

Secondary cyclogenesis

In the previous DNS, the late stage of the cyclogenesis was dominated by dissipative and diffusive effects. We have thus performed high Reynolds number LES at $Re = 10000$ to study the flow development subsequent to the occlusion process. The subgrid-scale structure function model with a cusp given by eq. 16 have been used. Here the grid is non-isotropic $\Delta x \neq \Delta y \neq \Delta z$. However, the structure function may be evaluated in each point with the help of the six closest, using the Kolmogorov's law for the structure function $F_2 \propto (\epsilon\, r)^{2/3}$. Is is obtained :

$$\begin{aligned}\bar{F}_2(\vec{x}, \Delta c) &= 1/6 \textstyle\sum_{i=1}^{3} \{\| \vec{u}(\vec{x}) - \vec{u}(\vec{x} + \Delta x_i \vec{e}_i) \|^2 \\ &\quad + \| \vec{u}(\vec{x}) - \vec{u}(\vec{x} - \Delta x_i \vec{e}_i) \|^2\} \, (\Delta c / \Delta x_i)^{2/3}\end{aligned}$$

where $\Delta c = (\Delta x_1 \, \Delta x_2 \, \Delta x_3)^{1/3}$

Figure 14 shows a time evolution of the vorticity contours of a cyclonic eddies. As compared with the DNS presented in the preceeding section, one may notice that the spiralling of the vorticity contours inside the core of the cyclonic eddies is much more pronounced. Due to viscous effects, the vorticity was indeed homogenized in the DNS. We have checked that the frontal region are much steeper in the LES indicating more energy near the wavenumber cut-off. The steepening of the fronts is associated with the appearance of a secondary instability resulting in a local intensification of the vertical vorticity. This instability seems to take place in regions where the local values of the Rossby and Froude number $Ro(\vec{x})$ and $Fr(\vec{x})$ verify the criterion $Ro(\vec{x})/Fr(\vec{x}) \leq 1.5$. The potential energy associated with local horizontal fronts is then converted into horizontal kinetic energy and gives rise to vertical vorticity intensification. It is important to notice that if the structure model without cusp is used excessive accumulation of energy is observed at the smallest scales eventually leading to numerical divergence. This demonstrates the importance of the cusp-like behaviour and the feasibility of the subgrid-scales prevously described for the LES of geophysical flows with quasi-twodimensional regions and sharp frontal regions.

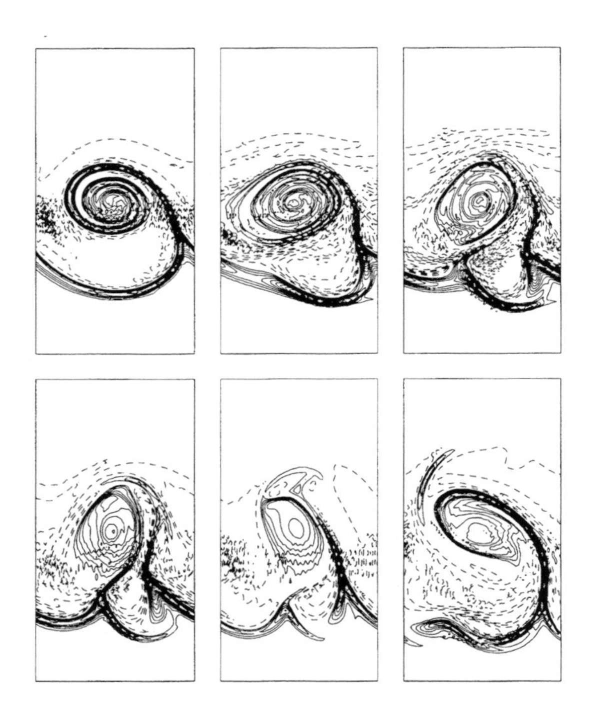

FIGURE 14. LES simulation at Re = 10000: time evolution of the vertical vorticity isocontours at $z = z_T$. Solid-lines cyclonic vorticity; dashed lines anticyclonic vorticity.

Acknowledgements : Some of the computations were carried out at the IDRIS (Institut du Développement et des Ressources Informatique Scientifique, Paris).

REFERENCES

1. Bartello, P., 1995. Geostrophic adjustement and inverse cascade in rotating stratified turbulence, *J. Atmos. Sci.*, **223**, pp. 521–536.
2. Bartello, P., Métais, O. & Lesieur, M., 1994. Coherent structures in rotating three-dimensional turbulence. *J. Fluid Mech.*, **273**, pp 1-29.
3. Basdevant, C. & Sadourny, R., 1983. Modélisation des échelles virtuelles dans la simulation numérique des écoulements bidimensionnels. *J. Mec. Theor. et Appl.*, Numéro Spécial, pp. 243-269.
4. Charney, J.G., 1971. Geostrophic turbulence. *J. Atmos. Sci.*, **28**, pp. 1087–1095.
5. Chollet, J.P. & Lesieur, M., 1981, Parameterization of small scales of three-dimensional isotropic turbulence utilizing spectral closures. *J. Atmos. Sci.*, **38**, pp. 2747–2757.
6. Comte, P., Lesieur, M. & Métais, O., 1995, "Vortex dynamics in numerical simulations of transitional and turbulent shear flows", proceedings of the workshop on *Small-Scale Structures in Three-Dimensional Hydro and Magnetohydrodynamic Turbulence*, Lecture Notes in Physics, Springer.
7. Craya, A., 1958. Contribution à l'analyse de la turbulence associée à des vitesses moyennes. *P. S. T. Ministère de l'Air*, Paris, 345 pp.
8. Gage, K.S. & Nastrom, G.D., 1986. Theoretical interpretation of atmospheric wavenumber spectra of wind and temperature observed by commercial aircraft during GASP. *J. Atmos. Sci.*, **43**, pp. 729–740.
9. Garnier, E., 1996. *Etude numérique des instabilités de jets beroclines*, PhD thesis, Natl. Polytech. Inst., Grenoble.
10. Instabilités primaire et secondaire dans un jet barocline. *C.R. Acad. Sci.*, Paris, Série II *b*, **t. 323**, pp. 95–101.
11. Garnier, E., Métais, O. & Lesieur, M., 1997. Synoptic and frontal-cyclone scale instabilities in baroclinic jet flows. *J. Atmos. Sci.*, *in press.*
12. Germano, M., Piomelli, U., Moin, P. & Cabot W., 1991. A dynamic subgrid-scale eddy viscosity model. *Phys. Fluids A.*, **3** (7), pp. 1760–1765.
13. Herring, J.R. & Métais, O., 1989. Numerical experiments in forced stably stratified turbulence. *J. Fluid Mech.*, **202**, pp. 97-115.
14. Johnston, J.P., Halleen, R.M., & Lezius, D.K., 1972. Effects of spanwise rotation on the structure of two-dimnesional fully developped turbulent channel flow. *J.Fluid Mech.*, **56**, pp. 533-557.

[7] Kim, K., 1983, "The effect of rotation on turbulence structure", In *Proc. 4th Symp. on Turbulent Shear Flows*, Karlsruhe, pp. 6.14–6.19.

15. Kraichnan, R.H., 1976. Eddy viscosity in two and three dimensions. *J. Atmos. Sci.*, **33**, pp. 1521-36.

16. Kristoffersen, R. & Andersson, H. I., 1993. Direct simulations of low-Reynolds-number turbulent flow in rotating channel. *J. Fluid Mech.*, **256**, pp. 163–197.
17. Lamballais, E., 1996. "Simulations numériques de la turbulence dans un canal plan tournant", PhD thesis, Natl. Polytech. Inst., Grenoble.
18. Lamballais, E., Lesieur, M. & Métais, O., 1996. Effects of Spanwise Rotation on the Vorticity Stretching in Transitional and Turbulent Channel Flow. *Int. J. Heat and Fluid Flow*, **17**, pp. 324–332.
19. Lamballais, E., Lesieur, M. & Métais, O., 1996. Influence d'une rotation d'entraînement sur les tourbillons cohérents dans un canal. *C.R. Acad. Sci.*, Paris, Série II *b*, **t. 323**, pp. 95–101.
20. Lamballais, E., Métais, O. & Lesieur, M., 1996, "Influence of a spanwise rotation upon the coherent-structure dynamics in a turbulent channel flow", in *Direct and Large-Eddy Simulation II*, J.P. Chollet, L. Kleiser and P.R. Voke, eds., Kluwer Academic Publishers, *in press*.
21. Lesieur M. 1997. *Turbulence in Fluids* (third revised edition), Kluwer Academic Publisher, 515 p.
22. Lele, S.K., 1992. Compact finite difference schemes with spectral-like resolution. *J. Comp. Phys.*, **103**, pp. 16–42.
23. Lesieur, M., 1997, "Coherent vortices in rotating flows", in *New Tools in Turbulence Modelling*, O. Métais & J. Ferziger, eds., Les Éditions de Physique; Springer-Verlag, 298 p.
24. Lesieur, M., & Métais, O., 1996. New trends in large-eddy simulations of turbulence. *Ann. Rev. Fluid Mech.*, **28**, pp. 45–82.
25. Lesieur, M., Yanase, S. & Métais, O., 1991. Stabilizing and destabilizing effects of a solid-body rotation upon quasi-two-dimensional shear layers. *Phys. Fluids A*, **3**, pp. 403-407.
26. Lilly, D.K., 1983. Stratified turbulence and the mesoscale variability of the atmosphere. *J. Atmos. Sci.*, **40**, pp. 749-761.
27. Métais, O., 1997. "Vortices in rotating and stratified turbulence", in *Proceedings of ICTAM Kyoto 1996*, T. Kambe, T. Tatsumi & E. Watanabe, eds., Elsevier Science Publishers B.V., pp. 87–103.
28. Métais, O., 1997. Numerical simulation of geophysical turbulence. in *Lecture Notes on Plankton and Turbulence, Sci. Mar.*, **61** (Supl. 1), pp. 75–91.
29. Métais, O., Bartello, P., Garnier, E., Riley, J.J. & Lesieur, M., 1995a. Inverse cascade in stably-stratified rotating turbulence. *Dynamics of Atmospheres and Oceans*, pp. 193–203.
30. Métais, O., Flores, C., Yanase, S., Riley, J.J. & Lesieur, M., 1995b. Rotating free-shear flows. Part 2: numerical simulations. *J. Fluid Mech.*, **293**, pp. 47-80.
31. Métais, O. & Herring, J.R., 1989. Numerical simulations of freely evolving turbulence in stably stratified fluids. *J. Fluid Mech.*, **202**, pp. 117-148.
32. Métais, O. & Lamballais, E., 1997, "Coherent vortices in rotating flows", in *New Tools in Turbulence Modelling*, O. Métais & J. Ferziger, eds., Les Éditions de Physique; Springer-Verlag, 298 p.
33. Métais O. & Lesieur M, 1995. Coherent vortices in thermally stratified and rotating turbulence, *Int. J. Heat and Fluid Flow*, **16**, pp. 316–326.

34. Métais O. & Lesieur M., 1992. Spectral large-eddy simulations of isotropic and stably-stratified turbulence. *J. Fluid Mech*, **239**, pp. 157-94.
35. Métais, O., Riley, J.J. & Lesieur, M., 1994. Numerical simulations of stably-stratified rotating turbulence. In *Stably-stratified flows- flow and dispersion over topography*, I.P. Castro and N.J. Rockliff eds., Oxford University Press, pp. 139-151.
36. Métais, O., Yanase, S., Flores, C., Bartello, P. & Lesieur, M., 1992. "'Reorganization of coherent vortices in shear layers under the action of solid-body rotation". Selected proceedings of the *Turbulent Shear Flows VIII*, Springer-Verlag, pp.415-430
37. Piomelli, U. & Liu, J., 1995. Large-eddy simulation of rotating channel flows using a localized dynamic model. *Phys. Fluids A*, **7** (4), pp. 839–848.
38. Riley, J.J., Metcalfe, R.W. & Weissman, M.A., 1981. "Direct numerical simulations of homogeneous turbulence in density stratified fluids". In: B.J. West (Editor), AIP Conf. Proc. N° 76, New York, Nonlinear Properties of Internal Waves, pp. 79-112.
39. Smagorinsky, J., 1963. General circulation experiments with the primitive equations. *Mon. Weath. Rev.* **91**, 3, pp. 99–164.
40. Tafti, D.K. & Vanka, S.P., 1991. A numerical study of the effects of spanwise rotation on turbulent channel flow. *Phys. Fluids A*, **3**(4), pp. 642–656.
41. Tarbouriech, L. & Renouard, D., 1996. Stabilisation et déstabilisation par la rotation d'un sillage plan turbulent. *C.R. Acad. Sci.*, Paris, Série II *b*, **t. 323**, pp. 95–101.
42. Yanase, S., Flores, C., Métais, O., & Riley, J.J. 1993 Rotating free shear flows Part 1: Linear stability analysis, *Phys. of Fluids. A.* **5** (11), pp. 2725–2737.

MILESTONES IN ROTATING SHALLOW WATER MODELING OF ROSSBY VORTICES, PLASMA DRIFT VORTICES, AND SPIRAL STRUCTURES IN GALAXIES

M.V. Nezlin, A.Yu. Rylov, K.B. Titishov, G.P. Chernikov

RRC "Kurchatov Institute", Institute of Nuclear Fusion,
Kurchatov Square 1, 123182 Moscow, Russia

Abstract. A review and the current status is given of laboratory experiments on the modeling, using rotating shallow water, of the largest and longest-lived vortical structures in planetary atmospheres, oceans, magnetized plasmas and spiral galaxies. Basic theoretical ideas, partly preceding the experiments mentioned and partly essentially inspired by the latter, are also presented.

CONTENTS

CP414, *Two-Dimensional Turbulence in Plasmas and Fluids:* Research Workshop
edited by R. L. Dewar and R. W. Griffiths

I INTRODUCTION

In this and the following sections, we consider Rossby vortices on "shallow water". Let us recall the main definitions and the main parameters determining the properties of these vortices.

The Rossby-Obukhov radius is defined by

$$r_{\mathrm{R}} \equiv (gH_0)^{1/2}/f_0 \,, \tag{1}$$

where H_0 is the shallow water thickness measured perpendicularly to its free surface (note that we shall always assume that the free surface does exist); g is the free-fall acceleration, f_0 is the Coriolis parameter:

$$f \equiv f_0 + \beta y \tag{2}$$

where y is the meridional coordinate with respect to some reference point,

$$\beta \equiv \frac{df}{dy} \tag{3}$$

is the parameter of the β-effect caused by the meridional gradient in the Coriolis parameter.

The Rossby speed is the maximal phase velocity of linear Rossby waves:

$$V_{\mathrm{R}} \equiv \beta r_{\mathrm{R}}^2 \tag{4}$$

at $H_0 = \text{const}$, or

$$V_{\mathrm{R}} = -g\frac{d}{dy}\left(\frac{H_0}{f}\right) = -\beta\left(1 - \frac{d\ln H_0}{d\ln f}\right) r_{\mathrm{R}}^2 \tag{5}$$

in the general case.

According to definition, the Rossby number, that is, the ratio of the Rossby wave frequency to the Coriolis parameter, is always much less than unity. It is the so-called Rossby regime.

In the Rossby regime, a cyclone (i.e., a vortex rotating in the direction of the planet's rotation) is a "well" on the shallow-water free surface, whereas an anticyclone is an elevation on the free surface shallow water.

A linear Rossby wave packet is a wavy structure suffering dispersive decay; the minimal characteristic time of that decay (when the packet size is equal to the Rossby radius) is

$$T_l = \frac{8r_R}{V_R} \tag{6}$$

(see, for instance, [1]).

The Rossby soliton is a nonlinear Rossby wave packet, in which the dispersive spread is compensated by the nonlinearity. According to definition, the Rossby soliton is a long-lived structure surviving much longer than the linear Rossby wave packet of the same size.

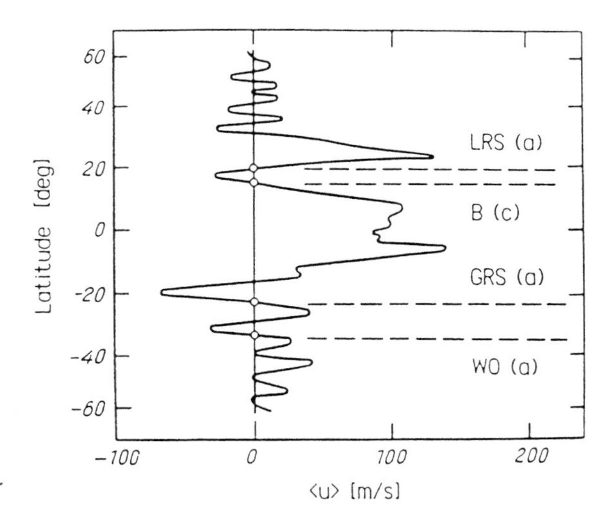

FIGURE 1. Zonal flows in the upper Jovian atmosphere: average velocity $\langle u \rangle$ of the wind blowing along the parallel (positive wind direction is eastwards) as a function of the planetary latitude. The position of major large-scale, long-lived vortices (GRS–Great Red Spot, LRS–Little Red Spot, WO–White Ovals, B–Barges) are indicated as well as their signs (a–anticyclone, c–cyclone). From the book [1] of one of the authors of this paper. Reprinted with kind permission of Springer-Verlag ©1993.

A good example demonstrating different properties of the Rossby structures are the largest giant planet vortices, the famous Great Red Spot of Jupiter (GRS) is among them — see Fig. 1. This vortex rotates with a period equal to one week, whereas the rotation period of Jupiter is about 10 hours. So, the Rossby number is very small compared to unity.

The vortices mentioned exist in a system exhibiting zonal flows and possess also the following other properties. Firstly, they are long-lived structures. For instance, the life time of a linear Rossby wave packet having the same size as the GRS is about a few months, whereas the life time of the actual GRS exceeds 300 years. The dimensions of the Jovian vortices are larger or of the order of the Rossby barotropic radius. For instance, that radius in the region of the GRS is about 6000 km, whereas the dimensions of the GRS are (12×25) thousand km. A highly essential property of the largest vortices of Jupiter and Saturn is their clear-cut cyclone-anticyclone asymmetry: almost all those vortices [with only one exception in Jupiter's atmosphere — the so called "Barges" (Fig. 1); and only one in Saturn's atmosphere — the so called "UV-Spot"] turn out to be anticyclones: their own rotation on the planet is directed contrary to the planetary rotation.

II FREE ROSSBY SOLITON CONCEPT AND THE FIRST SOLITON CONDITION

The longevity of the giant planet vortices and their cyclone-anticyclone asymmetry have inspired the development of the Rossby soliton concept. One of the first soliton theories of the GRS and like vortices was the theory [2], where the possibility of the existence of a Rossby soliton on the zonal shear flow with "solid lids" (i.e., without a free surface) was shown. According to [2], the Rossby soliton is a purely wavy eddy, which collides with like eddies elastically, i.e. it pierces them. This conclusion was in apparent contradiction with the real properties of the GRS: the latter merges at collisions with like structures. The problem of the cyclone-anticyclone asymmetry also remained open in the theory [2].

A vortical theory of the Rossby soliton was suggested in [3], also in an approximation of the "solid lid". According to [3], the GRS and like vortices could merge under their mutual collisions. However the question regarding the cyclonic-anticyclonic asymmetry was also not solved in [3].

A new approach to the problem of the Rossby soliton was suggested in the works [4–7]. In those works a model of the free Rossby soliton was suggested. According to [4–7], a Rossby soliton, in principle, may exist even without zonal flows and without the solid lid, that is, in the presence of a free surface of the "shallow water". The main statements of the theory [4–7] are the following.

1. The Rossby soliton is an anticyclone, i.e., an elevation of the free surface; a cyclone (i.e., a depression of the free surface) cannot be a Rossby soliton.

2. The Rossby soliton is a cylindrically symmetrical structure.

3. The Rossby soliton has a characteristic radius exceeding the Rossby-Obukhov radius

$$a > r_{\mathrm{R}} \, , \tag{7}$$

4. The Rossby soliton radius is:

$$a = \frac{\sqrt{3} r_{\mathrm{R}}}{\sqrt{h}} \tag{8}$$

   where h is a dimensionless elevation of the free surface (the ratio of the elevation to the unperturbed shallow water thickness, $H_0 = \mathrm{const}$).

5. The soliton has an universal profile [7]:

$$h = H_0 \cosh^{-\frac{4}{3}} \left(\frac{r}{a} \right)^{\frac{3}{4}} \tag{9}$$

6. The Rossby soliton propagates westwards, i.e., contrary to the planetary rotation, like the linear Rossby waves.

7. The propagation velocity of the Rossby soliton exceeds the Rossby speed:

$$V_{\mathrm{dr}} > V_{\mathrm{R}} \tag{10}$$

   Under this condition, it is out of the Cerenkov resonance with the linear Rossby waves and, therefore, it does not radiate those waves. (Otherwise, it could not be a soliton.) The condition (10) is the first necessary soliton condition. (As shown below, there exists also a second necessary soliton condition.)

8. The Rossby soliton is a purely wavy structure and collides with like structures elastically, i.e., the structures pierce each other freely.

9. A remarkable property of the Rossby soliton is its physical analogy with the drift wave soliton in the magnetized plasma: these solitons are described by the same equations. Namely, in geophysical fluid dynamics it is the generalized Charney-Obukhov equation for the dimensionless elevation (h), in plasma physics it is the generalized Hasegawa-Mima equation for the dimensionless plasma potential (i.e. potential over the electron temperature).

It is seen that the new soliton concept suggested is able to explain the longevity and the cyclone-anticyclone asymmetry of the GRS-like vortices. Therefore, it was a step forward, in spite of its purely wavy nature.

In [4–7], a generalized Charney-Obukhov equation describing the above formulated properties of the soliton under study was obtained. Namely,

$$\left(\Delta h - \frac{h}{r_R^2}\right)_t + \beta h_x + \beta h h_x + \frac{gH_0}{f_0} J(h, \Delta h) = 0 . \tag{11}$$

where the lower indices denote differentiation with respect to time (t) and the direction (x) of the wave propagation; J is the Jacobian. Compared to the known classical Charney-Obukhov equation (which is symmetrical with respect to cyclones and anticyclones — see, for instance, [1]), the equation (11) — due to the new term $\beta h h_x$ — contains the cyclone-anticyclone asymmetry effect and gives a solitonic solution only for the anticyclone (recall that the case H_0 = const is under consideration). It is the so called scalar nonlinearity, of the Korteweg-de Vries type, — the third term of the equation (11) — which is responsible for the cyclone-anticyclone asymmetry. Since, according to [4–7], the considered structure is axially symmetrical, the Jacobian in (11) is equal to zero. It means that the scalar nonlinearity is the single one equilibrating the dispersive spread of the structure. Therefore one may say that the properties of the soliton under consideration turn out to be similar to those of the first soliton in the history of science, which was discovered by Scott Russel in 1833 and later was described mathematically by Korteweg and de Vries.

III EXPERIMENTAL APPROACH TO CREATION OF ROSSBY SOLITONS

Since we knew the properties formulated above of the theoretically predicted Rossby soliton, our team (working in the Institute of Nuclear Fusion of the Russian Research Centre "Kurchatov Institute") decided to try to create it in the laboratory. (The initial staff of the team was: S.V. Antipov, M.V. Nezlin, E.N. Snezhkin, A.S. Trubnikov. Later, S.V.A. and A.S.T. left the team and were replaced by A.Yu. Rylov, G.P. Chernikov, and K.B. Titishov.)

It was clear from the very beginning that the traditional approach applied to studies of Rossby waves (vortices) was unsuitable for experiments with Rossby solitons. The situation is the following: A typical traditional device has a flat (or inclined) bottom. Such a device cannot be rotated sufficiently fast to satisfy the requirement (7), because otherwise the free surface of the rotating liquid — quite necessary according to the theory considered — will become so curved that the liquid will not be a shallow water system. Therefore, in all traditional devices, the Rossby radius essentially exceeds the setup radius and the condition (7) turns out to be unattainable.

Since the free surface of the rotating liquid acquires the parabolic shape, our setups are paraboloids. One of them ("the little one") has the working diameter 28 cm, the working heighth 14 cm and is rotated with a period 759 ms (therewith, the shallow water thickness $H_0 = 5$ mm is kept constant over all the free surface of the water). The second paraboloid ("the large one"), 72 cm in diameter and 36 cm in height, is rotated with a regulated period in the range of 800–900 ms. In particular, the value $H_0 = 1$ cm is kept constant at the rotation period 840 ms. (See [1] for details.)

IV ROSSBY SOLITONS IN THE LABORATORY AND THEIR UNEXPECTED PROPERTIES

In these experiments with parabolic vessels, we created for the first time an object which may be called a Rossby soliton. The experiments [1] have shown that the object possesses the following properties.

- It corresponds to the theoretical predictions (1), (3), (6), and (7). In particular, its characteristic diameter is about $(2.5 - 4)r_R$.
- However, it does not satisfy the theoretical predictions (2), (4), (5), and (8).

In particular, unlike the prediction (8), the two structures considered collide inelastically: they merge (or destroy each other), but never pierce each other [1]. Predictions (4) and (5) are in a clear-cut contradiction with the experiment [1], which shows that the object under consideration has a more or less arbitrary height profile (which is determined by the method of creation of the structure). And its characteristic size is independent of the amplitude [1].

A new fact of principal importance is that the object under study captures the particles of the medium and involves them in its drift motion along the parallel. Therewith, the maximal linear velocity on the rotation profile of the object exceeds the velocity of its propagation as a whole. At such conditions, the object's stream-lines form a separatrix, inside which there is a trapped fluid. In other words, it manifests the clear-cut properties of a "real" vortex. Strictly speaking, it is not a wave because there are no waves whose phase velocity is less than the velocities of the captured particles.

So, the object is rather dualistic, combining both wavy and vortical properties. It is a new object in nonlinear physics. We call it the solitary vortex.

The trapping of the medium has proved to be a quite necessary condition for the longevity of the vortex considered. From the theoretical viewpoint, it may be understood as follows. As shown in [8,9], the equation (11) contains also the additional term

$$\beta h_x \frac{2y}{R}, \tag{12}$$

where y is the meridional (radial) coordinate and R is the radius of "curvature" of the system. This term may cause the so called twisting of the structure considered. However, the structure may avoid the twisting effect provided its maximal rotation velocity exceeds the maximal phase velocity, namely [1]:

$$V_{\mathrm{rot}} > V_{\mathrm{R}} \tag{13}$$

Clearly, the condition (13) is just the condition of the medium trapping by the vortex (!). It has been shown for the first time in [1] (see also the references there) that (13) is a necessary condition for longevity of the vortex considered.

V FIRST TYPE OF THE CYCLONE-ANTICYCLONE ASYMMETRY

According to our experiments, in the "small" setup , cyclones do not survive for a measurable life time, whereas anticyclones survive longer than the linear Rossby wave packet [1]. In the "large" paraboloid, at a rather high thickness of the water layer (3–6 cm), the ratio of the life time of anticyclones to that of cyclones was equal to 2.3, whereas at thickness about (1.5–2) cm the ratio was about 4–5. Therewith, the life time of cyclones did not exceed that of the linear Rossby wave packet, but the life time of anticyclones was essentially larger and was equal to the molecular (table) viscous life time of the Rossby vortices. So, a clear-cut cyclone-anticyclone asymmetry was demonstrated in the experiments considered. It was in a good agreement with the analogous phenomenon observed by the spacecraft "Voyager-2" in the atmospheres of Jupiter and Saturn.

VI SELF-ORGANIZATION OF ROSSBY SOLITARY VORTICES

In the first stage of the experiments considered, there was an impression that the solitary vortices under study are formed as a result of compensation of the dispersive spreading (intrinsic in the linear wave packet) by the scalar nonlinearity (of the KdV type), which is presented by the third term in the generalized Charney-Obukhov equation (11). Therewith, it was assumed that the structure is axially symmetrical, so that the vector nonlinearity is absent (the Jacobian in (11) is equal to zero). The following facts supported such a supposition.

Firstly, the structure under study is an elevation of the shallow water free surface, like the classical Russel-Korteweg-de Vries soliton.

Secondly, as with the classical soliton, the characteristic size of the structure exceeds some scale (in this case, the Rossby-Obukhov radius, r_{R}).

Thirdly, the scalar nonlinearity produces the cyclone-anticyclone asymmetry and provides fulfillment of the soliton condition (10).

Let us elucidate the two last circumstances. The equation (11) shows that the phase velocity of a nonlinear Rossby structure (at $H_0 = \text{const}$) is

$$V_{\text{ph}} = V_{\text{R}}(1 + h) . \tag{14}$$

Therefore for fulfillment of the soliton condition (10), we must have

$$h > 0 , \tag{15}$$

implying an anticyclone. Unlike this case, a cyclone ($h < 0$) will be in Cerenkov resonance with the linear Rossby waves, will lose energy due to radiation of those waves, and therefore cannot be a soliton.

However, at the same time with the mentioned "pros", there were also very essential "cons", which consisted in the following.

The characteristic size of the solitary vortex turned out to be independent of its amplitude and essentially less than follows from the relationship (8). Therewith, the difference between the sizes mentioned was the larger, the less the vortex amplitude. The absence of a correspondence between the vortex size and its amplitude means that the scalar nonlinearity and the Rossby wave dispersion cannot equilibrate each other. For instance, this circumstance may be formulated as follows: if the vortex size is kept constant, whereas the vortex amplitude is increased, then what prevents the vortex overturning under the influence of the scalar nonlinearity? And, analogously, if the vortex amplitude is kept constant, whereas the vortex size is decreased, what prevents the vortex from spreading?

The answers to these questions are the following (see, for instance, [10,11] and references given there). Firstly, the initial supposition that the structure considered is axially symmetrical and the Jacobian in (11) equals zero, turns out to be wrong. Hence, we are forced to state that the structure is not circular and, hence, the Jacobian in (11) is not zero. And, on the contrary, the vector nonlinearity presented by the Jacobian in (11) plays a crucial role in the vortex dynamics, providing the necessary balance between the vortex dispersion and its nonlinearity. So, if the scalar nonlinearity turns out to be larger than dispersion, the vector nonlinearity "works" against the scalar one; in the contrary case, it works against dispersion.

More thorough consideration [12] shows the following vortex dynamics. In the process of its evolution, the structure under study is deformed, so that it may be considered as a combination of a monopolar anticyclonic vortex and a dipolar addition polarized along the meridian (i.e. perpendicularly to the vortex propagation direction). The direction of that polarization is determined by the sign of a imbalance between the scalar nonlinearity and the wave dispersion: if the scalar nonlinearity prevails, then a cyclone will be situated to the "south" of the monopole, and anticyclone, to the "north" of the monopole. Therewith, the dipole field causes the monopole to accelerate in the direction of its drift along the parallel, preventing the vortex from overturning.

In case of the opposite sign of the imbalance mentioned, the dipole polarization has the opposite sign, which prevents the vortex from spreading. (In regard to the connection between the vortex acceleration or deceleration and the sign of dipole polarization see the section 8 devoted to the vortex drifts.)

As a conclusion to this section, let us mention a number of a rather basic theoretical papers [13–15] and references cited there. In those works, partly inspired by the experiments by our team, the properties of the solitary Rossby vortices with trapped fluid were analyzed thoroughly and a very good agreement with the experiments described was obtained.

VII ON THE ADEQUACY OF ROSSBY VORTEX MODELING USING PARABOLIC VESSELS

The parabolic vessel differs essentially from the planet being modeled because the centrifugal force from the planetary rotation, firstly, is comparable with the gravitational force and, secondly, is a function of the meridional coordinate, y — Fig. 2. Accordingly, the effective acceleration, g^* (the geometric sum of the gravitational and centrifugal accelerations), perpendicular to the shallow-water free surface, is a function of the meridional coordinate:

$$g^*(y) = g/\cos\alpha(y) \ . \tag{16}$$

This leads to a rather fundamental effect [16]: in the generalized Charney-Obukhov equation,there appears one more quadratic nonlinearity proportional to dg^*/dy and exactly destroying the scalar nonlinearity. Therewith, the cyclone-anticyclone asymmetry disappears from the generalized Charney-Obukhov equation (!).

The clear-cut contradiction between such a conclusion and the experiments considered stimulated the second part of the theoretical work [16]. Unlike the first part, performed in the lowest approximation with respect to the Rossby number, the second part was made in the next approximation, in which the Euler equation was written with due account of the finiteness of the Rossby number. The new approximation gave a number of new terms in the generalized Charney-Obukhov equation [10,11]. Therewith, a new term proportional to the beta parameter, β, turned out to be approximately equal to the disappeared scalar nonlinearity, multiplied by the ratio $(r_{\rm R}/a)^2$. In other words, the scalar nonlinearity was effectively restored with proportionality coefficient equal to $(r_{\rm R}/a)^2$. It is necessary to underline that in our experiments the ratio mentioned is never small; in particular, in the experiments described below the ratio is of the order of unity. Therefore, the question regarding the adequacy of the parabolic vessel in relation to the real geometry of the objects being modeled has, in principle, been solved. We shall return to this question in the next section.

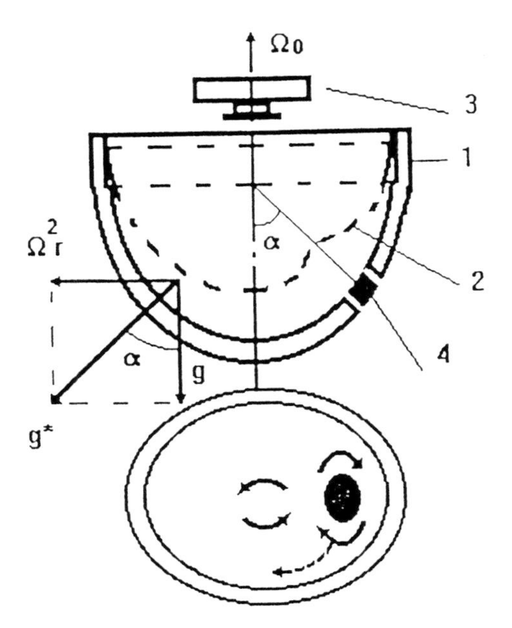

FIGURE 2. The equilibrium of a layer of fluid in a rotating paraboloid; also illustrating diagrammatically how anticyclonic Rossby vortices can be generated experimentally in shallow water that is in solid-body rotation. In the upper part of the diagram, (1) indicates a vessel with approximately paraboloidal bottom; (2) is the surface of the water, which is spread over the bottom of the paraboloid when it is in rotation; (3) is a camera corotating with the vessel; (4) is a vortex source. In the lower view, the solid arrows indicate the direction with which the vessel itself rotates and the anticyclonic direction with which the vortex rotates; the dashed arrow shows the direction in which the vortex drifts in the absence of a gradient in the fluid depth (or for positive sign of the meridional gradient); the vortex lags behind the global rotation of the system; α is the angle between the vessel rotation axis and the normal to the fluid surface. From the authors' paper [11].

VIII SECOND TYPE OF THE ROSSBY VORTEX CYCLONE-ANTICYCLONE ASYMMETRY

In the preceding sections of the paper, we have considered the properties of Rossby vorticies at constant shallow water thickness (H_0 = const). Now, let us consider the situation when H_0 has a considerable negative gradient in the meridional direction. (It means that the paraboloid is rotated slower than in the case H_0 = const, so that the shallow water thickness at the paraboloid pole is larger than at the periphery.) Then, the Rossby vortex propagation direction changes from westward to eastward, in accordance with the relationship (5). In this situation, the Rossby vortices manifest the following properties which seem to be rather paradoxical at first glance — see [10,17].

Firstly, a cyclone propagates faster eastwards, the larger its amplitude (that is, the thinner the water thickness in the cyclone (!)). Secondly, an anticyclone propagates faster eastwards, the less its amplitude (that is, again, the less the water thickness). These properties are contrary to those observed at $H_0 =$ const [1].

At some period of the paraboloid's rotation (depending on the vortex amplitude), the gradients in the Coriolis parameter and in the shallow water thickness equilibrate each other and, according to (5), the linear Rossby waves do not propagate. Therewith, the anticyclones of finite amplitude propagate faster westwards, the larger their amplitude, and cyclones of finite amplitude propagate faster eastwards, the larger their amplitude. In definite ranges of the vessel rotation periods, the eastward cyclone drift velocity and the westward anticyclone drift velocity exceed the Rossby speed, and the necessary soliton condition (10) is fulfilled. In particular, in the case of eastward vortex propagation, the solitonic condition may be fulfilled for cyclones only.

All of these paradoxical properties of the vortices under consideration may be successfully explained by means of two independent methods. Firstly, as shown above, the phase velocity of a nonlinear Rossby structure in the paraboloid, taking due account of the gradient in H_0, according to the generalized Charney-Obukhov equation [10,11], is expressed by the relationship:

$$V_{\mathrm{ph}} \simeq V_{\mathrm{R}} \left(1 - \frac{h \left(\frac{r_{\mathrm{R}}}{a} \right)^2}{1 - \frac{d \ln H_0}{d \ln f_0}} \right) . \tag{17}$$

At $d \ln H / d \ln f > 1$, the soliton condition (10) may be fulfilled only for $h < 0$, that is, only for cyclones. Therewith, the larger the cyclone amplitude (h), the faster its eastward velocity. And the larger the anticyclonic amplitude, the less, according to (17), its eastward propagation velocity. So, the relationship (17) explains all of the seeming paradoxes mentioned above.

The second — independent — method of interpretation of the paradoxes is connected with an integral relationship [16], derived on the basis of rather general considerations, which are quite independent of the Charney-Obukhov equation (see, for instance, [17] and [10,11]). The relationship gives the centre of mass velocity , V_c, for a solitary Rossby vortex in the parabolic device:

$$V_c = V_{\mathrm{R}} \left[1 + \frac{1}{g H_0} \frac{V^*}{V_{\mathrm{R}}} \frac{\int\int H v^2 dx dy}{\int\int \delta H dx dy} \right] , \tag{18}$$

where V^* is the Rossby speed (4) at $H_0 =$ const, and V_{R} is the total Rossby speed defined by (5); the value δH is the shallow water elevation, positive for an anticyclone and negative for a cyclone, v is the vortex's own rotation velocity, x is the coordinate along the parallel circle (azimuthal direction), and y, along the meridian (radial direction). By the way, it is easily seen from (18) that in the parabolic device the nonlinear β-effect (the scalar nonlinearity) is effectively restored, with a factor approximately equal to r_R^2/a^2 . If one takes into account that in the case considered the gradients in the shallow water thickness and in the Coriolis parameter have the same — negative — signs, then it is easy to see that the relationship (18) explains all of the paradoxes mentioned, independently of (17). Also, it is easy to see that the difference between the values V_c and V_R turns out to be equal to the difference between the phase velocity (17) and V_R. Finally, the eastward drift of the finite amplitude cyclones and the westward drift of the finite amplitude anticyclones at the "standing periods" (when the linear Rossby waves do not propagate) are also simply explained by the relationship (18).

It is useful to note for further consideration, that the centre of mass of a linear Rossby wave packet propagates with the Rossby speed, as may be seen from (18):

$$(V_c)\,\mathrm{lin} = V_R\ . \tag{19}$$

We call the cyclone-anticyclone asymmetry described in this section "inverted" (or "anomalous") asymmetry. This type of asymmetry should be realized, in particular, in the magnetized plasma (see below).

IX MERIDIONAL DRIFT OF ROSSBY VORTICES AND THE SECOND SOLITON CONDITION

Our experiments have shown that the Rossby vortices under study drift not only along the parallel circle, but along the meridian as well. The results of those experiments are presented in Figs. 3,4. In the figures, the drift velocities of cyclones and anticyclones along the parallel (V_x) and along the meridian (V_y), as well as the vortex life time τ, (the e-folding time of a decrease in the maximal linear rotation velocity on the vortex profile after turning off the local vortex source), are shown as functions of the paraboloid's rotation period. The eastward (in the sense of the vessel rotation) and northward (to the vessel periphery) directions have been defined as positive.

(It is necessary to note that the results presented in Figs. 3,4, have been obtained just recently and, in this sense, some of their details need to be checked. However, in total, they are quite definite.)

It is seen from Figs. 3,4 that the westward propagating cyclones deviate to the paraboloid pole (in the direction of the gradient in Coriolis parameter), while anticyclones deviate to the vessel periphery (counter to the gradient in Coriolis). This is a known fact from fluid dynamics. Indeed, according to the law of conservation of potential vorticity (the Ertel theorem),

$$\mathbf{e}_z \frac{(\mathrm{curl}\,\mathbf{v} + \mathbf{f})}{H} = \mathrm{const} \tag{20}$$

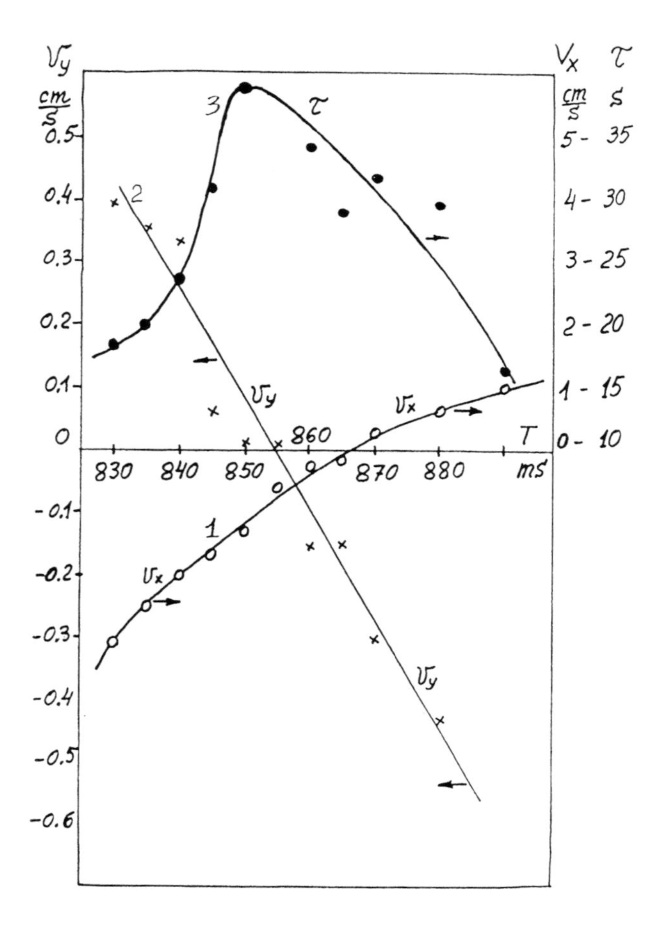

FIGURE 3. The anticyclonic Rossby vortex drift velocities: along the parallel (V_x, curve 1) and along the meridian (V_y, curve 2), as well as the vortex life time (curve 3), as the functions of the paraboloid rotation period, T.

(see, for instance, [1]), at H = const the above-mentioned deviations of the Rossby vortices from the parallel circle lead to loss of vortex energy.

When the direction of vortex propagation along the parallel is reversed (due to the presence of the negative meridional shallow water thickness gradient), the directions of the Rossby vortex drifts along the meridian are changed too. Namely, the cyclones deviate to the vessel periphery, and anticyclones, to the vessel pole. A very clear-cut regularity in Figs. 3,4 attracts one's attention: there exists a direct correlation between the value of meridional drift velocity (V_y) and the vortex life time τ: when V_y passes a minimum, τ passes its maximum. The correlation mentioned reveals the second soliton condition [recall that the first one is (10)]:

$$V_y = 0 \,. \tag{21}$$

It is worth mentioning that the equatorward meridional drift of the famous natural anticyclonic vortex "The Great Dark Spot of Neptune" with a velocity of about one degree per month turned out to be a cause of the disappearance of that beautiful natural phenomenon. (The vortex was discovered at 22 degrees of the Southern latitude by the spacecraft "Voyager-2" in 1989, as it left the Solar system [18].)

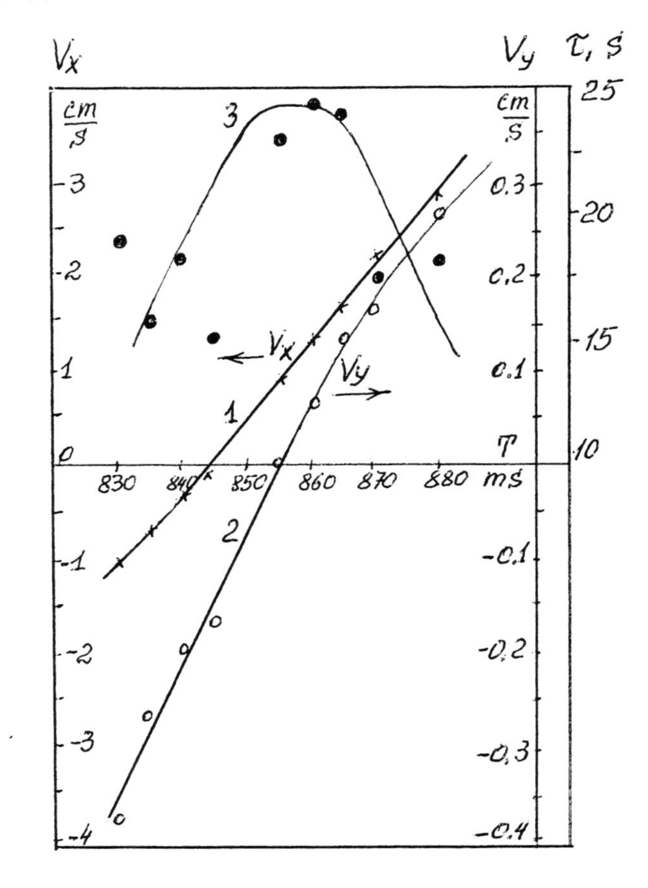

FIGURE 4. The cyclonic Rossby vortex drift velocities: along the parallel (V_x, curve 1) and along the meridian (V_y, curve 2), as well as the vortex life time (curve 3), as the functions of the paraboloid rotation period, T.

The condition (21) may be elucidated as follows — Fig. 5.

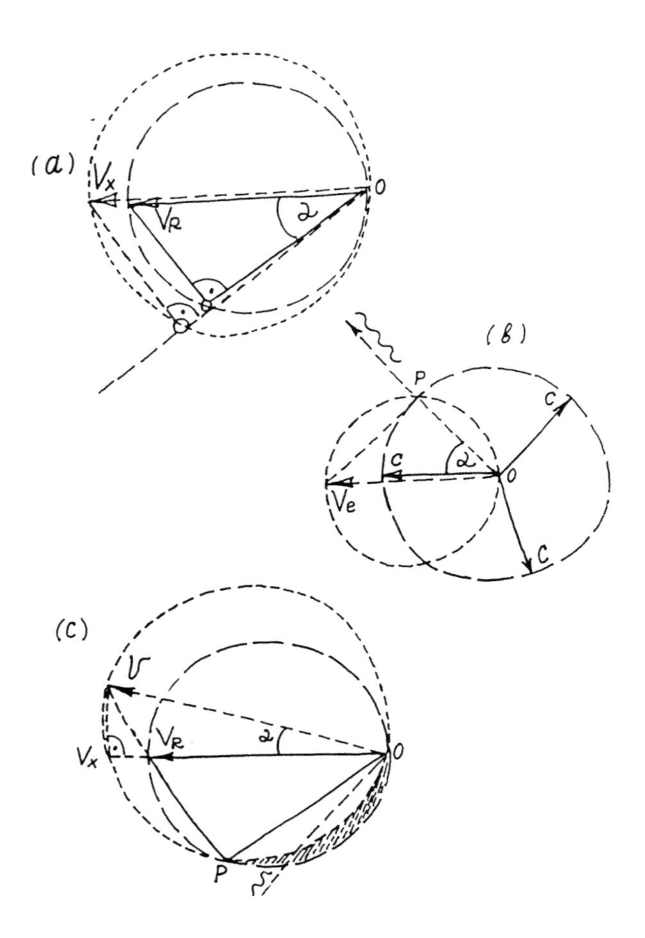

FIGURE 5. (a). An illustration showing the absence of Rossby wave radiation at any angle α in case when a Rossby vortex moves just along the parallel with a velocity, V_x, exceeding the Rossby speed, V_{R}.
(b). An illustration showing the Cerenkov radiation of an electron moving in a medium with the velocity, V_e, which is larger than the light velocity, c. The latter is isotropic, unlike the Rossby velocity. At the angle α, there is the Cerenkov resonance and light radiation.
(c). An illustration showing the Rossby wave radiation by a Rossby vortex moving at some angle α with respect to the parallel. The vortex velocity projection to the parallel direction exceeds the Rossby velocity. Nevertheless, there is a direction, indicated by the letter P, along which the projections of the vortex velocity and the Rossby velocity are equal. So, at all angles indicated by shading, there will be the Rossby wave radiation. It is an illustration of the second necessary soliton condition: $V_y = 0$.

The Rossby speed is maximal in the direction along the parallel. Its value along a given direction is proportional to the cosine of the angle of this direction with respect to the parallel. So, if one draws a circle whose diameter is equal to the westward Rossby speed (let H_0 be constant), then the Rossby speed along a given direction will be equal to the corresponding chord. If a vortex propagates at some angle to the parallel, then the projection of its velocity to the given direction will be equal to the corresponding chord of the circle whose diameter is equal to the vortex maximal velocity. It is seen from Fig. 5 that the circles mentioned intersect each other. It means that at all angles exceeding that corresponding to the intersection, there will be Rossby wave radiation, and the larger the vortex meridional velocity, the larger will be the radiation and the less will be the vortex life time. It is just what is shown in Figs. 3,4. In case of strong radiation, the vortex will not be a soliton. Therefore the second soliton condition is (21).

Let us now turn to an interpretation of the experimental data concerning the Rossby vortex meridional drifts and their directions.

The general idea forming the basis of our consideration is the following. In the process of drift along the parallel circle, the vortex, in principle, suffers a deformation caused by the dispersive and nonlinear effects. This deformation will cause appearance of a hydrostatic pressure gradient along the vortex trajectory. The latter, being crossed with the Coriolis parameter, will cause a gyroscopic effect — the advection flow and the vortex drift in the meridional direction. In principle, the physics of the gyroscopic effect mentioned is the same as that governing the very existence of the Rossby waves and their propagation along the parallel circle. Indeed, as is known [see, for instance, the expression (5)], the direction of the Rossby wave propagation is determined by the vector product $\mathbf{f} \times \mathrm{grad} f$, and the proportionality coefficient corresponds to the value of the Rossby speed defined by (5). So, finally, we may write

$$\mathbf{e}_x V_x = \left(\frac{gH}{f^3}\right) \mathbf{f} \times \mathrm{grad} f \, . \tag{22}$$

In accordance with the Ertel theorem, a gradient in the shallow water thickness will cause, physically, a drift just as that caused by the gradient in the Coriolis parameter. The only difference will be in the sequence of the multipliers in the vector product — since the Coriolis parameter is in the numerator of the potential vorticity (20), and the shallow water thickness, in the denominator. Thus, according to (22) and (5), we have

$$\mathbf{e}_y V_y = \left(\frac{g}{f^2}\right) \mathrm{grad} H \times \mathbf{f} \tag{23}$$

In the case considered, the term $\mathrm{grad} H$ is caused by the vortex deformation in the process of the vortex drift along the parallel circle.

(The authors mention with pleasure that the considered understanding of the gyroscopic effect under consideration as a general reason for the drifts (22) and (23) has appeared as a result of a discussion between one of us (M.V.N.) and Serguey Bazdenkov during this Workshop.)

Note that the drifts (22) and (23) are, physically, the same effects as the plasma drift in the crossed fields ($\mathbf{B} \times \mathrm{grad}B$) and ($\mathbf{B} \times \mathrm{grad}n$), where $\mathbf{B}$ is the magnetic field strength and $n = n(y)$ is the plasma density.

The relationships (22) and (23) may be substantiated also in an independent way. Let us take into account the following relationships:

$$H_0 = H_z \sin\phi \qquad \text{and} \qquad f = 2\Omega \sin\phi \, ,$$

where H_z is the shallow water thickness measured along the axis of the system rotation (and H_0 is measured perpendicularly to the shallow water free surface), $\sin\phi$ is the sine of the latitude, and Ω is the angular frequency of the system rotation ($\Omega = \mathrm{const}$). Hence, $H_0/f = H_z/2\Omega$, and the relationship (5) may be written as follows:

$$V_{\mathrm{R}} = \left(\frac{g}{f}\right) \frac{d}{dy} H_z \, , \tag{24}$$

or, in the vector form,

$$\mathbf{e}_x V_{\mathrm{R}} = \left(\frac{g}{f^2}\right) \mathrm{grad} H_z \times \mathbf{f}. \tag{25}$$

The meridional velocity will be expressed in the same way as (25), with only one difference: one has to take grad H_z not in the meridional direction (as in (25)), but in direction of the vortex drift. Then ,

$$\mathbf{e}_y V_y = \left(\frac{g}{f^2}\right) \mathrm{grad} H_z \times \mathbf{f} \, . \tag{25'}$$

From (22–25) we have (in order of magnitude):

$$\frac{V_y}{V_x} = \frac{khL}{a} \, , \tag{26}$$

where L is the characteristic size of the Coriolis parameter gradient, and k is the fraction of the initial "hill", h (in an anticyclone) or "well" (in a cyclone), which produces the perturbed pressure gradient under consideration. According to Figs. 3,4, the value of k is on the order of a few % , that is, on the order of magnitude, $k = h$. These estimates will be very useful for us in the next section. It is useful to note that the minimal value h necessary for the vortex formation is a/L [1]. Therefore, (26) means $V_y/V_x \geq khL/a$ (26).

Now, let us consider the concrete cases of the Rossby vortex meridional drift.

Case I: H_0 = const, the Rossby vortices propagate westwards.

(1,a) : Anticyclone, effect of the Rossby wave dispersion (the vortex amplitude is arbitrarily small, but finite).

The vortex will suffer dispersive spread, like a linear Rossby wave packet.

That is [19], its front part will become more smooth, and the back part, more steep. It means that deformation of the vortex will cause a perturbation of the anticyclonic "hill": there will appear a hydrostatic pressure gradient directed contrary to the vortex westward velocity. According to (23), this pressure gradient will cause the meridional drift of the anticyclone to the vessel periphery.

(1,b): Cyclone, effect of dispersion.

The front side of the vortex will become more smooth, and the back side more steep. Since the cyclone is a depression of the shallow water, the pressure gradient will be directed contrary to that in case of anticyclone. So, the pressure gradient will be directed ahead and will cause the cyclone meridional drift in the poleward direction.

(1,c); Anticyclone of a large amplitude, effect of scalar nonlinearity .

According to (14) and (18), the scalar nonlinearity will steepen the front side of the vortex and smooth the back side. The pressure gradient will be directed ahead, contrary to that caused by dispersion. It means that the effects of dispersion and nonlinearity may be mutually compensated, and thus the westward propagating anticyclone, in principle, may be a Rossby soliton.

(1,d): Cyclone of a large amplitude, effect of scalar nonlinearity.

The scalar nonlinearity will smooth the front side of the vortex and steepen its back side. Therefore, the pressure gradient will be directed, ahead, like when under the influence of wave dispersion. It means that in the case of a westward propagating cyclone the effects of dispersion and nonlinearity cannot be mutually compensated. Therefore, such a cyclone cannot be a Rossby soliton.

Case II: H_0 is not constant, the shallow water thickness has a considerable negative meridional gradient; the Rossby vortices propagate eastwards.

(2,a): Anticyclone, effect of dispersion.

The pressure gradient, as in case (1,a), will be directed contrary to the vortex propagation velocity. Since the latter has changed its sign, the meridional drift will be directed to the paraboloid pole.

(2,b): Cyclone, effect of dispersion.

The pressure gradient will be directed ahead, and the vortex will drift to the paraboloid periphery.

(2,c): Anticyclone of a large amplitude, effect of scalar nonlinearity.

According to the "paradoxical" result (16), (18), the vortex will propagate slower, the larger its amplitude. Hence, the pressure gradient will be directed contrary to the vortex propagation velocity, that is, in the same direction as under the influence of dispersion. It means that in the case of an eastwards propagating anticyclone the effects of dispersion and nonlinearity cannot be mutually compensated. In other words, such an anticyclone cannot be the Rossby soliton.

(2,d): Cyclone of large amplitude, effect of scalar nonlinearity.

According to (16) and (18), the cyclone propagates faster, the larger its amplitude. Therefore the pressure gradient will be directed contrary to its propagation velocity and will cause the meridional drift to the paraboloid pole. It means that the effects of dispersion and scalar nonlinearity, in principle, may be mutually compensated, and the eastwards propagating cyclone may be a Rossby soliton.

The reader may be easily convinced that the cases I and II are in a good agreement with the experimental data presented in Figs. 3,4.

As for the question to what extent the results presented in Figs. 3,4 definitely reveal the existence of a cyclonic Rossby soliton, this question, in principle, needs more detailed study. But, nevertheless, we believe those cyclonic vortices in Fig. 4, which have the maximal life time, are very near to the cyclonic Rossby solitons.

To conclude this section, let us make a remark concerning the connection between the polarization of the dipolar addition, which is formed during the Rossby solitary monopolar vortex self-organization (see Section 5), and the direction of the vortex acceleration. It is seen from (23) that if the cyclonic part of the dipole is situated from the "southern" side of the anticyclonic solitary monopole and, correspondingly, the anticyclonic part, from the northern side, then the monopole will be accelerated and therefore will not suffer nonlinear overturning. Analogously, if the dispersive effects are prevailing, then the pressure gradient caused by the vortex deformation will have the opposite sign, and the monopolar vortex will be decelerated. As shown in [12], the mentioned changes in the solitary vortex propagation velocity turn out to be very significant. The latter shows that the vortex drift in the crossed fields ($\text{grad}H \times \mathbf{f}$), induced due to the vortex self-deformation, is very essential.

X ON ANOMALOUS PLASMA DIFFUSION ACROSS A STRONG MAGNETIC FIELD

As is well known, the dynamics of drift vortices in magnetized plasmas, in which the gradients in density and the electron temperature are of the same sign, is strictly analogous to dynamics of Rossby vortices on rotating shallow water with the same signs of the gradients in the Coriolis parameter and water thickness [10,11].

As shown in the preceding section, in the case of a negative sign of the above-mentioned gradients in rotating shallow water, the Rossby solitons have cyclonic vorticity. It means that in magnetized plasma, with negative gradients in the density and the electron temperature (as in the widespread tokamaks and stellarators), the drift wave solitons will have cyclonic vorticity, that is, the vortex curl will be parallel to the magnetic field direction.

In [1], a mechanism for the anomalous plasma diffusion across the magnetic field was suggested, connected with the trapped-particle transport from one solitary vortex to another during vortex collisions. Here, we suggest a likely collisionless mechanism.

We proceed from the results of the preceding section. According to those results (see Figs. 3,4), the plasma drift cyclones should move to the periphery, i.e., across the magnetic field.

One may estimate the transverse plasma transport taking the ratio of the meridional/radial (V_y) and the azimuthal (V_x) velocities of the drift vortices (cyclones, since they move to the plasma periphery). To do that, let us recall some aspects of the above-mentioned analogy between Rossby vortices and drift vortices in magnetized plasma [10,11]. Namely,

- the plasma analogue of the vector Coriolis parameter ($\mathbf{f}$) is the vector ω_B directed along the magnetic field and having the absolute value equal to the ion Larmor frequency,

- the analog of the shallow water thickness (H_z) is the plasma density, (n),
- the analog of the Rossby speed (V_R) is the plasma drift velocity (V^*),
- the analog of the perturbed (grad$H$) caused by the vortex deformation is the electric field caused by the deformation of the vortex potential profile due to dispersive and nonlinear effects considered in the preceding section; that electric field is directed along the vortex trajectory and, being crossed with the magnetic field, causes the radial (outwards) drift plasma motion across the magnetic field,
- the physical mechanism driving drift waves (vortices) around the magnetic field and causing the transverse motion ("anomalous" plasma diffusion across the magnetic field) is the same as that considered in the preceding section [see relationships (22–26)].

Therewith, according to the mentioned analogy, in the plasma case, the velocity V_x is somewhat larger than the plasma drift velocity. The latter is:

$$V^* = \frac{cT_e}{eBL}, \tag{27}$$

where c is the velocity of light, T_e is the electron temperature, e is the electron charge, B is the magnetic field strength, L is the characteristic size of the plasma density gradient.

According to (22–26), the ratio V_y/V_x is of the order of :

$$\frac{V_y}{V_x} = \frac{khL}{a}, \tag{28}$$

where L is the characteristic size of the Coriolis parameter gradient (a counterpart of the above-mentioned size L in plasma), h is a counterpart of the dimensionless plasma potential, that is the potential over the electron temperature, and k is the fraction of the vortex pressure appearing as a result of the vortex deformation considered in the preceding section. It seems reasonable to take for the estimate, in order of magnitude:

$$k = h, \tag{29}$$

which corresponds to the experimental data presented in Figs. 3,4.

Then, taking into account that the minimal possible value of the dimensionless plasma potential in the drift vortex (necessary for plasma trapping) is [1]

$$\frac{a}{L}, \tag{30}$$

we find:

$$\frac{V_y}{V^*} = \frac{a}{L} \,. \tag{31}$$

where a is the drift vortex size.

The plasma radial flow density will be $nV_y = Dn/L$, where D is the plasma diffusion coefficient. Hence,

$$D = \left(\frac{cT_e}{eB}\right)\frac{a}{L} \,. \tag{32}$$

It is very interesting to compare the obtained value of D with the famous Bohm diffusion coefficient. The latter is

$$\left(\frac{1}{16}\right)\frac{cT_e}{eB} \,. \tag{33}$$

It is seen that the plasma diffusion under consideration will be of the order of the Bohm diffusion. For instance, if, in the context of (33), one assumes, that $V_y/V_x = 1/16$ (which is in agreement with the data of Figs. 3,4), then it follows immediately that the diffusion coefficient is just equal to the Bohm coefficient (33). Note that this corresponds well with the experimental data [20].

We may suggest the following method for an essential decrease of the plasma transport across the magnetic field. It is necessary to create an unstable plasma shear flow of anticyclonic vorticity in the region of plasma density gradient.

Such a flow, due to its anticyclonic vorticity, will destroy cyclones and thus prevent their fast drift outwards from the plasma.

In real tokamaks and stellarators, there exists a radial electric field directed into the plasma, such that the field has a maximum (in absolute value) at the radius of maximum density gradient [20]. It can be easily seen that the plasma flow produced by the crossed fields $(\mathbf{E}, \mathbf{B})$ between the mentioned $\mathbf{E}$-field maximum and the separatrix will have anticyclonic shear. Such a shear may destroy cyclones and thus improve the plasma confinement.

We assume that the above-mentioned mechanism may be responsible for the essential improvement of the plasma confinement as a result of the transition between the L- and H-modes in tokamaks and stellarators, because this transition is followed by an essential increase in the electric field strength of the above-mentioned (just suitable) direction and the radial profile.

XI GENERATION OF ROSSBY VORTICES BY UNSTABLE SHEAR FLOWS

All the results concerning this section have been published (see, for instance, [1,21]. Here, we only wish to mention the following points.

1. A shear flow may generate or destroy vortical structures. The first case takes place when both the flow and the vortex have vorticity of the same sign. The second case takes place when the vorticities are of opposite signs. This remark is made in the context of a likely mechanism (considered in the preceding section) for essential reduction of plasma diffusion across the magnetic field by means of destruction of the plasma drift cyclones by a sheared electric field having anticyclonic vorticity.

2. The cyclone — anticyclone asymmetry for the case $H_0 =$ const, considered in sections 4,5, has been additionally substantiated by the experiments of our team with unstable shear flows in rotating parabolic devices, in which the shallow water has a free surface.

3. In experiments involving the generation of chains of Rossby vortices (and spiral structures) a regularity has been revealed, which is common for creation of quite different structures and quite different experimental conditions. It is that, if a chain of some structures (having a given number of structures along the system perimeter) is generated, it means that the growth rate of the shear flow instability turns out to be maximal for a given chain mode (i.e. for a given number of structures along the system perimeter). In particular, if only one vortex along the whole system perimeter is generated [1], it means that the instability growth rate with respect to that mode ($m = 1$) is larger than that with respect to other modes. For details see [22].

XII MODELING SPIRAL STRUCTURES IN GALAXIES AND DISCOVERY OF INTERARM VORTICES

The material of this section has been published too [1,22]. We only wish to recall here the following.

In 1986, our team observed for the first time anticyclonic vortices in a system of spiral arms. The experiments were carried out by means of a setup consisted of two paraboloids with free surface shallow water. The latter were rotated differentially: there was an essential velocity jump between them, analogous to that existing in the most of the spiral galaxies. The shear flow regime was effectively supersonic.

The observed vortices have had the following properties: (1) they were anticyclones, (2) they were situated between the spiral arms, (3) their transverse dimensions were of the order of a few Rossby-Obukhov radii.

Basing on that observation, our team [23] has predicted the existence of analogous vortices in real galaxies.

XIII ASTRONOMICAL OBSERVATION OF INTERARM VORTICES PREDICTED ON THE BASIS OF LABORATORY EXPERIMENTS.

We recall that, according to recent publications, a special program based on the prediction mentioned in Section (12) was realized by the Russian and French astronomers and giant vortices having the predicted properties were discovered in real galaxies by direct astronomic observations.

The vortices discovered proved to have just the predicted properties. Namely, they turned out to be anticyclones situated between the spiral arms and having dimensions of the order of a few Rossby-Obukhov radii. For details, see, for instance, [22] and references given there.

XIV A COMMON MECHANISM GENERATING PLANETARY AND GALACTIC VORTICAL STRUCTURES.

Generation of the observed Rossby solitary vortices and spiral structures proves to be caused by the common mechanism determined by the centrifugal instability. The latter takes place when the inner part of the system is rotated faster than the periphery. The pecularity of the centrifugal instability observed is that it manifests itself in the regime, which is effectively supersonic: the equivalent Mach number [the velocity shift over the velocity $(gH)^{1/2}$] of the long shallow water waves reached the value about 10 and even more [1,22].

This instability differs essentially from the Kelvin-Helmholtz (KH) instability because it is kept even at the (equivalent) Mach numbers up to ten or even more, whereas the KH instability of the shallow water ceases at Mach numbers of about two (see [1] and refences given there).

XV CONCLUSION

Thus, our laboratory approach based on rapidly rotating shallow water experiments proved to be a rather fruitful method for investigation of planetary, galactic, and plasma structure dynamics.

XVI ACKNOWLEDGEMENTS

The authors express their gratitude to Prof. Robert Dewar and Dr. Ross Griffiths for the organization of an excellent and very productive Workshop. The authors are grateful to Serguey Bazdenkov for a fruitful discussion on gyroscopic phenomena in rotating fluids. This work has been implemented under financial support by the Russian Foundation of Basic Research (grant No. 96-05-64061), by the INTAS Foundation (grant No. 95-0988), and by the Russian Ministry of Science.

REFERENCES

1. M.V. Nezlin and E.N. Snezhkin:"Rossby Vortices, Spiral Structures, Solitons", Springer-Verlag, Heidelberg, 1993.
2. T. Maxworthy, L.G. Redekopp: Icarus **29**, 261 (1976); Science **210**, 1350 (1980).
3. R.Z. Sagdeev, V.D. Shapiro, V.I. Shevchenko: Sov. Astron. Lett. **7**, 279 (1981).
4. G.R. Flierl: POLYMODE News 62, 1 1979.
5. E.N. Mikhailova, N.B. Shapiro: Izv. Atm. Ocean. Phys. 16, 587 (1980).
6. V.I. Petviashvili: Sov. Phys. JETP Lett. **32**, 619 (1980).
7. J.G. Charney, G.R. Flierl: in "Evolution of Phys. Oceanography", ed. by B.A. Warren, C. Wunsch (MIT Press, Cambridge, 1981).
8. G.P. Williams, T. Yamagata: Journ. Atmos. Sci. **41**, 453 (1984).
9. N.N. Romanova, V.Yu. Tseitlin: Izv. Atm. Ocean. Phys. **20**, 85 (1984).
10. M.V. Nezlin, G.P. Chernikov: Plasma Physics Reports **21**, 922 (1995).
11. M.V. Nezlin, G.P. Chernikov, A.Yu. Rylov, K.B. Titishov: Chaos **6**, 309 (1996).
12. J. Nycander, G.G. Sutyrin: Dyn. Atmos. Oceans **16**, 473 (1992).
13. G.G. Sutyrin: Dokl. Akad. Nauk USSR, Earth Sci. **280**, 38 (1985).
14. G.G. Sutyrin, I.G. Yushina: Sov. Phys. Dokl. **33**, 179 (1988).
15. G.G. Sutyrin: Chaos **4**, 203 (1994).
16. J. Nycander: J. Fluid Mech. **254**, 561 (1993).
17. M.V. Nezlin, A.Yu. Rylov, K.B. Titishov, G.P. Chernikov: Izv. Atm. Ocean. Phys. **33** (3), 1997. In press
18. R.A. Kerr: Science **245**, 929 (1989).
19. G.R. Flierl: J. Phys .Oceanogr. **7**, 365 (1977).
20. M.G. Shats, D.L. Rudakov, B.D. Blackwell, G.G. Borg, R.L. Dewar, S.M. Hamberger, J. Howard, L.E. Sharp: Phys. Rev. Lett. **77**, 4190 (1996).
21. M.V. Nezlin: Chaos **4**, 187 (1994).
22. M.V. Nezlin: in "Physics of the Gaseous and Stellar Disks of the Galaxy", ed.by I.R. King, ASP Series, **66**, 136 (1994).
23. M.V. Nezlin, V.L. Polyuachenko, E.N. Snezhkin, A.S. Trubnikov, A.M. Fridman: Sov. Astron. Lett. **12**, 213 (1986).

Geostrophic Turbulence and Geophysical Circulations

Peter B. Rhines

School of Oceanography, University of Washington
Seattle, Washington USA 98915

Abstract. Basic vorticity principles give immediate insight into the behavior of turbulent flow of a classic, homogeneous fluid. Analogously, potential vorticity principles help to describe the large-scale flows of the Earth's oceans and atmosphere, and those of the other planets; system rotation and density stratification are dominant effects. In both cases the tracer-like properties of vorticity and potential vorticity yield insight and analysis. The Ertel-Rossby potential vorticity arises essentially from Kelvin's circulation theorem for fluid circuits lying on surfaces of constant potential density. Here we describe some of the dynamics of such fluids, notably the large-scale 'geography' of potential vorticity provided by the shape of the planet (or bounding container), waves that depend upon this geography, mean circulations that determine the geography, and geostrophic turbulence that actively help to shape this geography. The tracer-like property of potential vorticity encourages us to consider fluid circulations as being the product of boundary sources and sinks of potential vorticity, distributed by mean circulation and mixed (often homogenized) by geostrophic turbulence, with attendant back-effects on the mean circulation.

I INTRODUCTION

Among fluid dynamical studies aimed at oceans and atmospheres and possibly applicable to problems in plasma physics, wave/turbulence interactions and turbulence/mean circulation interactions stand out. We have seen over the past 25 years a growing understanding of wave/turbulence/circulation interrelationships, yielding for example some quantitative theory of the evolution of energy-containing quasi-geostrophic eddies, the interaction of eddies with time-averaged circulation, such as the development of zonal jets on Earth and the great outer planets, and the evolution of large regions of homogenized potential vorticity, both 'staircases' and 'gyres'. Together, these ideas begin to describe the complete problem of turbulent circulating fluids. By 'describe' we mean first at the diagnostic level to establish relationships between measurable eddy properties and the mean circulation they induce, and then at the

CP414, *Two-Dimensional Turbulence in Plasmas and Fluids:* Research Workshop
edited by R. L. Dewar and R. W. Griffiths

predictive level to give models, at least, of the circulation and eddies subject to prescribed boundary- and initial conditions. Two groups have been particularly active in this: stratospheric meteorologists, and dynamical oceanographers. Previous extended accounts and reviews in this area are available [1,2-6].

The outline of this paper is as follows: classical vorticity principles; potential vorticity, including its integral properties, 'impermeability' and 'invertibility'; the geography of the potential vorticity of the large-scale circulations (of oceans and atmospheres); brief discussion of the principle features of geostrophic turbulence; transport of potential vorticity by eddies/waves and its attendant relation to momentum transport, homogenization of potential vorticity as an often-occupied limiting state, sources and sinks, boundary 'reservoirs', and related breeds of geophysical turbulence such as rotating convection.

II CLASSICAL VORTICITY PRINCIPLES

Vorticity of an ordinary, Newtonian fluid, denoted $\vec{\omega}$, is a spin-like quantity, the curl of the vector velocity, $\vec{u}$, which has a particularly tangible nature. It is the vector associated with the anti-symmetric part of the rate-of-strain tensor $\partial u_i/\partial x_j$ (in that $\partial u_i/\partial x_j = -\varepsilon_{ijk}\omega_k$). It represents the angular momentum, $\vec{L}$, of an infinitesimal sphere of fluid about its center (times $2/I$, where I is the moment of inertia). Of course the 'sphere of fluid' ceases to remain a sphere for very long, so that angular momentum connections must be made with care. More generally the two properties are related through the identity

$$|\vec{r}|^2\vec{\omega} = -2\vec{r} \times \vec{u} + \nabla \times \left(|\vec{r}|^2\vec{u}\right)$$

or

$$\vec{L} = -\frac{1}{2}\iiint \rho|\vec{r}|^2\vec{\omega}\, dV$$

provided the mass-density ρ is a constant and velocity vanishes at the limits of volume integration; $\vec{r}$ is position in space. Angular momentum appears as the $2d$ spatial moment of the vorticity. The classic description of relative fluid motion near a point, e.g. [7], is completed by the symmetric part of the rate-of-strain tensor, which relates to a pure straining motion, and its trace, which is the rate of expansion or contraction of the fluid, $\nabla \cdot \vec{u}$.

In a constant-density fluid $\vec{\omega}$ is conserved following ideal (non-dissipative, unforced) fluid movement, in the sense of a vector tracer [8]. The curl of the momentum equation for an homogeneous (ρ = constant), incompressible fluid is

$$\frac{D\vec{\omega}}{Dt} = (\vec{\omega} \cdot \nabla)\,\vec{u} + \nu\nabla^2\vec{\omega} \tag{1}$$

ν being the kinematic viscosity and $D/Dt = \partial/\partial t + \vec{u} \cdot \vec{\omega}$ the usual time-rate of change following the fluid velocity. Inviscid flows ($\nu = 0$) thus obey conservation of the vector vorticity, in the sense that the first right-hand side term describes the stretching and tipping of the vorticity vector, as if it were a passive material 'arrow' marked with dye. If the infinitesimal arrow is initially drawn parallel to $\vec{\omega}$ it continues to represent vorticity for all time, or so long as it remains of infinitesimal length. Vortex lines are infinitesimal 'dye arrows' chained together. A tube comprising a 'bundle' of vortex lines has a 'strength', $\iint |\vec{\omega}| dA$, where the integration is across the tube, normal to the vorticity. This strength is conserved in absence of dissipation and density variations. This integral is also identical to the circulation integral about the periphery of the tube, at a given position along it, $\oint \vec{u} \cdot d\vec{s}$, where $d\vec{s}$ is arc length, and its permanence is just the same as is found in Kelvin's circulation theorem. While these remarks emphasize a tube made up of vortex lines, the Kelvin circulation/vortex flux integral can be evaluated for an arbitrary closed path in the fluid. The tangible quality of vortex lines is probably the most useful result of the fluid dynamics of turbulence. While vorticity is not a passive 'dye', we can visualize deformations of dye lines in complex fluid motions and begin to infer a persistent stretching of vortex lines; indeed, the permanence of vortex lines leads to important topological invariants (like knottedness) that constrain turbulent flows [9].

Reintroducing dissipation, we find the volume integrated dissipation rate for kinetic energy in this simple fluid by forming the scalar product of $\vec{u}$ with the momentum equation. The identity

$$\vec{u} \cdot \nabla^2 \vec{u} = \vec{u} \cdot \nabla (\nabla \cdot \vec{u}) - \vec{u} \cdot \nabla \times \vec{\omega}$$

leads to the following expression for this dissipation (simplifying for an incompressible fluid, $\nabla \cdot \vec{u} = 0$):

$$\nu \iiint \vec{u} \cdot \nabla^2 \vec{u} \, dV \equiv -\nu \iiint |\vec{\omega}|^2 dV$$

provided that at the outer surface bounding the integration (unit normal $\hat{n}$) the area integral

$$\nu \oint\!\!\oint \vec{\omega} \times \vec{u} \cdot \hat{n} \, dS \tag{2}$$

vanishes. Thus, immediately, the increase in the variance of the vorticity amplitude associated with vortex-line stretching is equated with dissipation, and this is the secret of energy destruction, no matter how small the viscosity [irrotational flows do of course dissipate energy, yet through the normally negligible term (2). The intuitive sense that lines marked with dye in a fluid experience a rapid increase in length can be developed in quantitative models; typically one argues that the relative separation velocity of two marked particles increases in proportion to their separation, so long as that difference-velocity is dominated by eddies much larger than the separation. If this is so, then one sees the length of a chain of such particles increasing like $\exp(t)$. Dissipation of kinetic energy, or enstrophy (which is $\frac{1}{2}\vec{\omega} \cdot \vec{\omega}$), and diffusive mixing of passive chemical constituents or tracers all rely on this quasi-exponential stretching of material lines (or vortex lines). It is the process by which one stirs cake batter and ultimately mixes it, by increasing the surface area between the 'chocolate' and 'vanilla' phases. Vorticity is related to small-scale energy (indeed the Fourier spectrum of vorticity with respect to wavenumber $\vec{k}$, $\vec{k} \cdot \vec{k}\ E(\vec{k})$ is a high-pass filter operation on the kinetic energy spectrum, E). Turbulent dissipation thus involves the production of small-scale energy of sufficient magnitude to carry out the explicit, molecular dissipation.

It is interesting to note at this point that, though we are dealing with simple fluid continua, some related descriptive results occur in ensembles of particles. Define the position vector marking a particle to be $X_i(t;\ X_i^0,\ t^0)$ where t is the time, and the superscripts indicate initial position and time. U_i is its time-derivative. The statistics of ensembles of particles (which may be marked particles of fluid) first described by Taylor [10], involve tensors like $U_i(t)U_j(t+\tau)$, which are known as Lagrangian velocity correlations. The product $\langle U_i X_j \rangle$ may be called a particle diffusivity, κ_{ij}. It has the properties that its symmetric part gives the rate of 'expansion' of the best-fit ellipsoid to the ensemble of particles

$$\frac{d}{dt}\langle X_i X_j \rangle = \kappa_{ij} + \kappa_{ji}$$

and its antisymmetric part describes the angular momentum of the ensemble:

$$-\varepsilon_{ijk} L_k = \kappa_{ij} - \kappa_{ji}$$

for unit masses. These relations remind us of the rate-of-strain tensor for continuum fluid, above. The antisymmetric part is sometimes called a 'skew' diffusivity, and it is prominent in waves which involve orbital particle motions. Analysis of movement of fluid particles gives us information about the 'Lagrangian'-mean circulation, $d/dt(X_i(t, X_i^0, t^0))$, and the dispersion by turbulence of ensembles of particles, but despite long interest the subject seems still in its infancy; and practical results are not numerous. Ideas of continuous tracer fields and fractal descriptions of chaotic mixing are a promising, related avenue.

Geophysical flows involve two important additions to the classic fluid: mean system rotation, at a rate $\vec{\Omega}$, and stratification of fluid mass-density. For the former, the vorticity equation above holds perfectly if the vorticity is replaced by absolute vorticity, $\vec{\omega}_a \equiv \vec{\omega} + 2\vec{\Omega}$ ('relative' plus 'planetary'). In cases of large-scale flows, the system rotation is usually much greater than the vertical component of relative vorticity, in which case the vortex stretching term is dominated by $(2\vec{\Omega} \cdot \nabla)\vec{u}$. The ratio $|\vec{\omega}|/|2\vec{\Omega}|$ has a typical magnitude

$$R_0 \equiv U/\Omega L \ll 1$$

where U is a typical horizontal velocity and L the horizontal length-scale of the flow; R_0 is known as the Rossby number. We visualize parallel lines of planetary vorticity threading through the fluid. These are unlike typical marked dye lines that, as G.I. Taylor [11] long ago remarked; in fact they provide an important 'stiffness' to the fluid, resisting both stretching, bending and tipping. The fundamental, quasi-exponential-with-time stretching of vortex lines, seen above in three-dimensional turbulence, cannot be expected to occur in these large-scale flows! Energy dissipation by small-R_0 turbulence cannot occur by cascade to small scale.

Density variations work on the vorticity through the new term $\nabla p \times \nabla \rho/\rho^2$ on the right-hand side of eqn. 1. Classical meteorologists would draw vertical cross-sections of pressure and density, and count the number of intersections between contour curves of the two fields. These 'solenoids' quantitatively describe the production of vorticity by density twisting. However, in these same large-scale flows, most of the twisting is balanced by a steady tipping of planetary vortex lines, $(2\vec{\Omega} \cdot \nabla)\vec{u}$, rather than a time-change of the relative vorticity. But, this twisting is acting almost exclusively on the horizontal vorticity, owing to the nearly vertical orientation of both ∇p and $\nabla \rho$. The vertical component of vorticity is subject to smaller but crucial changes induced by vertical stretching of the planetary vorticity. And, continuing the thread of argument about activation of diffusive processes by cascade to small length-scale, we see that density twisting cannot be expected to alter the limited vorticity production in a rapidly rotating fluid, at least insofar as the *vertical* vorticity is concerned.

Beyond its importance in the evolution of turbulence, vorticity plays a central role in force balances in fluids. At large Reynolds number the generation of vorticity is often dominated by what happens near fluid boundaries (where typically the velocity vanishes). The boundary layers on airplane wings, though thin, contain a total amount of vorticity roughly independent of their actual boundary-layer thickness. They act as sources of vorticity that is shed to the fluid interior to dominate a huge volume of fluid [7, p. 277]. The lift of the airfoil is proportional to the free-stream speed and the circulation about it. This circulation induced by the solid wing may be visualized as the net vorticity of a 'bound' vortex, which arises to locate the rear stagnation point at the trailing edge of the wing. While it is true that global integrals of the vorticity, and its flux through the boundaries of the fluid, often have little interest (the volume integrated vorticity vanishing, providing the volume extends out to a region of no motion), they serve to remind us that at the initiation of flight a 'starting' vortex is created with equal and opposite vorticity to that bound in the moving wing. This separating vortex pair possesses the downward fluid impulse (equal to the dipole moment of the vorticity distribution) increasing at a rate which is equal and opposite to the lift. As the separation between them increases the impulse is more and more perfectly vertical, and hence the 'induced drag' of a two-dimensional airfoil then vanishes. In three dimensions the two vorticies are connected by vortex lines emanating from the wing tips (recall that vorticity is solenoidal, $\nabla \cdot \vec{\omega} = 0$: vortex lines cannot terminate within the fluid).

As we move to consider potential vorticity, we will ask how these qualities of vector vorticity are transformed into the new setting of a rotating, density-stratified fluid.

III POTENTIAL VORTICITY (POVORTY)

Because fluid buoyancy produces horizontal vorticity, it useful to evaluate Kelvin's circulation theorem on a special fluid circuit, which lies on a surface of constant potential density, θ. In meteorology θ is the potential temperature, and the constant-θ surfaces are isentropic. In oceanography θ involves both potential temperature and salinity. Nonlinearities in the equation of state complicate the oceanic case, and these can add exotic instabilities and convection not visible with the present, approximate equation [12]. Potential density is defined as the fluid mass-density, after adjusting the pressure to a constant reference value. This being done, the contribution from the pressure/density correlation will vanish, for $\oint (\nabla p/\rho) \cdot d\vec{s} = 0$, integrated about such a closed circuit, if ρ is a unique function of p upon it. This picks out a special component of vorticity for our interest, and by manipulating Kelvin's theorem or equivalently taking the scalar product of eqn. (1) with $\nabla\theta$ we find

$$\begin{aligned} \frac{Dq}{Dt} = 0, \quad q &= \rho^{-1} \vec{\omega}_a \cdot \nabla\theta \\ &= \rho^{-1} \nabla \cdot (\vec{\omega}_a \theta) \end{aligned} \qquad (3)$$

the last identify following from the solenoidal nature of absolute vorticity. Here q is the Ertel-Rossby potential vorticity, which might more simply be called 'povorty'. With the warning that it is not yet widely accepted terminology, we henceforth use this abbreviation. In the more general case with externally imposed forces, thermodynamic forcing, viscous dissipation we write, now in flux form [13]

$$\begin{aligned} \frac{\partial}{\partial t}(\rho q) &= -\nabla \cdot \vec{J} \\ \vec{J} &= \rho \vec{u} q - \omega_a \Im - \vec{F} \times \nabla\theta \end{aligned} \qquad (4)$$

where

$$\frac{D\theta}{Dt} = \Im$$

expresses the thermodynamic forcing (from radiation, heat diffusion, or parametrized small-scale mixing of heat), and $\vec{F}$ is the force per unit mass acting on the fluid, incorporating momentum diffusion and externally imposed forces. The flux form of the conservation equation for q emphasizes that povorty cannot cross surfaces of constant potential density. For, in the expression for the its flux $\vec{J}$ the third term lies parallel to these surfaces, and the first and second terms combine to give a component across the surfaces at just the rate the surfaces themselves are moving (whether due to advection or due to thermodynamic forcing, which causes the θ=constant surfaces to move relative to the fluid). This, or the second expression for q given in eqn. 3, shows that povorty integrated between two adjacent surfaces of constant potential density, is altered only by flux along θ-surfaces. It is conservative even in the face of advection and thermodynamic forcing (the first two RHS terms), providing the momentum forcing F vanishes. For, indeed, it is just proportional to the absolute vorticity, $\vec{\omega}_a \equiv \vec{\omega} + 2\vec{\Omega}$. After a further integration over a tablet-shaped region bounded above and below by constant-θ surfaces, and at the sides by a curve that moves with the fluid, conservation of the absolute Kelvin circulation

$$\oint \vec{u} \cdot d\vec{s} + 2\Omega A$$

follows; A is the area of the tablet *projected upon the equatorial plane, perpendicular to* $\vec{\Omega}$. Thus, povorty integrated through a fluid volume between two surfaces of constant potential density, is the absolute Kelvin circulation, which is robustly conserved, even in the presence of diffusion and thermodynamic forcing. There are many premonitions of povorty in the early literature, from Lamb's discussion of rotating basins of water ([14], p. 333) to Rossby's development of a layered model of the density stratification ([15]), and Ertel's more complete derivation ([16]). In these it was early recognized that the overwhelming strength of the planetary vorticity can dominate the production of relative vorticity, under vertical stretching.

The fluid boundaries, the top and bottom of the ocean, however, are not impermeable to povorty. Specifying 'natural' boundary conditions like the heat flux through the boundaries and stress or velocity at the boundary does not simply constrain the flux of povorty. Indeed, in the trivial example of a motionless, temperature stratified fluid in an insulated box, heat diffusion will cause the surfaces of constant θ to escape the box, and with them the povorty (which will be zero at the isothermal end-state). This is somewhat the same as in the case of vorticity flux through a rigid boundary in a classical, non-rotating fluid.

We have now two fields of particular importance, the potential density θ and povorty, q (in addition to the classical velocity and pressure fields). Surfaces of constant potential density, θ, lie nearly horizontally (at large scale) in oceans and atmospheres, and there is no gravitational penalty if fluid moves along these surfaces; they become the central reference surfaces for large-scale dynamics. Their influence is measured by the buoyancy frequency,

$$N = (\frac{g}{\rho}\frac{\partial \rho}{\partial z} - \frac{g^2}{c^2})^{1/2} \approx (-\frac{g}{\rho}\frac{\partial \theta}{\partial z})^{\frac{1}{2}},$$

c being the speed of sound and z the vertical coordinate. Maps of trace-chemical concentrations on constant-θ surfaces often reveal circulation and mixing to great advantage. Covering these surfaces are contours of constant povorty, q, which further restrict the paths of circulation. For, a simple steady solution of eqn. (2) would be just

$$q = q(\psi)$$

where ψ is a stream function for the flow. Ideally this would be the Bernoulli function, but in many approximate cases we take it to be a quasi-two dimensional, non-divergent function such that $\vec{u} = \hat{z} \times \nabla\psi$, $\hat{z}$ being a vertical unit vector. In general, whether or not the flow is of this simple a nature, it is useful to make scatter-plots of q against ψ. Functional relations sometimes appear in parts of the flow (like regions of closed streamlines), and their inclination suggests stability properties of the flow. If such a plot is made from data along a curve in the fluid that terminates on boundaries, or is closed upon itself, the area enclosed by the $q(\psi)$ plot is proportional to the net advection of povorty across that curve [17].

Geophysical fluid dynamics deals with the situation where the horizontal divergence of the velocity is small, $O(R_0)$, compared with typical vorticity magnitude: yet is dynamically important to vortex stretching.

It is most important to appreciate the diverse nature of the various contributing parts that make up povorty, from the sphericity of the planet, from the large-scale topography of its surface (both 'solid Earth' terms), from the planetary-scale disposition of the constant-θ surfaces, from smaller scale tilt of these surfaces associated with rapid currents, and from the smaller scale relative vorticity. The large-scale terms provide a mean field which supports Rossby waves, topographic waves and flow instabilities. The smaller scale terms arise in turbulent interactions of the complexly eddying flows.

We divide the field into large-scale time-mean, $Q(x, y, z)$ and the remainder, $q'(x, y, z, t)$.

$$q = Q + q'$$
$$Q = \frac{2\vec{\Omega} \cdot \nabla\bar{\theta}}{\rho} + \frac{\bar{\omega} \cdot \nabla\bar{\theta}}{\rho} + O(a^2)$$

where the overbar is a time-average (though often it may be replaced by a zonal average round the planetary latitude circle, or an ensemble mean). The two contributions from the time-averaged flow are a 'stretching' or thickness, term 1, and the relative vorticity of the mean flow, term 2. In the thin sheets of fluid that are the oceans and atmosphere, the nearly horizontal disposition of large-scale θ-surfaces gives us for the stretching term,

$$Q \cong \frac{f}{\rho}\frac{\partial \theta}{\partial z} = \frac{\Delta\theta}{\rho}\frac{f}{h} \quad (f \equiv 2\Omega \sin(\mathit{latitude}))$$

Here z is the local vertical coordinate (radial on a sphere) and f is the local vertical component of $2\vec{\Omega}$, known as the Coriolis frequency; h is the thickness (vertically) of the layer between θ-surfaces differing by $\Delta\theta$. This term is indeed the dominant povorty field if the flow is of large-scale (compared to the Rossby deformation scale, NH/f where H is the fluid's total depth). This approximate form of q corresponds to the 'stiff' fluid described earlier, where planetary rotation is so strong as to prevent fluid from altering its thickness along lines parallel with $\vec{\Omega}$ (this is just a transliteration of conservation of f/h following the fluid). Thus constrained, the fluid still has freedom to advect, propagate, and suffer nonlinear steepening. Much of the large-scale time-mean circulation of the oceans is thought to be dominated by the 'thickness' povorty. But full-fledged geostrophic turbulence almost always involves the relative vorticity, and hence the full expression for q.

The scaling for geophysical flows that vary slowly in space and time, and are 'thin' in aspect ratio is summed up as: $R_0 \equiv U/fL \ll 1$, $fT \ll 1$, $H/L \ll 1$ where T is a typical time scale of the flow, H and L typical vertical and horizontal length scales of the flow and U a typical horizontal velocity. A great aid to further progress was the reduction of this problem to a single variable, the pressure or geostrophic streamfunction e.g., [18] from independent formulations made in the 1950s by Charney and Oboukhov.

$$\frac{D^g q_g}{Dt} = 0$$

$$q_g = \nabla_h^2 \psi + \left(\frac{f^2}{N^2}\frac{\partial \psi}{\partial z}\right) + f_0 + \beta y \tag{5}$$

Here a number of approximations have been made: the 'β-plane' or tangent plane approximation to spherical geometry (y being the north-south Cartesian coordinate); variation in the mean density field only in the vertical ($N = N(z)$ only), and the nearly 'geostrophic' balance between pressure gradient and Coriolis force that makes pressure an approximate stream function for the horizontal velocity. The 'thin-fluid-sheet' aspect ratio allows the advective operator $D^g q/Dt$ to follow only the horizontal motion of the fluid, being just $\partial q/\partial t + \partial(\psi, q)/\partial(x, y)$. ∇_h^2 is the horizontal Laplacian.

Having a single equation (plus boundary conditions) involving

horizontal velocity, $\hat{z} \times \nabla\psi$,

vertical velocity, $-\frac{f}{N^2}\frac{D^g}{Dt}\frac{\partial \psi}{\partial z}$, perturbation potential density, $-\frac{\rho f}{g}\frac{\partial \psi}{\partial z}$,

expressed in the single variable ψ, was a major simplification that has led to thousands of research papers on the dynamics of waves, turbulence and circulation.

The value of the povorty tracer is enhanced by its 'invertibility'. The resemblance of the operator in eqn. 5 to a (stretched) three-dimensional Laplacian suggests that something like a Poisson-equation inversion will take us from povorty to stream function, and hence to velocity and perturbation potential density. This turns out to be true in general, provided that the flow is in a balanced state, typically that of geostrophic balance [1], that the full povorty field can be inverted to give velocities and density fields providing that the distribution of mass with respect to density is known. but one must remember that motions with zero or constant povorty do exist. These are generalizations of common irrotational flows. An example is the Kelvin wave, essentially a long gravity wave turned into an evancescent mode trapped near an ocean boundary, or trapped at the equator. It has constant povorty and would be missed by a povorty 'inversion'.

Povorty flux can in the case of statistically steady flow be written [19]

$$\vec{J} = \nabla\theta \times \nabla B$$

where B is the Bernoulli function,

$$B = h + \frac{1}{2}|\vec{u}|^2 + \Phi$$

h is the enthalpy (internal energy + pressure/density), $h \approx c_p T$, and Φ is the geopotential field, $\Phi \approx gz$. Thus the potential-vorticity flux lies at the intersection of θ and B surfaces, essentially flowing with the circulation. As before, mixing and exotica like breaking of internal gravity waves can cause flow to penetrate θ-surfaces and q-contours.

While Schaer's interest was the vorticity production in small-scale flows past mountains, we have at the opposite extreme, for large-scale incompressible, adiabatic circulation, a Bernoulli function

$$B \approx \rho_0^{-1}(p + \rho g z)$$

which acts as a streamfunction for flow on surfaces of constant potential density. In the instance of steady flow of an ideal incompressible fluid (for which $\theta = \rho$, the density), and without dissipation or turbulence, one has an interesting relation between three key variables, B, θ and q [20]:

$$\begin{aligned} q &\equiv f\rho^{-1}\frac{\partial\theta}{\partial z} \\ &= fg/\left(\rho_0^2 B_{\theta\theta}\right) \end{aligned}$$

Here we take θ rather than z to be the vertical coordinate. The steady flow povorty problem, q is a function of ρ and B, looks like

$$\frac{\partial^2 B}{\partial \theta^2} = -fF(\theta, B)$$

for some function F that is determined by boundary conditions.

The interaction between the large- and small-scale povorty fields is at the heart of much geophysical circulation and transient motions, and it is not evident in the above formulas involving povorty flux. A representative problem is that of a uniform-density, uniform thickness layer, a two-dimensional flow governed by

$$\frac{\partial}{\partial t}\nabla^2\psi + \frac{\partial(\psi, \nabla^2\psi)}{\partial(x, y)} + \beta\frac{\partial\psi}{\partial x} = \nu\nabla^4\psi \,. \tag{6}$$

Here the povorty is $q = f_0 + \beta y + \nabla^2\psi$. The first two terms represent the time rate-of-change of the relative vorticity following the fluid, and the third is the advection of planetary vorticity.

IV SMALL-AMPLITUDE WAVES

Small amplitude motions (with small time-averaged flow also) balancing the first and third terms are Rossby waves. In these 'planetary' waves the contours of constant povorty, which would be latitude circles y=const. in absence of motion, are deformed slightly (following the motion of the fluid), waving with phase moving westward at rate β/k^2. Here $\vec{k}$ is the wave-vector of the sinusoidal disturbance field (of the form $\psi = \mathrm{Re}\,(\mathrm{i}\vec{\mathrm{k}} \cdot \vec{\mathrm{x}} - \mathrm{i}\omega \mathrm{t})$). Rossby waves are linearized, small-amplitude versions of the prominent weather systems of the atmosphere and energy containing eddies of the oceans. They are energized by release of gravitational potential energy from the sloping density surfaces, as well as by instability of kinetic-energy rich jet-like flows and direct generation by time-dependent forcing (such as wind-fields blowing on the ocean and seasonal thermodynamic forcing). There is a strong standing-wave component in the atmospheric westerly winds, expressing the forcing of Rossby waves by mountain topography. Other linearized balances are possible with this equation, when there are significant large-scale mean velocity fields, say a zonal flow $U(y)$. Then one has the Orr-Sommerfeld equation for instability of a parallel flow.

As Rossby waves access the gradually varying, large-scale povorty field, their frequency is greatest at long wavelength, atypical of the family of geophysical waves. The waves are *dispersive*, with phase speed depending on wavelength, *anisotropic* (only the velocity component normal to the mean contours of povorty yielding a restoring tendency), and their phase propagation always has a *westward* component, relative to the mean circulation (which itself may be moving eastward). With density stratification included, we find an infinity of vertical 'baroclinic' modes involving significant vertical motion. At scales longer than NH/f these modes become non-dispersive, purely westward propagating, and they are very active in transmitting peturbation density fields about the terrestrial oceans. In this limit they obey the 'stiff-column' physics in which the separation thickness of constant-θ surfaces, measured parallel to $\vec{\Omega}$, remains constant following the fluid.

While here described for the β-effect of the spherical planet, the related family of topographic waves exists with β replaced by $h\nabla(f/h)$, h being the layer depth; in the vorticity equation $J(\psi, q)$ becomes the more general $J(\psi, f/h)$.

V CASCADE BASICS

Some useful and fundamental ideas describing the effects of straining and vorticity of eddies in classical fluid flow has already been given above. Vortex-line stretching plays a central role in the energy cascade to small scales and dissipation. In geostrophic turbulence that vortex-stretching is prevented, and instead we notice that the contours of constant povorty, moving with the fluid, are prone to extending their length, quasi-exponentially with time. If this happens the gradient of povorty will increase with time, in direct proportion. Now, if an equation for the squared povorty is formed from eqn. 6, it has a dissipation term (from Navier-Stokes friction) proportional to the squared gradient of relative vorticity. Written in terms of the spectral density $E(k)$ of kinetic energy (with respect to the scalar wavenumber k),

$$\frac{\partial}{\partial t}\int_0^\infty E dk = -2\nu \int_0^\infty k^2 E dk$$

$$\frac{\partial}{\partial t}\int_0^\infty k^2 E dk = -2\nu \int_0^\infty k^4 E dk$$

The first equation describes the dissipation of total energy, proportional to the vorticity spectrum, k^2E. The second describes the dissipation of *enstrophy*, or squared vorticity, by the 4th moment of the energy spectrum, which is the spectrum of the vorticity *gradient*. Energy dissipation is bounded by its initial value, since vortex stretching is absent. Squared vorticity dissipates proportional to the squared spatial gradient of vorticity. It is this that is accelerated by turbulent stretching of the contours of constant (potential) vorticity. An initial, narrowly peaked spectrum evolving under these constraints cannot send more than a small fraction of its energy to large wavenumber. Set $\nu = 0$ temporarily; then the 'mass' and 'moment of inertia' of the distribution $E(k)$ are constant, and as the spectrum broadens the majority of energy must move toward small wavenumber. A way of writing this more clearly is that if

$$\frac{\partial}{\partial t}\int_0^\infty (k - k_1)^2 dk > 0$$

then

$$\frac{\partial}{\partial t} k_1^2 < -2\nu k_4^4 < 0$$

where

$$k_n = \int_0^\infty k^n E dk / \int_0^\infty E dk \, .$$

The centroid wavenumber k_1 decreases if the second moment of the E about k_1 should increase due to chaotic turbulent interactions. Starting with a narrow distribution of energy at wavenumber k_0, no more than one-half the total initial energy can reach wavenumber $2k_0$, for to balance all the remaining energy would have to move to $k = 0$. This is a variant of the arguments given by Fjortoft, Batchelor and Kraichnan ([21-23]). If energy is continuously generated by external forcing at wavenumbers near $k = k_0$, initial- tendency arguments suggest that it might be possible to form quasi-steady spectral distributions in which enstrophy was carried to the 'right' of k_0, to large wavenumber , while energy was carried to the 'left', to small wavenumber. Dimensional arguments then suggest spectral shapes proportional to $k^{-5/3}$ for the energy-carrying spectral 'pipeline', at wavenumbers below k_0 and k^{-3} for the enstrophy carrying pipeline, above.

Objections have often been raised to the detailed spectral shapes predicted in this inertial subrange theory, on the grounds (a) that with slopes as steep as k^{-3} the shear and strain at wavenumber k is not dominated by eddies of about that wavenumber (but rather by much larger eddies), and, (b), the dimensional and similarity arguments ignore the possibility of evolving coherent structures within the turbulence field. But nevertheless, the general tenets of the cascade have been demonstrated by many numerical simulations (beginning with Lilly's, [24]) .

For two-dimensional turbulence, plots of ψ and $\omega_z = \nabla^2\psi$ in actively evolving turbulence emphasize the straining of vorticity contours into long, thin tongues which, through the smoothing of the inverse Laplacian operator, induce large-scale flow fields. There is also at work the coalescence of pairs of vorticies of like sign, which occurs as they orbit about their center of vorticity. This is the spirit of the 'double cascade'. Fornberg, Basdevant *et al.* and McWilliams ([25-27]) demonstrate that as the cascade develops the coalescence process causes the vorticity field of pure two-dimensional turbulence to devolve into ever more sparsely separated strong, monopole vortex cores. This was not forseen by the similarity theory of the cascade, and modifies it significantly (producing non-Gaussian statistics and steeper-than-k^{-3} wavenumber spectra, though with a background field of continuous turbulence envisioned in the classical work). Manifestations of two-dimensional vortex interactions may be seen widely, for example in the 'roll-up' of a vortex sheet between two streams of differing velocity (e.g. [28]).

Geostrophic turbulence with density stratification adds interesting behavior connected with the variations of velocity now allowed in all three directions.

However, Charney ([29]) showed that the same basic integral invariants for energy and squared povorty prevent an energy cascade to large wavenumber, just as before. Violations can occur, associated with frontal development, but they do not characterize the turbulence as a whole. Rhines ([2]) demonstrates that the cascade sends energy from baroclinic (depth-varying) velocity fields to barotropic-mode (two-dimensional) velocity, with remarkable efficiency. A special case of this cascade is seen in the baroclinic instability of the westerly winds, and succeeding development toward effectively barotropic turbulence.

VI WAVE/TURBULENCE INTERACTION

It is convenient that the general, three-dimensional problem involves significant two-dimensional turbulence effects, as the latter is much simpler to understand. A particular problem of wide interest is the interaction of turbulence with the large-scale povorty field contributed by planetary sphericity, boundary topography, or large-scale mean circulation. The most basic case is β-plane turbulence, eqn. 6 ([30]). For a cascading turbulent spectrum with peak at wavenumber $k = k_m$ a similarity estimate the decrease in k_m with time is $k_m \sim (Ut)^{-1}$ where ρU^2 is the kinetic energy density. The characteristic time-scale of the turbulence is $\tau_{2DT} \sim (k_m U)^{-1} \sim t$. Meanwhile the Rossby wave time-scale, essentially the inverse frequency of waves at a given wavenumber, varies like $\tau_{RW} \sim (\beta/k_m)^{-1} \sim (\beta U t)^{-1}$. As the energy proceeds to larger scale the wave restoring effect, while perhaps small initially, becomes relatively stronger. We find

$$\tau_{RS} \sim \tau_{2DT}$$

as

$$t \to (\beta U)^{-1/2}$$

$$k_m \to k\beta \equiv (\beta/U)^{1/2}.$$

The phase of Fourier components becomes randomized by Rossby-wave radiation, slowing the persistent stretching of povorty contours that is the heart of the turbulent cascade. At the same time the pair interactions between vortices of like sign is upset, and formation of coherent, long-lived vortex monopoles no longer occurs.

The result of the turbulent cascade is now a field of propagating Rossby waves, a peculiar effect of large-scale vorticity dynamics (if one is used to thinking of waves breaking to make turbulence). Of course the latter process is still possible in less turbulent initial fields. The scale of the arrested cascade $k\beta^{-1} \sim (U/\beta)^{1/2}$ has been verified in many numerical experiments ([4,31]). It is accompanied by anisotropy of the velocity field, favoring zonal winds or currents (along lines of the large-scale povorty field).

Perhaps the most surprising effect is the generation of jets as a part of the developing cascade [4,32,33]. The elongation of eddies along mean povorty contours is accompanied (through the Rossby wave dispersion relation) by a decrease in their frequency. Essentially tubular 'plumes' of flow radiate westward, with very small frequency. They take on the appearance of slowly varying jets of nearly zonal flow. We find that particularly when there is a persistent energy source for the turbulence, the zonal jets become prominent, long-lived features. Their width, predicted to be $\sim k\beta^{-1}$, has been supported by these several sets of numerical simulations. Banded zonal circulations prominent in the great outer planets may be related to this effect, providing the static stability of the vertical density gradient is sufficient. Earth itself has jet streams that may partially relate to this Rossby wave/turbulence transition. The Jovian and other planetary circulations may instead driven by deep convection, induced by the large interior heat source and weak sunshine. Jets then can occur for rather different reasons ([34,35]), yet the steering by the mean povorty gradient is common to both theories.

The development of jet-like flows driven by turbulence is an example of the general problem of eddy-transport of momentum and povorty. The tracer-like nature of povorty suggests a mixing-length theory for its transport, and indeed increases the value of observations of other tracers that move with the fluid. The theory shows that the eddy-transport of povorty $-\hat{z}\times\langle q'\vec{u}'\rangle$ is an effective *force* on the Eulerian-mean fluid. Its principle result is a set of connections between the Lagrangian diffusivity tensor (described earlier), the mean povorty field, and the effective horizontal force exerted by turbulent eddies (or Rossby waves) ([3]). This explains, for example, the principle mechanism by which horizontal momentum is transported vertically in the atmosphere and oceans, the development of counter-currents at the edges of unstable jets, and the concentration of jets by radiating instabilities, and the generation of long, slowly varying Rossby wave 'plumes' by smaller-scale geostrophic turbulence.

As we have said, both Rossby waves and fully developed turbulence stir the large-scale povorty. In the former case we can work out more exactly the povorty flux that is available to induce mean circulation. The tilted wavecrests of a Rossby wave indicated a flux of westward momentum away from a wave source, connected with their transport of pseudo-momentum, $P \equiv Ek_x/\omega$, where E is now the spatial energy density, k_x the east-west wavenumber and ω the frequency measured by an observer moving with the mean velocity. The transport is equal to the P times the group velocity of the waves, and this in turn is equal to the north-south povorty flux, $\langle q'v'\rangle$. Various interesting expressions follow, for instance that the force exerted by the waves to accelerate a mean circulation, is related to the gradient of energy-density in a Rossby-wave packet:

$$\begin{aligned}-\hat{z}\times\langle q'\vec{u}'\rangle &= \nabla E \text{ reflected about the wavevector,}\\ &= -\nabla E + 2(\nabla E\cdot\vec{k})\vec{k}/\vec{k}\cdot\vec{k}\,.\end{aligned}$$

Similarly, small-amplitude instabilities of a zonal flow (both baroclinic- and barotropic) yield accelerations of the mean flow that relate to povorty transport by the growing waves. The very instability itself is best viewed as a positive-feedback instability of generalized Rossby waves ([36]).

Since forces and momentum are intimately tied up with povorty dynamics, it is not surprising that wave-drag in flows crossing mountainous topography can be connected with povorty transport. This gives us a framework to analyze the large-scale force balance of both oceans and atmospheres.

VII HOMOGENIZATION

A limiting state for strongly energized geostrophic turbulence is one of homogenized povorty, in which its gradient is 'expelled' to the edges of the domain. The effect is particularly important in regions of closed streamlines, for there the mean inward advection across a streamline vanishes, and even weak geostrophic turbulence can mix the region to a state of near uniformity ([37]). Because of the invertibility principle, this essentially gives a theory, or at least partial theory, of the general circulation ([38]).

Not only is povorty homogenized in single, closed gyres of circulation and cat's-eyes, but in vast open domains like the spherical atmosphere, geostrophic turbulence can work to create a stair-case of povorty, using its energy to widen the bands of homogenization so far as possible. This is important to the monopole vortex theories of such flows, and the remarkable life cycle of coalescence of vorticies discovered by Marcus [39] and others. In his case the environment was assumed to have uniform povorty, but we now see that producing the uniform strips of povorty is an important part of the problem. It is reminiscent of the layering of a stratified, non-rotating fluid driven either by heating or mechanical turbulence [40]. In that situation the potential energy is not available to mix away the entire stratification; rather, a *parfait* of thin mixed layers forms.

VIII BOUNDARY EFFECTS

Boundaries act on fluids in non-trivial ways. They may be the most significant sources of vorticity and povorty in the entire domain, particularly if stable density stratification isolates the fluid interior from direct forcing (either thermodynamic or mechanical). As describe in the discussion of airfoils above, the transport of fluid from near the boundary to the interior, though it involves only a small amount of anomalous fluid, can nevertheless have a global effect. Here we have avoided detailed discussion of boundary conditions for povorty, and the effects of the actual turbulent boundary layers in nature. Even in the simple, inviscid case the condition of vanishing normal velocity translates into a difficult, mixed condition, essentially $D\theta/Dt = 0$ at the boundary. Bretherton ([36]) suggested that the quasi-geostrophic case (eqn. 5) with horizontal top and bottom boundary, can be conceptually improved by adding delta-function contributions to q at the boundaries, say $z = 0, -H$. In this way a non-uniform potential density along the boundary reappears as a boundary 'charge', and the new boundary condition on $\tilde{q}$ is homogeneous. Now many aspects of povorty theory: instability criteria, eddy flux formulas for povorty and wave/mean-flow interaction calculations, are made simpler. We ([3]) have found that non-horizontal boundaries (bottom topography; mountains) can also be incorporated into the definition of povorty, and in doing so unifying the idea of Rossby waves, topographic waves and interior povorty gradients. In the quasi-geostrophic formulation we visualize a thin layer of constant θ appended to the fluid. This layer has little dynamical effect, but it formally simplifies the boundary condition and adds (through its sudden discontinuity in potential density) a new contribution

$$\tilde{q} = q + f_0 \left(h + \frac{f_0}{N^2} \frac{\partial \psi}{\partial z} \right) \delta(z + H)$$

such that $D\tilde{q}/Dt = 0$ with a homogeneous boundary condition for $\tilde{q}$ is the recast problem. Here $h(x, y)$ is the topographic height relative to the mean position of the lower boundary at $z = -H$. This addition expresses the variation in θ along the perturbed position of the boundary. It is a small-amplitude version of the boundary sheet

$$\vec{\omega}_a \cdot \hat{n} \theta \delta(\vec{x} - \vec{x}_b)$$

which cancels the volume integral of the interior povorty. $\delta(\cdot)$ is the delta function, $\vec{x}_b$ the position of the boundary, and $\hat{n}$ an inward normal vector at the boundary (yet unfortunately in this more general case the boundary contribution is not conserved following the fluid motion (I.M. Held, private communication). The boundary 'charge' or boundary 'reservoir' of povorty has great implications to interior dynamics. It is important only when there are variations of potential density along the boundary, either due to tilted density surfaces intersecting a horizontal boundary (Eady instability, outcropping of θ-surfaces at the ocean surface), or level potential density surfaces intersecting sloping topography (topographic waves, shedding of boundary povorty).

It is apparent that the intersection of constant potential density surfaces with the boundary of the fluid provides a region of large povorty (indeed, the 'wedges' of fluid between θ-surfaces suggests this, as f/h becomes singularly large at the boundary). This source or reservoir of povorty can be stripped off into the interior of the fluid. We have recently found it to play an important role in determining the ocean circulation, as seen in recent highly resolved, isopyncal numerical simulations [41, fig. 1], where both the sloping boundaries and the 'outcropped' density field at the sea surface provide the source. In the figure, a 100×100 grid-point by 8 isopycnal layer numerical model simulates the ocean circulation driven by wind. The domain is 2500km on a side, and the sloping bottom topography at the western boundary provides a source of high povorty. The circulation at this level is dominantly a clockwise gyre with a western boundary current. At other levels the ocean surface provides sources of high povorty while flow from low latitude provides a low-povorty signal. Geostrophic turbulence rapidly mixes these sources into the central 'plateau'.

In finite amplitude calculations for Eady's baroclinic instability, the same boundary source can be tapped [42], with major consequences to the initially zero-porvorty interior, plumes of high povorty erupting from boundaries into the interior.

Layer 4 Potential Vorticity at Day 8500

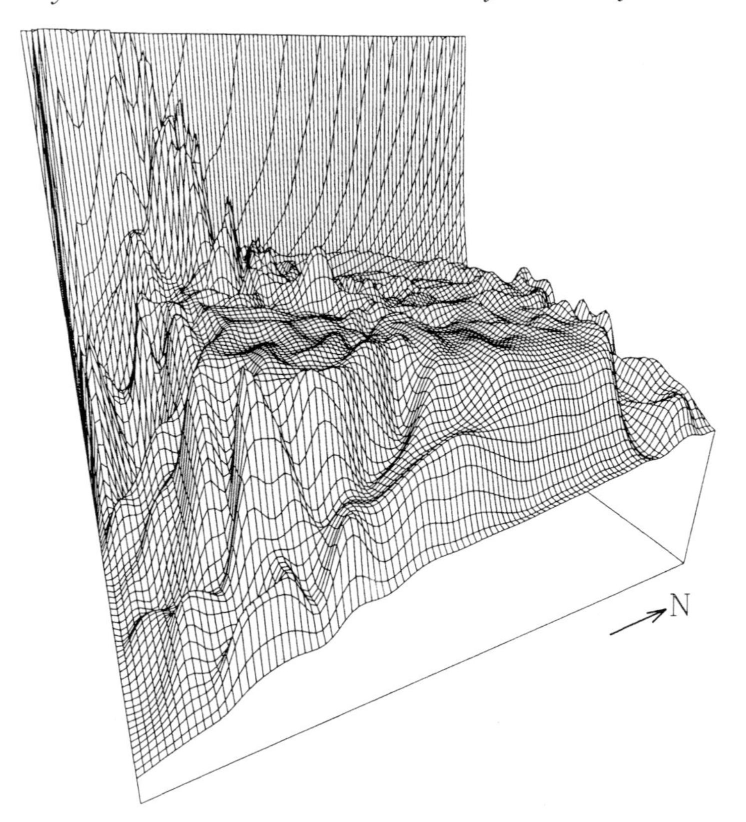

FIGURE 1. Potential vorticity ('povorty') field in the 4th layer of an isopycnal ocean simulation (Hallberg and Rhines [41], viewed from the southeast. A plateau of homogenized povorty fills the interior circulation gyre, and it is fed by a ridge of high 'boundary' povorty being stripped off the sloping western boundary by the jet-like circulation.

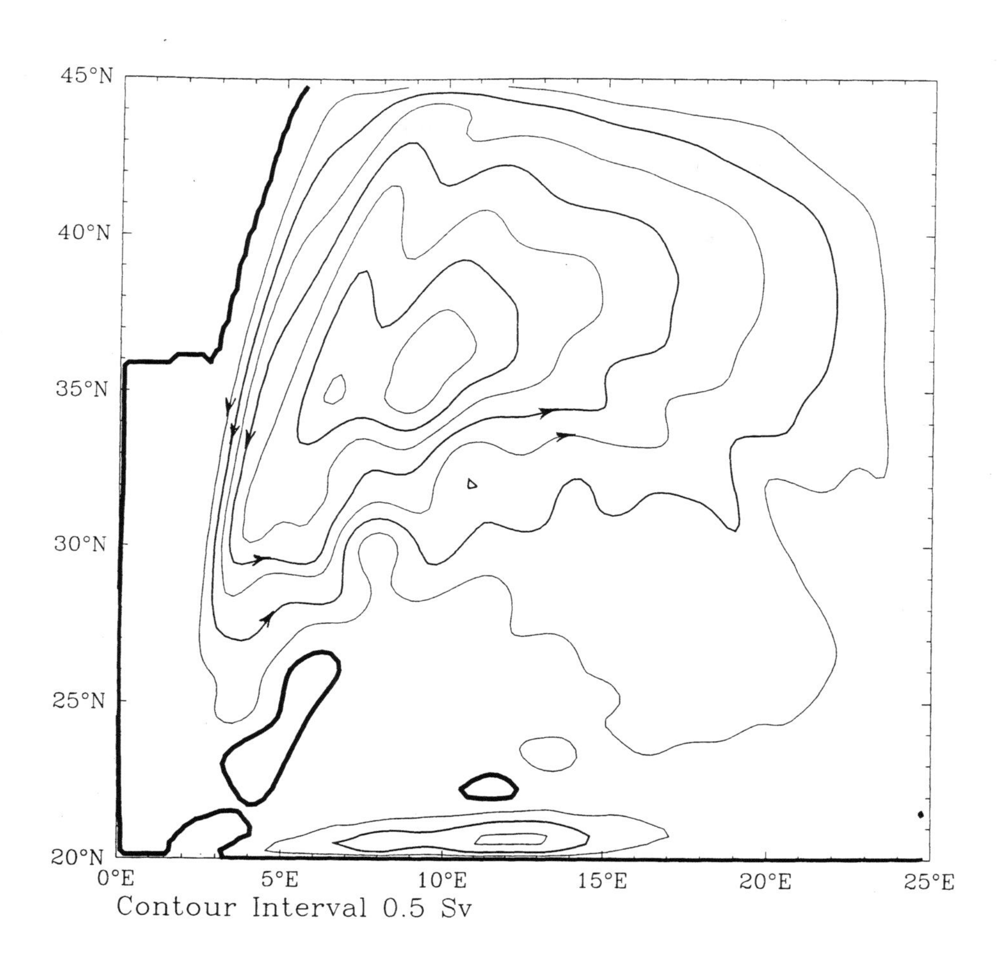

FIGURE 2. Transport streamfunction for the circulation corresponding to povorty field in figure 1. Averaged over 1000 days after the simulation has fully spun up. Intense geostrophic turbulence is superimposed upon this time-averaged streamfunction pattern. Although the depth-averaged circulation contains two gyres (cyconic in the north, anti-cyclonic in the south), at this density surface the circulation is almost entirely anti-cyclonic.

IX CONCLUSION

This short review of povorty dynamics relating to geostrophic turbulence and circulation in oceans and atmospheres has aimed at establishing the connection with classical vorticity conservation, yet with the stark contrast from the impossibility of massive vortex stretching in large-scale rotating stratified flows. Embedded small-scale turbulence is still at work in these flows, but it is unable to communicate fully with the great energy-containing eddies: the 1000km-scale weather of the atmosphere and 100 km-scale eddies of the oceans cannot dissipate energy as fast as they otherwise would. Turbulent boundary layers and thin, isolated patches of turbulence occur. The atmospheric boundary layer, with its mechanical- and cumulus-convective turbulence, fills 100s to 1000s of m of the 10km thick troposphere, so it is not really 'thin'. In the oceans, however, boundary layers are typically 10 to 100m thick, compared with the 3 to 5 km depth of the fluid. Dissipative damping of the large-scale circulation is correspondingly different: the weather systems of the atmosphere have an e-folding time for kinetic energy of 3 to 5 days (a little greater than the L/U inertial time scale) while oceanic circulation is damped much more slowly, over times of 100s to 1000s of days.

The tracer property of vorticity and povorty has enormous implications to flow instability, turbulence, its interaction with general circulation, and boundary sources and sinks of circulation. Analytical progress and intuitive understanding both profit from this relationship with observable, advected patterns in fluids. Flows with stably stratified density fields often have sources of povorty in only small regions, and at boundaries. The influx, advection, and mixing of povorty in these circumstances bears a resemblance to the classical problem of vorticity generation in ordinary fluids at large Reynolds number. Analysis of models and observations in this light is promising.

ACKNOWLEDGMENTS

I am grateful to Parker MacCready, Bob Hallberg and Isaac Held for discussions about vorticity and povorty, and to the National Science Foundation and Office of Naval Research for funding my research on geostrophic turbulence.

REFERENCES

1. Hoskins, B.J., McIntyre, M.E., and Roberston, A.W., *Q. Jour. Roy. Meteorol. Soc.* **111**, 877-946 (1985).
2. Rhines, P.B. in *The Sea*, vol VI, E.D. Goldberg Ed., Wiley-Interscience, 1977, pp. 189-318.
3. Rhines, P.B., *Ann. Revs. Fluid Mech.*, **11**, 404-441 (1979).
4. Rhines, P.B., *Ann. Revs. Fluid Mech.*, **18**, 433-497 (1986).

5. Rhines, P.B. and Holland, W.R., *Dyn. Atmos. and Oceans*, **3**, 289-325 (1979).
6. Rhines, P.B. in *Modelling Oceanic Climate Interactions*, J. Willebrand and D. Anderson Eds., Springer Verlag, 1993.
7. Batchelor, G.K., *An Introduction to Fluid Dynamics*, Cambridge Univ. Press (1967), 615 pp.
8. Lighthill, M.J., *An Informal Introduction to Theoretical Fluid Mechanics*, Oxford University Press, 1986, 260pp.
9. Moffatt, H.K., *J. Fluid Mech.*, **36**, 117-129 (1969).
10. Taylor, G.I., *Proc. London Math. Soc.*, **20**, 196-212 (1921) and in *The Scientific Papers of G.I. Taylor*, **2**, G.K. Batchelor Ed., Cambridge, 1960, pp. 172-184.
11. Taylor, G.I., *Reports and Memoranda of the Adv. Comm. for Aeronautics*, **345** (1917) and in *The Scientific Papers of G.I. Taylor*, **2**, G.K. Batchelor Ed., Cambridge, 1960, pp. 69-78.
12. McDougall, T.J., *Progress in Oceanography*, **20**, 185-221 (1988).
13. Haynes, P.H and McIntyre, M.E., *J. Atmos. Sci.*, **47**, 2021-2031 (1990)
14. Lamb, H., *Hydrodynamics*, 6th Ed., Cambridge Univ. Press (1932), 738 pp.
15. Rossby, C.G., *Quart. J. Royal Meteorological Soc.*, **66**, Suppl., 68-87 (1940).
16. Ertel, H., *Meteorolog. Zeitschrift*, **59**, 271-281 (1942).
17. Read, P.L., Rhines, P.B. and White, A.A., *J. Atmos. Sci.*, **43**, 3226-3240.
18. Gill, A.E., *Atmosphere-Ocean Dynamics*, Academic Press (1984), 662 pp.
19. Schaer, C., *J. Atmos. Sci.*, **50**, 1437-1428 (1993).
20. Killworth, P.D., *J. Phys. Oceanogr.*, **17**, 1925-1943.
21. Fjortoft, R., *Tellus*, **5**, 225-230 (1953).
22. Batchelor, G.K., *Phys. Fluids*, **12**, 233-238 (1969).
23. Kraichnan, R., *Phys. Fluids*, **10**, 1417-1423.
24. Lilly, D.K., *Geophys. Fluid Dyn.*, **3**, 289-319.
25. Fornberg, B., *J. Comp. Phys.*, **25**, 1-31 (1977).
26. Basdevant,C., Legras, B., Sadourny, R., and Beland, M., *J. Atmospheric Sci.*, **38**, 2305-2326 (1981).
27. McWilliams, J.C., *J. Fluid Mech.*, **146**, 21-43 (1984).
28. Aref, H. , *Ann. Revs. Fluid Mech.* **15**, 345-389 (1983).
29. Charney, J., *J. Atmos. Sci.*, **28**, 1087-1095 (1971).
30. Rhines, P.B., *J. Fluid Mech.*, **69**, 417-443 (1975).
31. Vallis, G. and Maltrud, M., *J. Phys. Oceanogr.*, **23**, 1346-1362.
32. Williams, G.P., *Climate Dyn.*, **2**, 205-260 (1988).
33. Panetta, R.L., *J. Atmos. Sci.*, **50**, 2073-2106.
34. Busse, F., *Icarus*, **29**, 255-260 (1976).
35. Condie, S. A. and Rhines, P.B., *Nature*, **367**, 711-713 (1994).
36. Bretherton, F.P., *Quart. J. Royal Meteorol. Soc.*, **92**, 335-345 (1966).
37. Rhines, P.B. and Young, W.R., *J. Fluid Mech.*, **122**, 347-367 (1982).
38. Rhines, P.B. and Young, W.R., *J. Marine Res.*, **40**, suppl., 559-595 (1982).
39. Marcus, P.S., *Nature*, **331**, 693-696.
40. Turner, J.S., *Buoyancy Effects in Fluids*, Cambridge Univ. Press (1974), 366 pp.
41. Hallberg, R.W. and Rhines, P.B., *11th Conf. Atmos. and Oceanic Fluid Dyn*,

Abstract volume, Amer. Meterolog. Soc. (1997), pp. J9-J12.
42. Garner, S.T., Nakamura, N., and Held, I.M., *J. Atmos. Sci.*, **49**, 1984-1996 (1992).

TURBULENCE AND STRUCTURES IN DUSTY PLASMAS

V.N. TSYTOVICH

General Physics Institute, Russian Academy of Science,
Vavilova str. 38, Moscow, 117942, Russia,
Tel.: 7 (095) 135-02-47, Fax: 7 (095) 1350270,
E-mails: tsyt@ewm.gpi.ru, tsytov@tp.lpi.ac.ru

Abstract. The field of dusty plasmas is a new field that is rapidly developing and has many applications in laboratory, space and astrophysical contexts. The basic properties of dusty plasmas are reviewed and the effect of dust on drift waves and vortices is described.

I DUSTY PLASMA AS A NEW STATE OF MATTER.

Dusty plasma has become a fashionable subject for research, driven mainly by several important applications. Interest in the field of dusty-plasma physics is driven by the industrial applications of etching and plasma deposition problems and the possibility of the production of new materials, by new space research related to recent space missions and the discovery of many new aspects of dust in space including the complicated structure of planetary rings, star formation, creation of dust in supernovae explosions and finally by pollution problems (dust in lower ionosphere and lower atmosphere). The field is also important in Controlled Thermonuclear Research (CTR).

It is often said that plasma is a new state of matter, though, since plasma consists mainly of freely moving particles, it is in a sense a gaseous state. Of course plasma is the ionized state of a gas, with free electrons and ions, which introduces many collective properties so it is reasonable to call it a new state of matter. But from the point of view of particle motions, the possible states of matter can have either fixed (or almost fixed) positions of particles (the solid and liquid states), or free particle movements (the gaseous state). In this sense plasma is the ionized gaseous state, so in a way astrophysicists are correct in calling plasma "hot gas", although this name sounds naive from the modern point of view of plasma physics. The problem whether other states of

CP414, *Two-Dimensional Turbulence in Plasmas and Fluids:* Research Workshop
edited by R. L. Dewar and R. W. Griffiths

matter can exist is an important general physical problem.

In usual matter the only possible particle binding has a quantum nature, since the exchange interactions are related to the overlapping of the wave functions. This interaction can produce particle attraction, all chemical bindings, and such states as the solid state of matter. Dust in plasmas provides a new dust-dust attraction, which is classical in nature and therefore can be considered as an elementary interaction of a new type of matter, including dust crystals and dust liquids. In this sense dust in plasmas can form very different structures, with properties quite different from usual matter, and one can speak about a new state of matter. This opens new possibilities which one can describe as the start of a super-chemistry, in which the elementary interacting entities will be not atoms, but dust particles in plasmas. Thus it is important to understand why the dust particles in a plasma can attract each other regardless of the Coulomb repulsion associated with their large charges of equal sign.

The key point for understanding dusty plasma as a new state of matter is the openness of the system — plasma with dust cannot survive in the absence of an external source of electrons and ions, or in the absence of a flux of them from the region where the dust is absent. The dust cloud is usually automatically charged in such a way as to create such fluxes. The reason for the presence of fluxes of plasma particles on the dust particles is the dust-charging process, which, even after reaching the equilibrium dust charge, does not vanish when the current onto a dust particle vanishes. The openness of dusty plasma systems and the presence of plasma-particle fluxes, both onto each dust particle and onto the dust cloud as a whole, creates additional forces between the dust particles, which are usually attractive. This effect can be explained, in a hand-waving fashion, as a pressure produced by the shadowing of plasma flux on one of the dust particles by another dust particle.

Another fundamental property of the dusty plasma is that it is a highly dissipative system, the dissipation being caused by absorption of plasma fluxes on dust particles. The rate of formation of self-organized structures in open systems is usually determined by the degree of dissipation, which in equilibrium is compensated by external sources. In a dusty plasma, a big flux of energy into the system creates possibilities for the development of self-organized structures. In fact the dusty plasma is a system extremely well adapted for the formation of structures — the structures that can be created in the absence of dust will be modified by the presence of dust, but this is less important than the fact that, in addition, new structures can be created with no analogy to known plasma structures. Thus the variety of structures that can be created in dusty plasmas is much greater than in usual plasmas.

Since plasma particles play the role of "food" for these structures, the competition for "food" can be of importance for the evolution of the structures. There appears also a possibility of modeling biological structures using dusty plasma — a topic already discussed in the current literature. The question of the existence of these structures in space plasma has also been raised, as well as the problem of storing information in the structures. Regions in space where the conditions are appropriate for the formation of dust in plasmas may be appropriate places to search for extraterrestrial life.

This means that dusty-plasma problems are of general importance in present physics — not only is dusty plasma an example of a new physical object, but *structures* in dusty plasma play a specially important role, being a necessary component of such a system. Turbulence in dusty plasma is the state of interaction of these structures.

II DUST-DUST INTERACTIONS

The difference between dust particles and other charged particles is that their charges are very large and are not fixed, being dependent on the surroundings. This is because the ion and electron fluxes creating the dust charges depend on local plasma parameters and the plasma-particle distribution functions. The presence of a neighboring dust particle changes the plasma fluxes, thus changing the charges on the dust particles, and therefore changes their interactions.

Even for fixed charges one should expect the presence of new forces between dust particles. The interactions could be different for the case where the distance between dust particles is less that the Debye screening length and in the case where it is larger than the Debye screening length. The first case is usually found in etching experiments and the second case is usually found in plasma-dust crystal experiments. For the latter case the Coulomb interaction is screened very much and the interactions due to mutual shadowing of plasma fluxes are most important.

Only the shadowing of the neutral plasma particle bombardment can compete with electrostatic forces for distances less than the Debye screening distance. This bombardment force is usually an attractive force, as we will demonstrate later, and can lead to dust agglomeration — an effect usually observed in etching experiments. In the conditions for plasma dust-crystal experiments the bombardment by neutrals is almost equal to or less than the plasma flux.

But these forces are always dominant for distances larger than the Debye screening distance. These non-Coulomb forces have a nature similar to the usual electrostatic forces, which, according to the general concept of forces, are due to exchange of photons between the interacting particles (or for electrostatic forces due to exchange of pseudo-photons). In this picture the electrostatic forces arise due to the process where one particle emits a photon and another absorbs it. This leads to an exchange of momentum and therefore leads to the existence of forces.

If a single spherical dust particle is present in a plasma, the bombardment by plasma particles does not transfer momentum (assuming the shape of dust particle also to be spherical). However, in the presence of another dust particle at a certain finite distance from the first one, the fluxes on the first particle are shadowed by the second dust particle and vice-versa.

Since the total spherical flux does not produce any force one can subtract it, and what is left is the absence of flux in a certain angular interval due to the shadowing effect, which is the equivalent of emission of this flux by one of the dust particles so as to compensate the part of the spherical flux in the shadowed region.

This flux acts on another dust particle in the same way as the flux of emitted photons and transfers momentum. It is easy to see that these forces are determined by the solid angle of the shadow, and therefore for distances larger than the size of dust particles the forces are inversely proportional to the square of the distance between the dust particles, as it is for Coulomb forces.

It is clear also that these bombardment forces should be attractive, and should increase proportionally to the surface area of the dust particles, i.e. proportionally to the square of their radius. This conclusion is independent of whether the bombardment is produced by neutral plasma particles or by charged plasma particles (electrons and ions). This also means that for large dust sizes the attraction forces can dominate, which seems to be the physical reason for the possibility of dust agglomeration in the presence of large Coulomb repulsions.

The presence of plasma fluxes on a single dust particle changes the shielding of its electrostatic field since the shadowing effect creates an additional charge distribution around the dust particles, with a potential proportional to the solid angle of the shadow. This leads to the presence of an electrostatic non-screened repulsion between dust particles, inversely proportional to the cube of the distance between dust particles (the derivative of an additional potential which is inversely proportional to the square of the distance). This repulsion is small compared with the attraction for large distances since the attraction force is inversely proportional to the square of the inter-dust distance, while the repulsion force is proportional to the cube of the inter-dust distance. This gives a molecular type of potential for the interaction of two dust particles in plasmas. Schematically the possible potentials of pair dust-dust interactions in dusty plasmas is shown on Fig. 1

III DUST ATTRACTION LENGTH

In the absence of dust, a plasma has several characteristic lengths such as Debye screening length, Larmor radius, or the scale length of density inhomogeneities. Dust-plasma physics need to introduce an additional characteristic length related to the dust sizes. We will call it the *Dust Fundamental Length* (DFL), since in the case all other characteristic lengths are unimportant (ab-

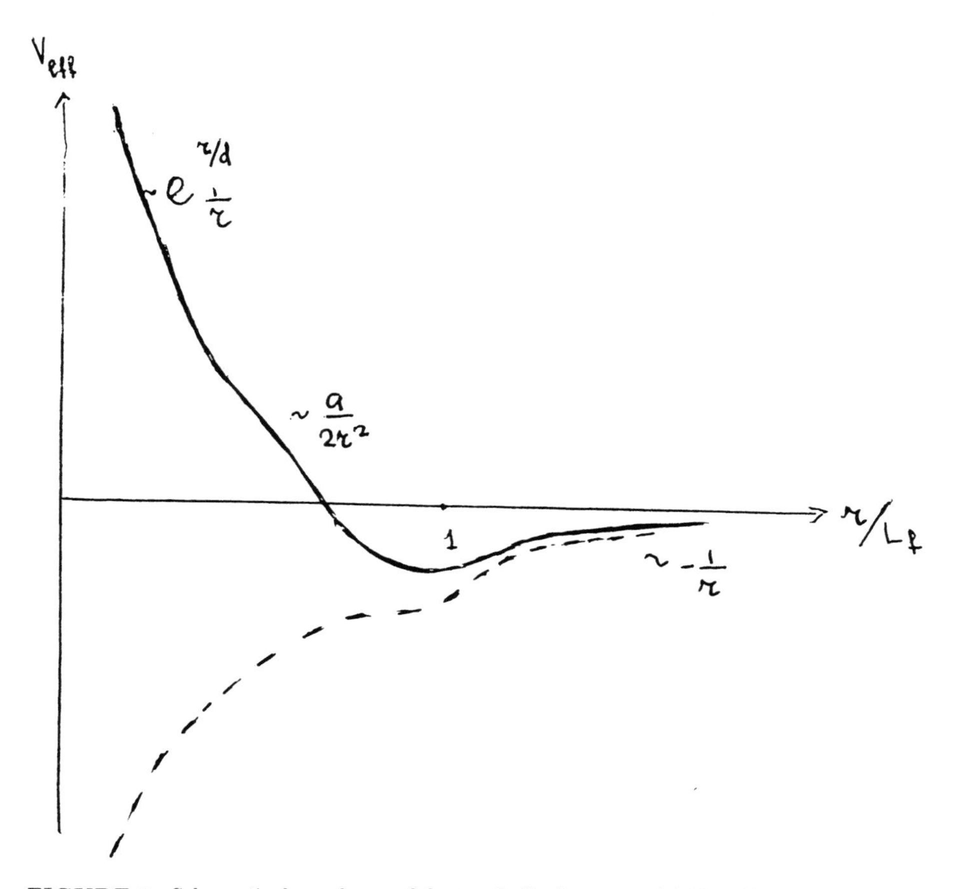

FIGURE 1. Schematic dependence of the total effective potential V_{eff} of interaction of two dust particles on the inter-dust distance r. The potential V_{eff} includes both the electrostatic potential and the potential of the bombardment forces. The solid line shows the behavior of the potential on the inter-dust distance in the case where a bound state of two dust particles can be formed. The dotted line shows the dependence of the potential on the inter-dust distance in the case corresponding to dominance of attraction, leading to dust agglomeration. The conversion from one curve to another is determined by the degree of plasma ionization determining the relative density of neutrals.

sence of external electric and magnetic field and homogeneity of plasma in absence of dust) most of the characteristics of dust-plasma structures can be expressed through this DFL.

This new fundamental length is

$$L_{\mathrm{f}} \equiv d\frac{d}{a_0} , \qquad (1)$$

where d is the Debye screening length and a_0 the size of dust particles. The Dust Attraction Length plays in dusty plasma a role similar to atomic size in atomic physics.

A dust molecule — the molecule formed from two equally charged dust particles — has a size equal to or of the order of L_{f}, the lattice size of crystals is of the order of L_{f}, and even the plasma-particle mean free path in dusty plasma is of the same order. Also the structures themselves have a size of the order of L_{f} etc. (for examples see below).

One of the most important points is that equations self-consistently describing self-organized structures in dusty plasmas, when expressed in a dimensionless length measured in units of L_{f}, do not have any small or large parameters. These equations include all the attraction and repulsion forces mentioned above. This is true only in the case such lengths as Larmor radius or Rossby radius do not play any important role for the structure. In cases when they do play a role the situation is more complicated, but the L_{f} and its ratio to other lengths plays a crucial role for the structures. This is another reason to call L_{f} a Fundamental Length.

The new attraction forces are similar to gravity and have the same dependence with respect to inter-dust distance as gravity forces. For the attraction produced by charged plasma particle fluxes one can express the forces in the form

$$\mathbf{F}_{\mathrm{attr}} = -\frac{\mathbf{r}}{r^3} Z_{\mathrm{d}}^2 e^2 \frac{a_0}{L_{\mathrm{f}}} \eta_a , \qquad (2)$$

where η_a is a numerical coefficient which depends on the plasma parameters and is of order unity for the simplest conditions (see below).

For the repulsion forces $\mathbf{F}_{\mathrm{rep}}$, the expression a_0/r should be substituted in (2) for a_0/L_{f} and η_r should be substituted for η_a, where η_r is the numerical coefficient characterizing the repulsion forces. It is easy to see that for numerical coefficients η_a and η_r of the order of 1, the repulsion and attraction forces balance each other for $r = L_{\mathrm{f}}$. The forces induced by neutrals contain a factor equal to the inverse degree of ionization. The total potential of these forces is either of molecular type or pure attraction in the case where the neutral bombardment force dominate, leading to dust agglomeration (see Fig. 1).

In 1776 Lesage proposed a model of Newtonian gravity supposing that each particle absorbs ether, and that the shadowing of this absorption creates the gravitational force. The special theory of relativity, by excluding the ether, showed that such model is nonrealistic and it was forgotten. However, dust in plasma absorbs electrons and ions and this creates real fluxes which lead to dust attraction. Thus dust attraction is a realization of Lesage's model of gravity.

This analogy shows that there should exist a deep similarity between dust attraction and gravity, although gravity is universal while the attraction we are discussing acts only on dust particles, but not on electrons and ions. Nevertheless an instability similar to gravitational instability and caused by dust attraction should exist in dusty plasmas. We write down the dispersion relation for gravitational instability together with the dust-attraction electrostatic instability to point out this similarity. The equation for gravitational instability has the well known form

$$\omega^2 = k^2 v_s^2 - Gnm \tag{3}$$

where G is the gravitational constant, n is the density of gravitating matter, m is the mass of gravitating particles and v_s is the speed of sound. If the dust particles give the main contribution to the mass density in dusty plasma, then $n \approx n_d$, where n_d is the dust density and $m = m_d$ where m_d is the mass of dust particles.

In this case

$$\omega^2 = k^2 v_s^2 - G n_d m_d \,. \tag{4}$$

Instability occurs when ω^2 is negative and determines the Jeans critical length. The dispersion relation for the dust-attraction instability has the form

$$\omega^2 = k^2 v_{sd}^2 - \omega_{pd}^2 \eta \frac{a_0}{L_f} \,, \tag{5}$$

where v_{sd} is the dust sound velocity ($v_{sd}^2 \equiv \omega_{pd}^2 d_i^2$) and ω_{pd} is the dust plasma frequency ($\omega_{pd}^2 = 4\pi n_d Z_d^2 e^2 / m_d$). From (5) it follows that the critical length for development of the electrostatic dust-attraction instability is of the order of the fundamental length L_f.

This is another reason to call this length a fundamental length. The reason why this length is relatively small is that the dust sound velocity entering in (5) is much less than the usual sound velocity entering in (4). By comparing the last terms of (4) and (5) we find that the effective gravity introduced by dust attraction can be described by G_{eff}

$$G_{eff} = \frac{4\pi Z_d^2 e^2}{m_d^2} \eta \frac{a_0}{L_f} \,. \tag{6}$$

The development of dust attraction is the starting point for formation of self-organized structures in dusty plasmas.

IV ELEMENTARY RELATIONS IN DUSTY-PLASMA PHYSICS

We will give here the simplest relations in dusty plasma physics

1. **Dust size a_0 and Debye length d**

$$a_0 \ll d\,; \quad \frac{1}{d^2} = \frac{1}{d_{\rm i}^2} + \frac{1}{d_{\rm e}^2}\,, \tag{7}$$

$d \approx d_{\rm i}$ for $T_{\rm e} \gg T_{\rm i}$. We can introduce the dimensionless dust size a

$$a \equiv \frac{a_0}{d_{\rm i}}\,. \tag{8}$$

2. **Charging of dust**

Floating potential arguments give an estimate of the dust charge in units of the electron charge e $Z_{\rm d}e^2/a_0 \sim T_{\rm e}$. We will use the dimensionless charge z defined by

$$z \equiv \frac{Z_{\rm d}e^2}{a_0 T_{\rm e}}\,. \tag{9}$$

3. **Temperature ratio**

$$\tau \equiv \frac{T_{\rm i}}{T_{\rm e}}\,, \tag{10}$$

where $\tau \ll 1$ in laboratory experiments, $\tau \approx 1$ in astrophysical applications.

4. **Dimensionless dust density**

$$P \equiv \frac{n_{\rm d} Z_{\rm d}}{n_{\rm e}}\,. \tag{11}$$

In plasma etching experiments P is of the order of 10, in plasma crystal experiments it is of the order of 1 and in astrophysical conditions it varies from 10 to 10^{-3}. The condition of quasi-neutrality of the background dusty plasma reads

$$n_{0,\rm i} = n_{0,\rm e}(1+P)\,. \tag{12}$$

5. **Charging cross-sections**

Subscript ch is used for charging cross-sections. For electrons

$$\sigma_{\rm ch}^{\rm e} = \pi a_0^2 \left(1 - \frac{2Z_{\rm d}e^2}{m_{\rm e}v^2}\right)\,. \tag{13}$$

For ions

$$\sigma^{\mathrm{i}}_{\mathrm{ch}} = \pi a_0^2 \left(1 + \frac{2Z_{\mathrm{d}}e^2}{m_{\mathrm{i}}v^2}\right) . \tag{14}$$

The impact parameter for electrons is $p_{\mathrm{e}} < a_0$, and for ions is $p_{\mathrm{i}} > a_0$. For $\tau \ll 1$, $p_{\mathrm{i}} \gg a_0$. The plasma particle fluxes on dust particles are

$$\frac{\langle \sigma^{\mathrm{e}}_{\mathrm{ch}} v \rangle}{\pi a_0^2 v_{\mathrm{Te}}} = 2\sqrt{\frac{2}{\pi}} \exp(-z) , \tag{15}$$

$$\frac{\langle \sigma^{\mathrm{i}}_{\mathrm{ch}} v \rangle}{\pi a_0^2 v_{\mathrm{Ti}}} = 2\sqrt{\frac{2}{\pi}} \left(1 + \frac{\tau}{z}\right) . \tag{16}$$

6. **Coulomb elastic collision cross-sections**

The usual estimate $e^2 Z_{\mathrm{d}}/p_{\mathrm{i,e}} \approx T_{\mathrm{i,e}}, \sigma_{id,ed} \approx \pi p_{\mathrm{i,e}}^2$ gives two important results (subscript C is used for Coulomb cross-sections)

$$\frac{\sigma_{\mathrm{C}}^{\mathrm{ed,id}}}{\sigma_{\mathrm{C}}^{\mathrm{ee,ii}}} \approx Z_{\mathrm{d}}^2 \gg 1 , \tag{17}$$

$$\sigma_{\mathrm{C}}^{\mathrm{ed}} \approx \pi a_0^2 z^2 \ln(d/a_0) , \tag{18}$$

$$\sigma_{\mathrm{C}}^{\mathrm{id}} \approx \pi a_0^2 \frac{z^2}{\tau^2} \ln(d/a_0) . \tag{19}$$

For $\tau \ll 1$ we find $\sigma_{\mathrm{C}}^{\mathrm{id}} \gg \{\sigma_{\mathrm{C}}^{\mathrm{ed}}, \sigma^{\mathrm{i}}_{\mathrm{ch}}\}$ and $\sigma_{\mathrm{C}}^{\mathrm{id}} \approx (z/\tau)\sigma^{\mathrm{i}}_{\mathrm{ch}}$.

7. **Charging frequency**

The current balance equation

$$\begin{aligned}\frac{\delta Z_{\mathrm{d}}}{dt} &= n^{\mathrm{e}}\langle \sigma^{\mathrm{e}}_{ch}\rangle - n^{\mathrm{i}}\langle \sigma^{\mathrm{i}}_{\mathrm{ch}}\rangle \approx -\pi a_0^2 v_{\mathrm{Ti}} n^{\mathrm{i}} \delta \left(\frac{z}{\tau}\right) \\ &= -\pi n^{\mathrm{i}} a_0^2 v_{\mathrm{Ti}} \frac{\delta Z_{\mathrm{d}} e^2}{a_0 T_{\mathrm{i}}} = -\nu_{\mathrm{ch}} \delta Z_{\mathrm{d}} ;\end{aligned} \tag{20}$$

$$\exp(-z) = \frac{1+P}{\sqrt{\tau\mu}}(\tau + z) ; \quad \mu = \frac{m_{\mathrm{i}}}{m_{\mathrm{e}}} . \tag{21}$$

The charging frequency ν_{ch} is approximately given by

$$\nu_{\mathrm{ch}} \approx \frac{a_0}{d_{\mathrm{i}}^2} v_{\mathrm{Ti}} = \frac{v_{\mathrm{Ti}}}{L_{\mathrm{f}}} . \tag{22}$$

An important estimate is

$$\frac{\nu_{\rm ch}}{\nu_{\rm ii}} \approx Z_{\rm d}\frac{\tau}{z}\ . \tag{23}$$

The exact expression for the charging frequency for the equilibrium dust charge is:

$$\nu_{\rm ch} = \frac{\omega_{pi}}{\sqrt{2\pi}}\frac{a_0}{d_{\rm i}}(1+\tau+z) = \frac{v_{\rm Ti}}{\sqrt{2\pi}L_{\rm f}}(1+\tau+z)\ . \tag{24}$$

8. **Frequencies of plasma particles collisions with dust**

Only charging collisions enter the continuity equation. The collision frequency in the continuity equation is denoted as $\bar{\nu}_{\rm e,i}$.

$$\bar{\nu}_{\rm e,i} \approx n^d \langle \sigma_{\rm ch}^{\rm e,i} \rangle v_{\rm Te,Ti}\ , \tag{25}$$

$$\bar{\nu}_{\rm i} \approx \pi a^2 N^{\rm d}\frac{z}{\tau}v_{\rm Ti} \approx \frac{P\nu_{\rm ch}}{1+P}\ . \tag{26}$$

This collision frequency describes the rate of particle recombination on dust. An exact expression for these frequencies for the equilibrium dust charge is

$$\bar{\nu}_{\rm e} = \nu_{\rm ch}\frac{P}{z}\frac{(\tau+z)}{(1+\tau+z)}\ ; \tag{27}$$

$$\bar{\nu}_{\rm i} = \frac{\bar{\nu}_{\rm e}}{1+P}\ . \tag{28}$$

The frequencies describing the momentum transfer from plasma particles to dust particles are usually larger, since both the charging and the Coulomb collisions contribute and the Coulomb frequency is larger. Denote these frequencies as $\tilde{\nu}_{\rm e,i}$. The exact expressions are

$$\tilde{\nu}_{\rm e} = \nu_{\rm ch}\frac{P(\tau+z)}{z(1+\tau+z)}\left(4+z+\frac{2z^2}{3}e^z\ln\frac{d}{a}\right)\ , \tag{29}$$

$$\tilde{\nu}_{\rm i} = \nu_{\rm ch}\frac{P}{(1+P)z(1+\tau+z)}\left(z+\frac{4}{3}\tau+\frac{2z^2}{3\tau}\ln\frac{d}{a}\right)\ . \tag{30}$$

For $\tau \ll 1$ we have

$$\tilde{\nu}_{\rm i} \approx (z/\tau)\bar{\nu}_{\rm i} \approx \nu_{\rm ch}Pz/\tau \approx (\nu_{\rm ii}Z_{\rm d}\tau/z)P(z/\tau) = \nu_{\rm ii}PZ_{\rm d} \tag{31}$$

9. **Mean free path of plasma particles**

The mean free path for the charging process is

$$L_{\mathrm{ch}} \approx \frac{\bar{\nu}_{\mathrm{i}}}{v_{\mathrm{Ti}}} \approx L_{\mathrm{f}} \; , \tag{32}$$

and for Coulomb collisions is

$$L_{C,\mathrm{i}} = L_{\mathrm{f}} \frac{\tau}{z} \tag{33}$$

10. **Dust sound waves**

The ion pressure against dust inertia describes the simplest short-wave-length ($k \gg 1/Lf$) dust sound waves

$$\frac{1}{k^2 d_{\mathrm{i}}^2} = \frac{\omega_{\mathrm{pd}}^2}{\omega^2} \; , \tag{34}$$

$$\omega = k v_{\mathrm{sd}} \; ; \quad v_{\mathrm{sd}} = \sqrt{\frac{P Z_{\mathrm{d}} T_{\mathrm{i}}}{m_{\mathrm{d}}(1+P)}} \; . \tag{35}$$

This expression can be compared with that for ion-sound waves

$$\omega = k v_{\mathrm{s}} \; ; \quad v_{\mathrm{s}} = \sqrt{\frac{Z_{\mathrm{i}} T_{\mathrm{e}}}{m_{\mathrm{i}}}} \; . \tag{36}$$

The expression (34) is valid only for $\tau \ll 1$. In the more general case

$$v_{\mathrm{sd}} = \frac{P Z_{\mathrm{d}} T_{\mathrm{e}}}{m_{\mathrm{d}}} \frac{\tau}{\tau + 1 + P} \; . \tag{37}$$

The long wavelength ($k \ll 1/L_{\mathrm{f}}$) dust sound waves are determined for $\tau \ll 1$ by the electron temperature (not the ion temperature as in (35)). For arbitrary τ the long-wavelength dust sound velocity is

$$\omega = k v_{\mathrm{sd}} \; ; \quad v_{\mathrm{sd}} = \frac{Z_{\mathrm{d}} T_{\mathrm{e}}}{m_{\mathrm{d}}} \frac{P(\tau + z) + (1+P)(1+\tau+z)}{\tau + z + 1 + P} \; . \tag{38}$$

11. **Dust temperature**

The thermal dust velocity is $v_{Td} = \sqrt{T_{\mathrm{d}}/m_{\mathrm{d}}}$. A dimensionless temperature can be defined as

$$\tau_{\mathrm{d}} = \frac{v_{Td}}{v_{\mathrm{sd}}} = \sqrt{\frac{T_{\mathrm{d}}(1+P)}{P Z_{\mathrm{d}} T_{\mathrm{i}}}} \; . \tag{39}$$

The last expression is written for the short-wave-length dust sound speed.

12. **Electrostatic energy accumulated on dust particles**

$$\frac{Z_d^2 e^2 N_d}{a N_e T_e} = Pz \; . \tag{40}$$

For usual parameters $z \approx 2$, $P \approx 1$ the energy accumulated on dust particles is rather big and can drive instabilities.

13. **Factor Γ**

The inter-dust distance is

$$\Delta = (4\pi N_d/3)^{-1/3} \; . \tag{41}$$

A useful relation is

$$\frac{a_0}{\Delta} = \left(\frac{P\tau}{(1+P)3z}\right)^{1/3} a^{2/3} \; . \tag{42}$$

Then

$$\tau_d^2 = \frac{a^{2/3}}{\Gamma}\left(\frac{(1+P)3z}{P\tau}\right)^{2/3} \; . \tag{43}$$

The ratio of Coulomb potential energy to thermal dust energy is denoted as Γ

$$\Gamma = \frac{e^2 Z_d^2}{\Delta T_d} \; . \tag{44}$$

V DUST-PLASMA STRUCTURES

The high dissipation rate in dusty plasma makes all self-organizing processes very fast to reproduce and to model the competition of dust structures for plasma as "food" for their survival. This opens the possibility of modeling biological systems and the evolution of self-organizing structures.

The question has already been asked whether such self-organized dust structures can exist in space, since both dust and plasma is very common in space. In particular the question has been raised whether dust-plasma crystals can already exist in space. But the question is indeed more general since the variety of self-organized structures could be large and the type of interactions between them could be very complicated. General questions arise about the natural evolution of these structures in space and the possibility of the natural storage of information in this evolution. If these questions can be answered positively it will be obvious that the present search for extraterrestrial intelligence was made in the completely wrong direction.

Plasma structures can usually be formed in the absence of dust. Any structures of this kind could have for example the electric potential inside the structure less (negative) or larger (positive) than the surrounding plasma potential. Suppose the difference of the potential is negative and is ϕ. Then the trapping energy for a dust particle will be $Z_d\phi$. It is well known that dust particles are well trapped in electrostatic potential wells due to their large charges. They are known to be trapped by the floating potentials of the walls of discharge chambers. The same is true for potential structures in plasma — they become traps for dust particles. If many dust particles are trapped in the structure, then they start to change the structure itself.

First of all they create an additional negative potential and absorb the plasma particles. The other plasma particles outside the structure will start to move toward the structure, creating the flux of plasma to the structure. Soon the structure becomes a dissipative structure and should be treated as a self-organized structure. This scenario is only one of the possible ways to create dust-plasma structures. In some laboratory experiments the plasma is created by volume ionization, as in the case of etching experiments. These structures after formation will also be dust-plasma self-organized structures.

Another possibility for creation of dust-plasma structures is an increase of instability caused by the presence of dust particles in the plasma. Since dust introduces a high rate of dissipation, all negative-energy modes, such as drift modes or beam modes, become much more unstable in the presence of dust in plasma.

As soon as the dust-plasma structure is formed it should create plasma fluxes toward the structure. One can foresee the consequences of the presence of these fluxes. First of all they transfer energy, momentum and angular momentum to the structure. Conservation of angular momentum implies that the structure in general should have some rotation, i.e. it should be a vortex structure. The simplest structure of this kind is the two-dimensional (2D) vortex structure. Therefore one can expect that the formation of vortex structures in dusty plasma should be typical phenomenon.

If the flux does not transfer angular momentum, we can ask as to the fate of the particle fluxes and momentum fluxes. Current conservation implies that outside the structure the flux of electrons is equal to the flux of ions (for stationary structures). The main part of the momentum is transferred to the structure by ions. In the case of symmetry (planar or spherical) the total momentum transferred could be zero (momentum fluxes transferred from different sides compensate each other). Then such a structure will be stationary, in the sense that it does not gain momentum.

Ions and electrons are absorbed by dust particles. But the collision frequency for ion-scattering by dust particles is larger than the collision frequency for their absorption by dust particles. This statement was illustrated above in detail. Thus the ions will be more often scattered than absorbed. Which means that the ions will be accumulated in the dust structure. Thus the structure not only traps dust particles but also accumulates ions. The pressure of the accumulated ions will work against the ram ion pressure of the ion flux towards the structure.

Thus one can foresee that dust-plasma structures will not only be structures where dust exists, but also structures where the plasma ion density is increased. Both dust and plasma densities should be distributed in the structure self-consistently. This structure can survive only in the presence of plasma outside the structure to create the fluxes toward the structure, or in presence of a constant source of external ionization in the structure. In both cases plasma serves as a "food" for the structure.

We shall later consider 2D vortex structures in dusty plasma, including drift waves, and discuss the problem of combined vortices in which both the neutral and charged components form a vortex which, through collisions of neutrals with dust particles, is bound to a drift vortex, forming a new type of vortex.

Only stationary dust-plasma structures can be described by a system of nonlinear equations which, in the dimensionless form (when the sizes are measured in the fundamental length introduced above), has no small parameter, and in this system we can describe universal structures. The latter fact can be foreseen from the properties mentioned in dusty plasma where the plasma particle mean free path in dusty plasma is equal to the fundamental length times the parameter P. Thus the stationary dust-plasma structures will be described by a universal equation not containing a small parameter (and in which all terms are of the same order of magnitude) if, on average, the parameter P is of the order of 1 in the structure. We will show that, for examples we will consider, this is indeed the case.

To illustrate this statement we can write down the equation of change of dust momenta in dimensionless form assuming that all variables depend only on the space variable $\mathbf{r}'/L_{\mathrm{f}}$, denoted in the subsequent equation as $\mathbf{r}$. As an example we can write the expression for the 1D momentum equation for dust particles taking into account only the attraction and the repulsion forces in the dimensionless coordinates x introduced here

$$\frac{u_{\mathrm{d}}}{\sqrt{2}v_{\mathrm{Ti}}}\frac{\partial}{\partial x}\left(\frac{u_{\mathrm{d}}}{\sqrt{2}v_{\mathrm{Ti}}}\right) = \eta_a Z_{\mathrm{d}}^2\frac{\partial U_a}{\partial x} + \eta_r Z_{\mathrm{d}}^2\int_{-\infty}^{+\infty}\frac{n_{\mathrm{d}}(x')}{n_{0,\mathrm{i}}}\frac{1}{2(x-x')}\,dx' , \quad (45)$$

$$\frac{\partial^2}{\partial x^2}U_a = -\frac{n_{\mathrm{d}}}{n_{0,\mathrm{i}}} . \quad (46)$$

We have written this expression to show that, indeed, the attraction forces can be described by a Poisson-type of equation, that the additional repulsion forces (the electrostatic repulsion is not included in the RHS of (45)) always have an integral form, and that there is, in dimensional variables, no small parameter — all the terms can be of the same order of magnitude.

One important point should be mentioned, related to thé dust attraction, which causes the contraction of dust clouds and helps the formation of dust-plasma structures. This attraction will work on large time scales, when the dust particles will be able to response to disturbances in their density. But for stationary dust structures the presence of this interaction is very important. If the size of the structure is larger than the plasma particle mean free path, the attraction creates a surface layer with a surface tension similar to a kind of "skin" separating the structure from the surrounding plasma.

Of course the structures can be non-stationary or can be varying in time so fast that the dust particles will be not able to response to these changes. Then these structures will be non-universal. The point is that dust in plasma opens many new possibilities to form self-organized structures, the variety of these structures in dusty plasma being much larger than in other systems.

The problem of turbulence in dusty plasmas involves coexistence, competition and interaction between different dust-plasma structures and is far more complicated than the problems of hydrodynamic turbulence or plasma turbulence. It will be the subject of many future investigations. At present, only a few first steps in understanding turbulence in dusty plasma have been taken, using the weak turbulence approach when dust-plasma structures do not appear, or do not play a significant role. In this case the dusty plasma also opens up new possibilities due to the appearance of new modes, such as dust-sound waves.

The presence of two sound branches in dusty plasma makes the weak turbulence qualitatively different from that where only one sound wave branch can exist. The point is that the interactions between two sound branches creates the possibility of forming universal Kolmogorov decay cascades and the formation of universal decay turbulence spectra with universal spectra $W_k \propto 1/k^{3/2}$. But the large energy input usually necessary to support dusty plasma makes the region of validity of this spectrum rather narrow, and outside this regime, dusty-plasma turbulence can convert to the state of interacting self-organized dusty-plasma structures.

VI HASEGAWA–MIMA AND HASEGAWA–WAKATANI EQUATIONS IN DUSTY PLASMAS

Drift vortices are well-known examples of 2D nonlinear structures in the absence of dust. We start with the simple question as to whether the presence of dust will modify the excitation of these vortices, the linear approach when the vortices are treated as drift waves being the simplest approach to reach some answers.

The qualitative results can be foreseen without calculations. Drift waves are negative energy waves, and any dissipation leads to their excitation. On the other hand it was already mentioned that the presence of dust makes the dusty plasma system much more dissipative, which obviously shows that the drift waves in dusty plasma can be more unstable and the threshold of instability will be much lower in presence of dust. Both indeed appear to be the most probable effects introduced by the presence of dust in plasma.

One can find not only linear instability but also can describe nonlinear vortices in dusty plasma by generalizing the Hasegawa–Mima and Hasegawa–Wakatani equations for nonlinear drift waves in dusty plasma. In this approach one can first consider the dust immovable and distributed uniformly, but inhomogeneously. One should then take into account movement of the dust, and the modification of its distribution to form self-consistent dust-plasma structures.

Drift waves exist in the presence, for example, of density inhomogeneity and in the presence of magnetic field. These two factors should be added in the balance equations written above. The magnetic field is described by the Lorentz force in the momentum equations for electrons and ions, while the density inhomogeneity is described by the pressure term. Balance of them leads to particle diamagnetic drift. For electrons we get the diamagnetic electron drift velocity

$$\mathbf{u}_{\mathrm{e},0} = -\frac{v_{\mathrm{Te}}^2}{\Omega_{\mathrm{e}} n_{0,\mathrm{e}}} \mathbf{z} \times \nabla n_{0,\mathrm{e}} = -\frac{v_{\mathrm{s}}^2}{\Omega_{\mathrm{i}} n_{0,\mathrm{e}}} \mathbf{z} \times \nabla n_{0,\mathrm{e}} \tag{47}$$

where $\mathbf{z}$ is the direction of the magnetic field, Ω_{e} is the electron cyclotron frequency, Ω_{i} is the ion cyclotron frequency and $v_{\mathrm{s}} = \sqrt{T_{\mathrm{e}} Z_{\mathrm{i}}/m_{\mathrm{i}}}$ is the ion-sound velocity.

We denote the direction of the density gradient by $\mathbf{x}$ and the direction of the drift wave propagation by $\mathbf{y}$. The characteristic length of electron-density inhomogeneity is denoted by L_n:

$$L_n^{-1} = \frac{1}{n_{0,\mathrm{e}}} \frac{\partial n_{0,\mathrm{e}}}{\partial x} \tag{48}$$

Then the (47) can be written as

$$u_{\mathrm{e},0,\mathrm{y}} = -v_{\mathrm{s}} \frac{\rho_{\mathrm{s}}}{L_n} ; \quad \rho_{\mathrm{s}} = \frac{v_{\mathrm{s}}}{\Omega_{\mathrm{i}}} \tag{49}$$

Here ρ_{s} denotes the ion Larmor radius for electron temperature. The fundamental length L_{f} is determined when the dust-dust interaction is taken into account. This occurs in the absence of magnetic field. In the presence of magnetic field and a density inhomogeneity, the length L_n is another important length. The fundamental length L_{f} will not play any role for immovable dust, which is the case we consider here. The natural small parameter in this case is the ratio of ion Larmor radius for electron temperature to the length L_n, namely

$$\epsilon \equiv \frac{\rho_s}{L_n} \tag{50}$$

The drift approximation corresponds to

$$\epsilon \ll 1 \, . \tag{51}$$

We can expand in this parameter, including linear and nonlinear effects up to quadratic nonlinearities. Although the dust is immovable, the *charge* on the dust can change.

We have the following small values in which we expand leaving only the quadratic nonlinearities:

$$\frac{\delta Z_{\rm d}}{Z_{\rm d}} \, ; \quad \frac{\delta n_{\rm e}}{n_{\rm e,0}} \, ; \quad \frac{e\phi}{T_{\rm e}} \, , \tag{52}$$

where $Z_{\rm d}$ is the equilibrium value of the dust charge, $n_{\rm i,0}$ is the equilibrium value of the ion density, $n_{\rm e,0}$ is the equilibrium value of the electron density, ϕ is the electrostatic potential of the drift mode and we denote the deviations of the corresponding variables from their equilibrium values with the symbol δ.

All equilibrium value can be inhomogeneous in space (we consider here the presence of the dependence on coordinate x only). Due to quasi-neutrality of the equilibrium state the ion equilibrium density can be expressed through the electron equilibrium density and the parameter $P = n_{\rm d} Z_{\rm d}/n_{\rm e,0}$:

$$n_{\rm i,0} = n_{\rm e,0}(1 + P) \tag{53}$$

The same is true for $\delta n_{\rm i}/n_{\rm i,0}$ if one assumes that in deviations from equilibrium the quasi-neutrality is maintained (this is the case where the wave length is larger than the Debye length). Indeed we have then

$$\delta n_{\rm i} \doteq \delta n_{\rm e} + \delta Z_{\rm d} n_{\rm d} \, ; \quad \frac{\delta n_{\rm i}}{n_{\rm i,0}} = \frac{\delta n_{\rm e}}{n_{\rm e,0}} \frac{1}{1+P} + \frac{\delta Z_{\rm d}}{Z_{\rm d}} \frac{P}{1+P} \tag{54}$$

In (54) variations of dust density are not taken into account, since in this section we consider the dust to be immovable. Equation (54) also gives an explanation why, in the list (52) of the small variables in which the expansion is made, the value $\delta n_{\rm i}/n_{\rm i,0}$ is not listed as an independent variable. In our expansion we take into account all possible linear and quadratic combinations of the variables (52).

The procedure is standard, but the new feature is that the new hydrodynamics should be used and new types of nonlinearities and linear terms appear due to variations of the dust charge $\delta Z_{\rm d}$. For the latter we can use the dust-charging equation. It appears that the nonlinear terms in the charging equation are absent

$$\frac{\partial}{\partial t}\frac{\delta Z_{\rm d}}{Z_{\rm d}} = -\nu_{\rm ch}\frac{\delta Z_{\rm d}}{Z_{\rm d}} + \frac{1+P}{P}\bar{\nu}_{\rm i}\left[\frac{\delta n_{\rm e}}{n_{\rm e,0}} - \frac{\delta n_{\rm i}}{n_{\rm i,0}}\right] . \tag{55}$$

The final result of the derivation for nonlinear drift modes in dusty plasma, taking into account only the quadratic nonlinearities in the expansion, we present in a form using new dimensionless variables which take into account the small parameter ϵ

$$t \to \Omega_{\rm i}\frac{\rho_{\rm s}}{L_n}t\ ; \quad x \to \frac{x}{\rho_{\rm s}}\ ; \quad y \to \frac{y}{\rho_{\rm s}}\ ; \quad n \to \frac{\delta n_{\rm e}}{n_{\rm e,0}}\frac{L_n}{\rho_{\rm s}}\ ;$$

$$\zeta \to \frac{\delta z_{\rm d}}{Z_{\rm d}}\frac{L_n}{\rho_{\rm s}}\ ; \quad \psi \to \frac{e\phi}{T_{\rm e}}\frac{L_n}{\rho_{\rm s}}\ ; \quad s = \frac{PL_{\rm d}^{-1}}{(1+P)(1+P-PL_{\rm d}^{-1})} . \tag{56}$$

These equations serve as a generalization of the known Hasegawa–Mima and Hasegawa–Wakatani equation for dusty plasmas

$$\frac{\partial}{\partial t}\nabla^2\psi - s\frac{\partial\psi}{\partial y} - \frac{c_{\rm e}}{1+P}(\psi - n) = \{\nabla^2\psi, \psi\} - \frac{P}{1+P}\{\zeta, \psi\}\ ; \tag{57}$$

$$L_{\rm d}^{-1} = \frac{\partial \ln n_{\rm d} Z_{\rm d}}{\partial \ln n_{\rm i,0}} ,$$

$$\frac{\partial n}{\partial t} + \alpha(n - z\zeta) + \frac{\partial\psi}{\partial y} - c_{\rm e}(\psi - n) = \{n, \psi\}\ ; \tag{58}$$

$$\frac{\partial\zeta}{\partial t} + \beta\zeta = \alpha(1+P)n , \tag{59}$$

where the notation $\{\cdot,\cdot\}$ is used for Poisson brackets

$$\{A, B\} = \frac{\partial A}{\partial x}\frac{\partial B}{\partial y} - \frac{\partial A}{\partial y}\frac{\partial B}{\partial x} . \tag{60}$$

The coefficient $c_{\rm e}$ describes the electron non-adiabaticity along the magnetic field lines due to collisions of electrons with dust particles. The usual binary collisions appearing in Hasegawa–Wakatani equations are here neglected within the framework of the new hydrodynamics, since they are to be assumed much less than the collisions of electrons with dust particles

$$c_{\rm e} = \frac{v_{\rm Te}^2 k_z^2}{\tilde{\nu}_{\rm e}\Omega_{\rm i}}\frac{L_n}{\rho_{\rm s}} \tag{61}$$

where k_z is the wave vector component along the magnetic field. Of course, for the case where $k_z = 0$, there is no dissipation along the magnetic field lines and therefore the instability can vanish. The most important point is that the $\tilde{\nu}_e$ is in dusty plasmas usually much larger than the binary collision frequency and for the same other parameters c_e is, by several orders of magnitude, less than in the absence of dust. The smaller c_e, the larger is the non-adiabaticity, and the lower the threshold of the instability.

If necessary, in the system of nonlinear equations (48–50) it is easy to take into account the effect of ion non-adiabaticity, which is usually much smaller than the electron non-adiabaticity.

The coefficients α and β describe the influence of charging and Coulomb collisions

$$\alpha = \frac{\bar{\nu}_e}{\Omega_i}\frac{L_n}{\rho_s} \; ; \quad \beta = \alpha(1+P)\left(1 + \frac{\nu_{ch}}{\bar{\nu}_i}\right) . \tag{62}$$

These equations generalize the known system of Hasegawa–Wakatani equations, which do not take into account dust but take into account the viscosity and diffusion terms caused by binary collisions. To convert our system to the Hasegawa–Wakatani system, one needs first to take into account the effects caused by binary collisions and then go to the limit $P \to 0$.

VII EXCITATION OF 2D DRIFT VORTICES IN DUSTY PLASMAS

The system of nonlinear equation obtained describes, in the linear limit, the modification of the drift instability in dusty plasmas. The linear system of equations is

$$(i\omega k^2 - sik_y)\psi - \frac{c_e}{1+P}(\psi - n) = 0 \; ; \quad s = \frac{PL_d^{-1}}{(1+P)(1+P-PL_d^{-1})} , \tag{63}$$

$$(-i\omega + \alpha + c_e)n - \alpha z\zeta + (ik_y - c_e)\psi = 0 , \tag{64}$$

$$(-i\omega + \beta)\zeta = \alpha(1+P)n . \tag{65}$$

This equation leads to the dispersion relation for $\bar{\omega} = \omega/\alpha$

$$i\bar{\omega}k^2 - is\frac{k_y}{\alpha} - \frac{c_e}{(1+P)\alpha} - \frac{c_e(i\frac{k_y}{\alpha} - \frac{c_e}{\alpha})}{\alpha(1+P)\left[-i\bar{\omega} + 1 + \frac{c_e}{\alpha} - \frac{z(1+P)}{-i\bar{\omega}+\beta/\alpha}\right]} = 0 . \tag{66}$$

This form of the equation is useful for obtaining limiting cases. The equation depends on two parameters c_e/α and k_y/α. For $c_e \gg \alpha$ we get

$$i\bar{\omega}\left(k^2+\frac{1}{1+P}\right)-\frac{k_y}{\alpha}\left(s+\frac{1}{1+P}\right)+\frac{z}{\left(-i\bar{\omega}+\frac{\beta}{\alpha}\right)}=0\,. \tag{67}$$

In this limit, as in the Hasegawa–Mima approach, the parameter c_e does not enter. We mention that here both coefficients c_e and α are related to the influence of dust — the coefficient c_e describes the non-adiabaticity caused by dust and the coefficient α describes the charging processes. (It is completely new, although the coefficient c_e is decreased by the presence of dust, usually by several orders of magnitude — approximately $Z_d P$ times). By assuming $c_e \gg \alpha$ we consider the case where the non-adiabaticity introduced by dust is much less than the effect of dust charging (the non-adiabaticity being larger when c_e is smaller).

The quadratic equation (67) is simple to solve analytically. But for the sake of deeper understanding we consider the limiting cases. First of all we can mention that in general the ratio β/α is of the order of 1. For the special case of small dust densities $P \ll 1$ this ratio is large:

$$\frac{\beta}{\alpha}\approx\frac{z(1+\tau+z)}{P(\tau+z)} \tag{68}$$

Then we find that the drift waves are always unstable - the instability is caused by dust only. Taking into account $P \ll 1, s \ll 1$ we find

$$\omega\approx\frac{k_y}{1+k^2}+i\frac{\alpha^2 z}{\beta} \tag{69}$$

The real part of drift frequency is the approximately the same as in absence of dust but the growth rate is of the order of $\bar{\omega}P \approx \nu_{\mathrm{ch}}P^2 \approx \nu_{\mathrm{i,i}}Z_{\mathrm{d}}P^2$

For P of the order of 1 the ratio β/α is of the order of 1, and the drift waves become so unstable that the real and imaginary part of the frequency is of the same order of magnitude. This happens if

$$\frac{k_y}{\alpha}(s+\frac{1}{1+P})\ll\frac{\beta}{\alpha}\left(k^2+\frac{1}{1+P}\right)\,. \tag{70}$$

Then there always exists an unstable root

$$\bar{\omega}=\frac{i}{2}\left(\sqrt{\frac{\beta^2}{\alpha^2}+\frac{4z}{k^2+\frac{1}{1+P}}}-\frac{\beta}{\alpha}\right)\,. \tag{71}$$

For the opposite inequality to (71) there is no instability: one of the roots describes simply the relaxation of dust charges ($\omega \approx -i\beta$) and another root describes a stable dust-drift wave

$$\omega\approx\frac{k_y\left(s+\frac{1}{1+P}\right)}{k^2+\frac{1}{1+P}}-i\frac{3\beta z\left(k^2+\frac{1}{1+P}\right)\alpha^2}{k_y^2\left(s+\frac{1}{1+P}\right)^2}\,. \tag{72}$$

An important feature is the change of the real part of the drift frequency in the presence of dust described by the first term of (72).

We can consider the case of small and large α values and both components of the wave vector of the same order of magnitude (P of the order or larger than 1). According to (71) in the unstable region the growth rate is of the order of α for $k \ll 1$ and decreases with the wave number for $k \gg 1$. For $\alpha \gg 1$ this growth rate is very large, the condition (70) for $k \ll 1$ means $k_y \ll \alpha$ and is always fulfilled since $\alpha \gg 1$; for $k \gg 1$ the condition (71) means $k \gg 1/\alpha$ and is also always fulfilled. For $\alpha \ll 1$ the instability growth rate is again of the order of α but for $k \ll 1$ the condition (70) gives the restriction $k_y \ll 1$ and for $k \gg 1$ (where the growth rate decreases with k and thus is much less than its maximum value) the condition (59) gives $k \gg 1/\alpha$.

For $c_e \ll \alpha$ the non-adiabaticity introduced by dust dominates. Assuming that both components of the wave vector are of the same order of magnitude we find that for k of the order of 1 and $\alpha \ll 1$ the instability is absent but the real part of the frequency is much changed by dust

$$\omega = \frac{sk_y}{k^2} - i\frac{c_e}{1+P} \,. \tag{73}$$

The instability appears for $\alpha \gg 1$

$$\omega = \frac{sk_y}{k^2} + i\frac{c_e(P(z+\tau) + z(P+1+z+\tau))}{(1+P)zP(\tau+z)} \tag{74}$$

We mention that the condition $c_e \ll \alpha$ reads $k_z^2 v_{Te}^2 \ll \bar{\nu}_e \tilde{\nu}_e \propto P^2$ and means the parameter P should be larger than a certain value (depending of course on the wave number component k_z, which is another independent variable).

VIII EFFECT OF DUST-DUST ATTRACTION, DUST-DUST REPULSION AND DUST-NEUTRAL COLLISIONS

Long-lived dust-plasma structures have time to redistribute the dust particles in the structure. One expects that dust will be concentrated in the structure and the structure will become negatively charged as a result of the fluxes of electrons and ions. At the present time, the program of investigation of such 2D dust-plasma structures can only be formulated. The first step is to generalize the nonlinear equation given above to take into account the dust-dust interactions, including the dust-dust attraction and dust-dust repulsion.

The results of this step are given below. The next step will be to solve these nonlinear equations for 2D structures. This step is only on the way to be performed since it needs a lot of numerical calculations and a lot of computer time, which must be left for future research. At least the equations needed for the numerics are at now written explicitly, and they can be used in different numerical approaches.

A problem related to the dust redistribution in the structures is the problem of dust-neutral interaction. The amount of neutral non-ionized component in many circumstances is not small. Thus in the lower ionosphere where the dust is observed and the drift structures are observed as well the degree of ionization is low and dust neutral collisions can be important in transferring momentum to the dust particles. Since the neutrals have no charges the cross-section of dust interaction with neutrals is just the geometrical cross-section of dust particles.

Thus the collision frequency of neutrals with dust can be written in a simple manner. In addition to neutral-dust collisions, collisions of neutrals with ions can be of importance in exchange of energy and momentum between the neutrals and ions. The comparison of the rate of ion-neutral collisions with dust-neutral collisions show that there exist broad range of parameters where the dust-neutral collisions dominate.

The next step can be to derive the nonlinear equations in which the dust-neutral collisions are taken into account, together with dust-dust interactions. The result of such an investigation is given below for the limit where the effective dust-neutral collision frequency is larger than the inverse time of the nonlinear process. The next step of investigation of dust-plasma structures using these equations is left for future research.

For such structures one is interested in the possibility of the merging into one structure of the properties of two types of vortices. If the neutrals have no interaction with dust, their motion, as for any subsonic motion in neutral gas, will be described as vortex motion or as a superposition of many vortices.

In dusty plasmas where the interactions of dust particles with neutrals is not taken into account, the structures will be drift vortices, modified by dust. In the presence of dust-neutral interaction one cannot separate these vortices, and the nonlinear structure appearing will simultaneously have the properties of neutral gas vortices and drift vortices.

In the lower ionosphere and upper atmosphere such structures can probably really be created since the conditions for their appearance exist there. The drift structures are much influenced by the degree of ionization and solar activity. The link to the usual vortex motion can create some motions in the upper atmosphere which depend on the degree of dust present, and therefore on the degree of existing pollution. A detailed analysis is given in the paper with S.V. Vladimirov which is published in these proceedings.

ACKNOWLEDGMENTS

I give my acknowledgments to the organizers of the workshop and especially to Prof. Robert Dewar for hospitality and discussions. The author thanks very much S. Vladimirov for fruitful collaboration and discussions and careful reading of the manuscript.

The author appreciates discussions at the workshop with W. Horton, S. Hamberger, T. Sheridan and other participants. The author is grateful to D. Melrose for hospitality during his stay in Sydney. The support of Sydney University, which made the author's visit to Australia possible, is gratefully acknowledged.

BIBLIOGRAPHY

This text is written according to the lectures given by the author at the workshop, and therefore has a lecture style, which is why no references in the text were given. The advantage of such a representation is that then the reader does not need to look up the original literature to have a description of the problem, and can use the text of the lectures directly. Thus this kind of representation is appropriate to the workshop style.

However, it is desirable to give references to works on related topics, as well as references to the papers in which the author has been involved and to which not much attention was paid in these lectures. It is desirable to mention the similarities and differences in approaches used by different authors and in the following we give such a critique.

1. *Dust attraction.*

   After the author's seminar in the General Physics Institute (Moscow) about the diminishing of the electrostatic dust-dust energy with an estimation of the role of fluxes on this interaction at distances less than the Debye distance, [1], the paper [2] appeared with the same results concerning the forces due to fluxes, but for distances larger than the Debye distances.

   The most important work in this field is the numerical simulations, which proved the existence of these forces [3] and the analytical expressions obtained in [4]. Thus, comparing [1,2] with [3,4] we conclude that the latter contain a more developed approach and results appropriate for comparison with experimental data.

   In paper [2] the repulsion force was missed, which was taken into account in [3,4]. Thus the molecular type of potential was obtained only in [3,4] and the conclusion that a new state of matter can exist in dusty plasma can be made only using the results of [3,4].

Recently the paper [5] was published discussing the problem of [1], it agrees with [1] in the conclusion that the electrostatic energy decreases when the distance between the dust particles decreases. Then in [5] the arguments of [6] were repeated, which are valid only for thermal equilibrium. But dusty plasma is an open and highly non-equilibrium state and these arguments do not work, [7].

The problem of the openness of the dusty plasma system was much emphasized in the recent review by the author, [8]. Concerning the thermophoretic forces one should mention the need to investigate them close to the wall [9]. Concerning the influence of the ion drift close to the wall on the dust-dust interaction, it is necessary to mention the work [10] on dust in the sheath. The general kinetic theory of a single dust particle close to the wall was formulated in [11].

2. *Problems of dust-plasma crystals.*

The first theoretical statement that, in dusty plasma, a Coulomb crystal can be obtained was made in [12] using the strongly correlated plasma approach of [13]. The criterion for crystallization was obtained in [13] by using Monte-Carlo simulations of strongly coupled systems, and is $\Gamma > 170$.

It is also not possible to improve the relation [13] by taking into account only the shielding of the Coulomb potential [14], but for distances larger than the Debye radius a new long-range, un-shielded repulsion appears, as well as the attraction of dust particles discussed above. A criterion for crystallization can be obtained by finding the binding energy as a balance of repulsion and attraction at distances larger than the Debye radius and comparing it with the dust kinetic energy.

This criterion seems to more appropriate for interpretation of existing experiments than the criteria [13,14]. The present experiments can be listed as [15]- [24]. Most of them were performed is the plasma sheath where both electric field and ion drift exists. The investigation of dust behavior in the sheath is only starting [10].

The paper [25] discusses the appearance of attraction as a wake field behind the dust particle produced by ion drift in the sheath. The consideration [25] can be compared with [26]. The appearance of a wake field is natural because of the particle longitudinal wave Cerenkov emission.

In [10] it was mentioned that the ions, according to the observations of [27], should have a large thermal spread along the drift, which causes attraction in the plane parallel to the wall. On the other hand the shielding should be considered as nonlinear, which still waits to be considered.
The author appreciates very much the discussion of these problems with S. Vladimirov and T. Sheridan at this workshop.

Concerning the priority of experimental discovery of the dust plasma crystals it should be given to Ref. [17]. The first publication on this subject was made at the Toulouse conference in 1994 [15], and in the lecture given by G. Morfill in Bochum at the Conference on Phenonemena in Ionized Gases in 1994 [16].

The first investigation of melting was made in [20], but the detailed investigation of this phenomenon was made in [18,19]. The author appreciates very much discussions with J. Allen and his colleagues concerning new phenomena in dust crystals observed at Oxford university [15], and a discussion with J. Peel on his measurements of dust charges in crystals [16].

3. *General dusty plasma theory.*

The general Klimontovich-type equation with charge as an independent variable was first discussed with O. Havnes and T. Aslaksen. The only publication at that time was [28] on the linear properties of collective plasma modes in dusty plasmas, showing the new effects appearing for variable charges of dust particles. The only results obtained up to the present time are the appearance of dust-dust attraction and repulsion, the distribution of charges of dust particles, which appear in certain cases to be Gaussian, and the limit of fast plasma-particle collisions with dust particles, where the usual coefficient of wave damping is proved to have the additional factor PZ_{d} as compared with the damping in the absence of dust [29].

Fluctuations in dust charges were also the subject of the paper [30], the results being very similar to those obtained in [11]. The problem of very low frequency fluctuations is of special interest for the theory of the pair correlation function in dusty plasma [31].

Already in [28] it was mentioned that, in practice, the charging frequency is rather large, which was the first step toward realizing that there should exist a broad range of frequencies less than the charging frequency for which a new hydrodynamics will be valid.

However this conclusion was not emphasized in [28]. The whole ideology and the equations of the New Hydrodynamics were given in [32], where all effective collision frequencies used above were calculated. The general new hydrodynamic dispersion equation and new plasma modes were investigated in [4,33,34].

4. *Dust in space.*

The importance of dust in space was realized many years ago. The treatment was elementary, using single particles without a self-consistent treatment of all components, and without taking into account the openness of the dusty plasma system [35,36].

The standard point of view on star formation is given in an excellent review [37] — although written 20 years ago it does not needed to be updated. In general star formation should be described as a self-organized process in which both gravitational effects, radiation and plasma effects contribute to the process.

Some aspects of the present point of view are given in the review by the author [8]. Concerning the planetary ring problem, some of the effects related to dust charging were already taken into account. The thickness of the ring was the subject of discussion in [38,39], as was the balance of gravitational attraction and Coulomb repulsion of dust particles. The gravitational aspect of planetary rings has been developed in much more detail (see the review [40]).

REFERENCES

1. Tsytovich V.N. *Comments on Plasma Physics and Controlled Fusion* **15** 349 (1994)
2. Ignatov A.M. *Plasma Physics (Russ. Journ.)* **27** 323 (1996)
3. Khodataev Ya.K., Bingham R., Tarakanov V.P. and Tsytovich V.N. *Journ. Plasma Phys.* **49** 224 (1996)
4. Tsytovich V.N., Khodataev Ya.K. and Bingham R. *Comments on Plasma Physics and Controlled Fusion* **17** 221 (1996)
5. Hamaguchi S. *Plasma Physics and Controlled Fusion* **18** 45 (1997)
6. Landau L.D. and Lifshits E.M. *Electrodynamics of Continuous Media* (Pergamon Press 1983)
7. Vladimirov S.V. and Tsytovich V.N. *Comments on Plasma Physics and Controlled Fusion* **18** (1997)
8. Tsytovich V.N. *Physics Uspekhi* **40** 53 (1997)
9. Havnes O., Nitter T., Tsytovich V., Morfill G and Hartquist T. *Plasma Sources, Science and Technology* **3** 448 (1994)
10. Benkadda S., Tsytovich V., Preprint Institut Méditerranéen de Technologie, Equipe Turbulent Plasma Marseille, France (1997)
11. Sitenko A.G., Zagorodny A.G. et al *Plasma Phys and Controlled Fusion* **38** A105 (1996)
12. Ikezi H. *Phys. Fluids* **29** 1764 (1986)
13. Ichimaru S. *Phys Rep.* **34** 1 (1982); *Basic Principles of Plasma Physics* (Benjamin Inc., Reading, Massachusetts, 1973)
14. Robbins M.O., Kerner K., Grest G.S. *J. Chem. Phys.* **88** 3286 (1988)
15. Thomas H. and Morfill G. *Proceedings of NATO Advanced Research Workshop on Formation, Transport and Consequences of Particles in Plasmas, Chateau de Bonas, Castera-Verduzan, France* (1993)
16. Morfill G. Proceedings XX Int. Conf on Phen. in Ionized Gases (Bochum August 1994) v. Inv. Papers, 3, (1994)
17. Thomas H., Morfill G et al *Phys. Rev. Lett.* **73** 622 (1994)

18. Thomas H. and Morfill G. *Nature* **379** 806 (1996)
19. Thomas H. and Morfill G, *J. Vac. Sci. Technol.* **A 14** 501 (1996); Morfill G. and Thomas H. *J. Vac. Sci. Technol.* A **14** 490 (1996)
20. Chu J.H. and Lin I. *Physica* , **A 205** 183 (1994)
21. Melzer A. et al *Phys.lett.*, **A 191** 301 (1994)
22. Fortov V., Nefedov A. et al *Phys. Let.* **A 219** 89 (1996)
23. Allen J. et al *Proc Ann. Plasma Phys. Conf*, Perth, Scotland (1996)
24. Hayashi Y. and Tachibana S.Jpn. Appl. Phys. **33** L804 (1994)
25. Vladimirov S.V. and Ishihara O. *Physics Plasmas* **3** 444 (1996); Ishihara O. and Vladimirov S.V. *Phys. Plasmas* **4** 69 (1997)
26. Vladimirov S.V. and Tsytovich V.N. *Comments in Plasma Physics and Controlled Fusion* (1997)
27. Bache G., Cheringrer L and Doveil F. *Phys. Plasmas* **2** 1 (1995)
28. Tsytovich V.N. and Havnes O. *Comments on Plasma physics and Controlled Fusion* **15** 267 (1994)
29. de Angelis U., Bingham R, Ponomarev A., Shukla P, and Tsytovich V. *Phys Fluids* (1993)
30. Cui C. and Goree J. "Fluctuations of the charge dust grains in a plasma" *IEEE Trans* (1993); Goree J. *Plasma sources, Science and Technology* **3** 400 (1994)
31. Tsytovich V.N., de Angelis U., Bingham R. and Resendes D J. Plasma Phys. (1997)
32. Benkadda S., Gabai P., Tsytovich V.N. and Verga A. *Phys. Rev E* **53** 2717 (1996)
33. Resendes D., Bingham R. and Tsytovich V.N. *Journ. Plasma Phys.* **49** 458 (1996)
34. Resendes D and Tsytovich V *Plasma Physics (Russ. Journ.)* **27** 34 (1996)
35. Spitzer L. *Physical Processes in the Interstellar Media* (Wiley, New York 1975)
36. Kaplan S.A. and Pikel'ner S.B. *Physics Interstellar Media* (Benjamin Press: N.-Y. 1980) *Fisika Mejsvezdnoi sredy* (Moscow: Nauka, 1979)
37. Kaplan S A and Pikel'ner S B *Annual Review of Astronomy and Astrophysics* **12** 113 (1974)
38. Havnes O, Morfill G.E. and Goertz C.K. *Journal of Geophysical Research* **89** 10999 (1984)
39. Aslaksen T.K. and Havnes O. *Journal of Geophysical Research* **97** 19175 (1992)
40. Gor'kavyi N N and Fridman A M *Physics of Planetary Rings* (Moscow: Nauka, 1994)

Resistive Interchange Turbulence

Masahiro Wakatani

Graduate School of Energy Science,
Kyoto University, Gokasho, Uji 611

Abstract. An overview for linear and nonlinear behavior of the resistive interchange mode and resistive drift wave is described. One specific point of view is interaction between resistive interchange turbulence and sheared flow induced by electric field in magnetically confined plasmas. Since the sheared flow produced by Reynolds stress suppresses the turbulent level, this mechanism may explain confinement improvement based on the L (Low mode) to H (High mode) transition and the ELMs (Edge Localized Modes) in tokamaks and stellarators.

INTRODUCTION

This is a lecture note edited from the three talks at the Workshop on Two-Dimensional Turbulence in Plasmas and Fluids held at Australian National University in June, 1997. The contents include several aspects of resistive interchange modes; linear instability and nonlinear evolution of unstable mode in the case of single helicity and multiple helicities, turbulent behavior and interaction with shear flow given externally or generated by Reynolds stress. The object of these studies is to understand roles of the resistive interchange mode in magnetically confined plasmas in tokamaks and stellarators. The materials for this lecture note are mainly taken from our works at Plasma Physics Laboratory, Kyoto University and collaborations with W. Horton (Institute for Fusion Studies, University of Texas) and A. Hasegawa (Bell Laboratories and Osaka University).

It is also expected that the interaction between resistive interchange mode and sheared plasma flow due to Reynolds stress gives an insight for the recent topic of confinement improvement based on L (Low mode) to H (High mode) transition (1) in both tokamaks and stellarators. A modeling for ELMs (Edge Localized Modes) is also possible based on the nonlinear interchange mode interacting with the self-generated sheared plasma flow (2).

CP414, *Two-Dimensional Turbulence in Plasmas and Fluids:* Research Workshop
edited by R. L. Dewar and R. W. Griffiths

LINEAR RESISTIVE INTERCHANGE MODE

The comprehensive linear stability theory of resistive interchange mode or g mode was given by Furth, Killeen, Rosenbluth (3). They showed three types of resistive MHD instabilities; tearing mode, g mode and rippling mode. For tokamak experiments tearing modes are the most intensively studied, since the magnetic island with mode numbers $(m, n) = (2, 1)$ are often observed and it is consistent with the tearing mode theory, where $m(n)$ is a poloidal (toroidal) mode number (4).

The rippling mode was studied for understanding edge turbulence observed in relatively small tokamaks with fairly low temperature edge plasmas (5). However, there is no clear experimental evidence of the rippling mode in such tokamaks.

The resistive interchange mode becomes unstable only for finite beta plasmas, since this mode is pressure-driven. Also plasma current is not essential for destabilizing the resistive interchange mode. Thus this instability seems more important in stellarators than tokamaks.

The ideal interchange mode is the most popular pressure-driven instability in magnetically confined plasmas. When there is no magnetic shear in the magnetic confinement system, the stability against the ideal interchange mode depends on magnetic well (or hill). For ideal interchange modes in sheared magnetic fields, the magnetic shear is effective for stability. The famous stability condition considering all effects in toroidal geometry is Mercier criterion (6). In the limit of cylindrical plasma, Mercier criterion reduces to Suydam condition (7) which compares the destabilizing term proportional to pressure gradient to the magnetic shear term.

Furth, Killeen, Rosenbluth showed that the resistive interchange mode becomes unstable, when ideal interchange modes are stable in sheared magnetic fields. Since resistivity reduces the magnetic shear stabilization, resistive interchange modes are easily destabilized particularly in magnetic hill configurations. Thus for suppressing resistive interchange modes, the magnetic well is more effective. From this linear stability point of view, heliotron plasmas are appropriate for studying behavior of the resistive interchange modes, since the magnetic shear is finite and the magnetic hill is inevitable in the edge region where resistivity is not negligible. For typical plasmas at the edge of Heliotron E (8), the magnetic Reynolds number S (more correctly Lundquist number) is on the order of 10^6. It is a ratio between resistive diffusion time for minor radius to Alfvén transit time for major radius.

Here we explain briefly the linear growth rate of resistive interchange mode in the limit of cylindrical plasma. For the resistive MHD equations we assume the ordering

$$\gamma_g \, (r - r_s)^2 \sim \eta \, , \tag{1}$$

where γ_g is a growth rate, r_s is a radius of resonant surface satisfying $q = m/n$ and

η is a resistivity. The safety factor is denoted by q. Since stable interchange modes belong to shear Alfvén branch,

$$\omega^2 = k_{||}^2 \, V_A^2 \tag{2}$$

is also valid for the resistive interchange mode, which suggests the ordering $\gamma_g^2 \sim (r - r_s)^2$. Here V_A is Alfvén velocity and parallel wavenumber $k_{||}$ vanishes at $r = r_s$. Thus

$$\gamma_g \sim (r - r_s) \sim \eta^{1/3} \tag{3}$$

is obtained from these relations. By solving the eigenvalue equation adequately, the growth rate of the most unstable mode is obtained as

$$Q^{3/2} = D_s \ , \tag{4}$$

where

$$Q = \gamma_g \left(\frac{\eta \, n^2 \, q'^2 \, B_{0\theta}^2}{\mu_0 \, \rho_0 \, r_s^2} \right)^{-1/3} , \tag{5}$$

$$D_s = -\frac{2 \, \mu_0 \, P'}{r \, B_{0z}^2} \left(\frac{q}{q'} \right)^2 \Big|_{r=r_s} , \tag{6}$$

and $q' = dq/dr$, $P' = dP/dr$, P is a pressure, $B_{0\theta}$ (B_{0z}) is a poloidal (toroidal) magnetic field in the cylindrical model. It is noted that the growth rate γ_g is estimated as

$$\gamma_g \propto \eta^{1/3} \tag{7}$$

from eqs.(4) and (5), which is consistent with the ordering (3).

REDUCED MHD EQUATIONS FOR HELIOTRON PLASMAS

Magnetic confinement systems with magnetic hill region are appropriate for studying resistive interchange modes. There were several experiments showing fluctuations corresponding to resistive interchange modes. Here we pick up the high beta plasma experiment in Heliotron E (9) which showed pressure-driven fluctuations in magnetic field, soft-X ray, electron cyclotron emission (ECE) and line averaged density, etc. Since currentless plasmas produced by the combination of ECRH and NBI heating were confined in Heliotron E, the observed fluctuations were interpreted to be driven by the resistive interchange mode. The reconnection of magnetic field line is possible by the tearing or resistive interchange mode. However, the plasma current is not necessarily required for the latter mode.

It is known that reduced MHD model of resistive MHD can be derived for currentless stellarator/heliotron plasmas. The equation of motion is reduced to the form of vorticity equation in the toroidal coordinates $(r,\ \theta,\ \zeta)$,

$$\rho \frac{d}{dt} \nabla_{\perp}^{2} \phi = - \frac{1}{\mu_0} \mathbf{B} \cdot \boldsymbol{\nabla} \nabla_{\perp}^{2} \Psi + (\boldsymbol{\nabla} \Omega \times \boldsymbol{\nabla} P) \cdot \hat{\zeta} \tag{8}$$

which is obtained by using the condition of charge neutrality $\boldsymbol{\nabla} \cdot \mathbf{J} = 0$, where $\mathbf{B} = \boldsymbol{\nabla}\Psi \times \hat{\zeta} + B_0\hat{\zeta}$ is a magnetic field vector, $\mathbf{J}$ is a current density, ρ is a mass density, P is a pressure. The velocity perpendicular to magnetic field line is given by $\mathbf{v}_{\perp} = \boldsymbol{\nabla}\phi \times \hat{z}$ and the plasma current along the toroidal direction is given by $J_{\zeta} = \mathbf{J} \cdot \hat{\zeta} = -\nabla_{\perp}^{2}\Psi/\mu_0$, where Ψ is a poloidal flux function and μ_0 is a vacuum permeability. An average curvature of magnetic field line is given by $\boldsymbol{\nabla}\Omega$, where $\Omega = \overline{\mathbf{B}^2}/B_0^2$ and $\overline{\mathbf{B}^2}$ is the average magnetic field over one field period in the toroidal direction of heliotron plasma. Thus $\overline{\mathbf{B}^2}$ is a function of r and θ, and B_0 is a magnetic field at the magnetic axis. Hereafter it is assumed that $\Omega \gg r/R_0$ which may be valid for Heliotron E plasmas. Then the toroidal curvature is negligible compared to the average curvature produced by helical magnetic fields.

Ohm's law is reduced to

$$\frac{\partial \Psi}{\partial t} = \mathbf{B} \cdot \boldsymbol{\nabla} \phi + \frac{\eta}{\mu_0} \nabla_{\perp}^{2} \Psi \, . \tag{9}$$

With the pressure evolution equation

$$\frac{\partial P}{\partial t} + (\boldsymbol{\nabla} \phi \times \hat{z}) \cdot \nabla P = 0 \, , \tag{10}$$

the reduced MHD equations are closed (10). It is noted that, when Ω is replaced with the toroidal curvature $r\cos\theta/R_0$, the reduced MHD equations (8), (9), (10) are applicable to tokamaks (11). In this case the equilibrium poloidal flux function Ψ_0 and the rotational transform $\iota(\Psi_0)$ are produced by toroidal plasma current instead of helical magnetic fields.

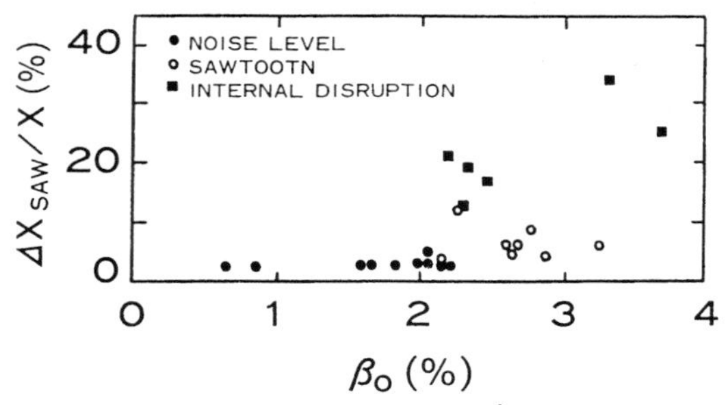

FIGURE 1. Central-chord soft X-ray sawtooth amplitude ($\Delta X\mathrm{saw}/X$) as a function of central $\beta(\beta_0)$ for discharges with moderately peaked pressure profiles. For $\beta_0 < 2\%$, only weak noise fluctuations are observed (Ref.9).

NONLINEAR RESISTIVE INTERCHANGE MODE

As shown in Fig.1 the pressure driven MHD fluctuations are observed when central beta value $\beta(0)$ exceeds about $(1.5 - 2)\%$ (9). This property is consistent with the linear stability of resistive interchange mode with $(m, n) = (1, 1)$ in currentless finite beta plasmas of Heliotron E (12). When η is assumed zero, the ideal interchange mode with $(m, n) = (1, 1)$ has the beta limit at $\beta(0) \simeq (1.5 - 2.0)\%$ which depends on the pressure profile. When η is finite, the resistive interchange mode with the resonant surface at $\iota(r_s) = n/m = 1$ is destabilized even for negligibly small beta $\beta(0) \simeq 0\%$, although growth rate decreases according to $\gamma_g \propto (P'(r_s))^{2/3}$. Roughly, when the growth rate exceeds $0.01R_0/V_A$, the nonlinear behavior of resistive interchange mode becomes clear (13). The mode structure of the soft-X ray fluctuation is shown in Fig.2 (b), which is localized near the $\iota = 1$ surface. At the nonlinear stage the internal disruption occurs as shown in Fig.2 (a) (9). This behavior is similar to the internal disruption observed in the central region of tokamak plasma, although there is no net plasma current in the Heliotron E. It is also noted that the results in Fig.2 seems consistent with the nonlinear behavior of resistive interchange mode shown by solving the reduced MHD equations (8), (9), (10) in the limit of straight plasma column (12). Thus we believe that the low mode number resistive interchange instabilities occur in the Heliotron E and generate the pressure-driven internal disruption. These results suggest high mode number resistive interchange modes are easily destabilized and resistive interchange turbulence may exist in the edge region with the magnetic hill in Heliotron E.

SHEAR FLOW EFFECT ON RESISTIVE INTERCHANGE MODE

In the experiments obtaining the H mode, the velocity shear due to the radial

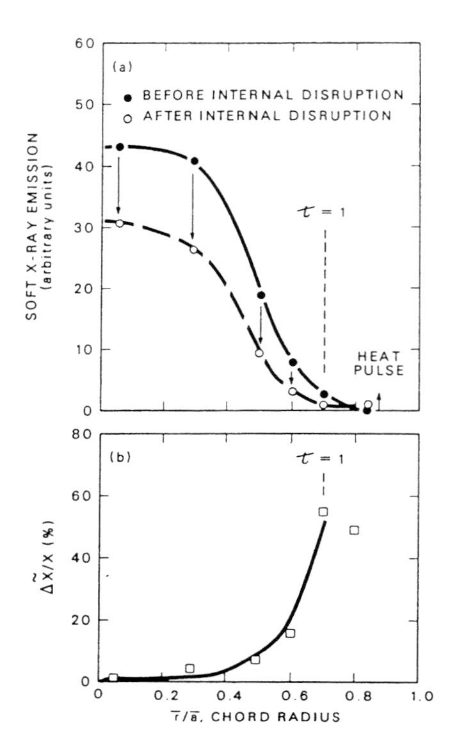

FIGURE 2. Profiles of soft X-ray signal amplitude as a function of detector chord radius: (a) total signal amplitude before and after large sawtooth (internal disruption) and (b) normalized sinusoidal precursor oscillation amplitude ($\Delta \tilde{X}/X$) just before internal disruption (Ref.9).

electric field is enhanced compared to the L mode. Thus it is interesting to study the shear flow effect on resistive interchange modes. In order to understand the shear flow effect qualitatively, we use reduced MHD equations for a simple slab plasma with pressure gradient, magnetic shear and velocity shear in the x direction;

$$\frac{\rho c}{B_0}\left(\frac{\partial}{\partial t} - \mu \nabla_\perp^2 + \frac{c}{B_0}\hat{z} \times \nabla \Phi \cdot \nabla\right)\nabla_\perp^2 \Phi = \frac{B_0}{c\eta}\nabla_\parallel^2 \tilde{\Phi} - \frac{d\Omega}{dx}\frac{\partial \tilde{P}}{\partial y} \tag{11}$$

$$\left(\frac{\partial}{\partial t} - \chi \nabla_\perp^2 + \frac{c}{B_0}\hat{z} \times \nabla \Phi \cdot \nabla\right)\tilde{P} = \frac{c}{B_0}\frac{dP_0}{dx}\frac{\partial \tilde{\Phi}}{\partial y}, \tag{12}$$

where $P = P_0(x) + \tilde{P}$ and $\Phi = \Phi_0(x) + \tilde{\Phi}$ with fluctuations $\tilde{P}$ and $\tilde{\Phi}$ (17). Viscosity μ and thermal diffusivity χ are included in the dissipative terms. It is noted that

ϕ in eq.(8) is equal to $-c\Phi/B_0$. The sheared magnetic field is shown as $\mathbf{B} = (0,\ B_y(x),\ B_0)$ and $|B_y(x)| \ll B_0$. For the time evolution of background poloidal flow $v_E = c(d\Phi_0/dx)/B_0$,

$$\frac{\partial v_E}{\partial t} + \gamma_d\, v_E = -\frac{\partial\langle\tilde{v_x}\tilde{v_y}\rangle}{\partial x} \tag{13}$$

is obtained from eq.(11) by adding the second term in LHS of eq.(13). Here γ_d is a damping rate and $\tilde{\mathbf{v}} = \hat{z} \times \nabla\tilde{\Phi}$. It is noted that $-\langle\tilde{v_x}\tilde{v_y}\rangle$ is Reynolds stress, which makes it possible to generate v_E from velocity fluctuations. It is noted that only electrostatic fluctuations can be described by eqs.(11), (12) and (13) (14).

After linearizing eqs.(11) and (12) for $\mu = \chi = 0$, and assuming $\tilde{\Phi}(x,y,t) = \Phi_k(x)\exp(-i\omega t + iky)$ for the mode structure, the eigenmode equation

$$\left[\frac{d^2}{dx^2}\, k^2 - \frac{kv_E''}{\omega - kv_E} - \frac{k^2}{(\omega - kv_E)^2} - \frac{ik^2x^2}{(\omega - kv_E)}\right]\Phi_k(x) = 0 \tag{14}$$

is obtained (15). It is noted that $x = 0$ denotes the position of resonant surface in the slab plasma with magnetic shear. Figure 3 shows the linear growth rate as a function of flow velocity with the profile $v_E = v_0\tanh(x/L_E)$. Here the growth rate is normalized with γ_g and the velocity v_0 is normalized as $v_0 k/\gamma_g$. The parameter in Fig.3 is $\mu = \Delta/L_E$, where Δ is the width of linear eigenmode in the case of $v_0 = 0$. Figure 3 shows that the significant stabilizing effect appears, when the conditions $\Delta < L_E$ and $kv_0 > \gamma_g(L_E/\Delta)$ are satisfied. For example, for $k \simeq 0.2(\mathrm{cm}^{-1})$ and $\gamma_g \simeq 10^5\,\mathrm{sec}^{-1}$, $kv_0/\gamma_g \sim 3$ corresponds to $v_0 \sim 1.5 \times 10^6\mathrm{cm/sec}$. This velocity was often observed in the H mode of tokamak plasmas. It is noted that the Kelvin-Helmholtz instability appears in the resistive magnetized plasma for $kv_0 \gg \gamma_g$ in Fig.3.

MODEL EQUATIONS FOR RESISTIVE DRIFT WAVES

It is well-known that electrostatic drift waves are also destabilized by the resistivity, which is called resistive drift waves. Linear theory of the resistive drift wave has been developed by including details of plasma physics based on the kinetic model (16). However, two fluids model is useful for studying nonlinear behavior of resistive drift waves, since the wave particle interaction is not essential for the nonlinear dynamics.

A simple model equations can be derived for density fluctuation $\tilde{n}$ and electrostatic potential fluctuation $\tilde{\Phi}$ by assuming (i) two dimensional motion for ion fluid, $v_{\|i} \simeq 0$, (ii) cold ions, $T_i \simeq 0$, and (iii) isothermal electrons $T_e = const.$ It is noted that Ohm's law for the resistive drift wave is different from that for the resistive interchange mode, since $\nabla_{\|}P_e$ is kept in Ohm's law.

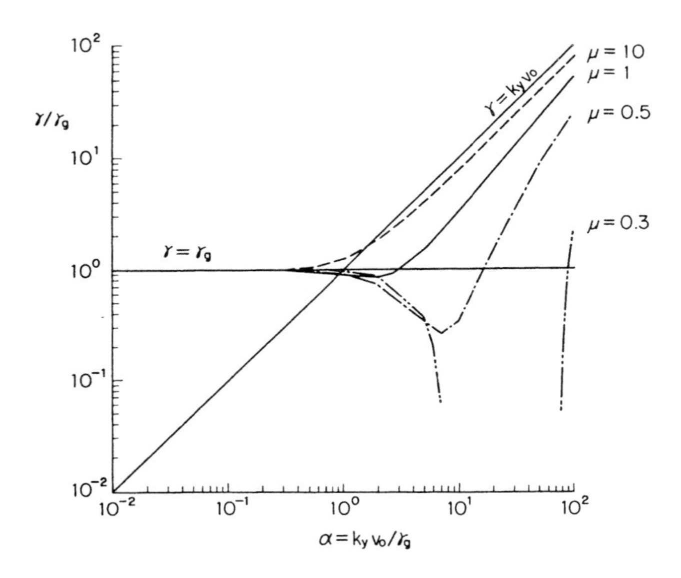

FIGURE 3. Growth rate γ/γ_g of resistive interchange mode versus $\alpha = kv_0/\gamma_g$ for $\mu = \Delta/L_E =$ 0.3, 0.5, 1 and 10 (Ref.15)

For resistive drift wave turbulence, the vorticity equation

$$\frac{d}{dt}\frac{\nabla_\perp^2 \tilde{\Phi}}{B_0 \Omega_i} = \frac{1}{en_0}\frac{\partial J_z}{\partial z}, \tag{15}$$

which is obtained from the ion momentum equation. The electron continuity equation is shown as

$$\frac{dn}{dt} = \frac{1}{e}\frac{\partial J_z}{\partial z}, \tag{16}$$

where $n = n_0 + \tilde{n}$. From Ohm's law the plasma current along the magnetic field line J_z is given by

$$J_z = \frac{T_e}{e\eta}\frac{\partial}{\partial z}\left(\frac{\tilde{n}}{n_0} - \frac{e\tilde{\Phi}}{T_e}\right), \tag{17}$$

and we obtain a closed model equations for $\tilde{n}$ and $\tilde{\Phi}$ (17). It is noted that these equations may include a sheared plasmas flow by assuming $\Phi = \Phi_0(x) + \tilde{\Phi}(x, y, z, t)$; where x denotes the direction of inhomogeneity. It is also possible to derive the model equations for a cylindrical plasma with magnetic shear (18).

Here we note that Hasegawa-Mima equation for drift wave turbulence may be obtained by subtracting eq.(16) from eq.(15), and assuming the Boltzmann relation $\tilde{n}/n_0 = e\tilde{\Phi}/T_e$ (19).

One significant point for the model equations (15), (16), (17) is that they include linear instabilities. For studying nonlinear behavior, dissipative terms such as viscosity and diffusivity are added in eqs.(15), (16), respectively. Thus the resistive drift wave turbulence is described with both the source and sink self-consistently. It seems interesting to study spectrum cascade by solving them numerically. In the wave number space, when there are two invariants, the dual cascade occurs. Usually the invariant with higher power of wavenumber goes to the small scale. The model equations (15), (16), (17) has two invariants. One is energy

$$\int \left[\frac{1}{2}(\tilde{n})^2 + \frac{\rho_s^2}{a^2}(\nabla_\perp \tilde{\Phi})^2\right] dV \tag{18}$$

and the other is enstrophy

$$\int \left[\frac{1}{2}\left(\left(\frac{\rho_s}{a}\right)^2 \nabla_\perp^2 \tilde{\Phi} - \tilde{n}\right)^2\right] dV , \tag{19}$$

which are conserved in the limit of no dissipation, where $\rho_s = C_s/\Omega_i$, C_s is an ion acoustic velocity and a is a plasma size. It is expected that the energy cascades inversely and the enstrophy cascades normally. These are confirmed by numerical calculations for a cylindrical plasma model (20).

During the numerical study of eqs.(15), (16), (17), it was found that a global electric field is produced from the resistive electrostatic turbulence as shown in Fig.4 (21). The mechanism to generate the radial electric field in the cylindrical plasma is Reynolds stress of the turbulent velocity field due to the resistive drift waves (22). In Fig.4 there is no global electric potential $\Phi_0(r)$ initially. During the development of the resistive electrostatic turbulence $\Phi_0(r)$ also grows and becomes positive in the central region and negative in the outside region. According to the boundary condition, $\Phi_0(a) = 0$ is always satisfied. As a next step we need to study whether the generated poloidal shear flow $v_\theta = (d\Phi_0/dr)/B_0$ is effective to suppress the radial transport or not, and what parameter is important to enhance the poloidal shear flow.

TRANSPORT DRIVE BY RESISTIVE INTERCHANGE TURBULENCE

Usually it is difficult to estimate turbulent transport analytically. However, three methods for deriving turbulent diffusivity, mixing length theory, scale invariance, two-point renormalized theory, give the same result for the resistive interchange turbulence.

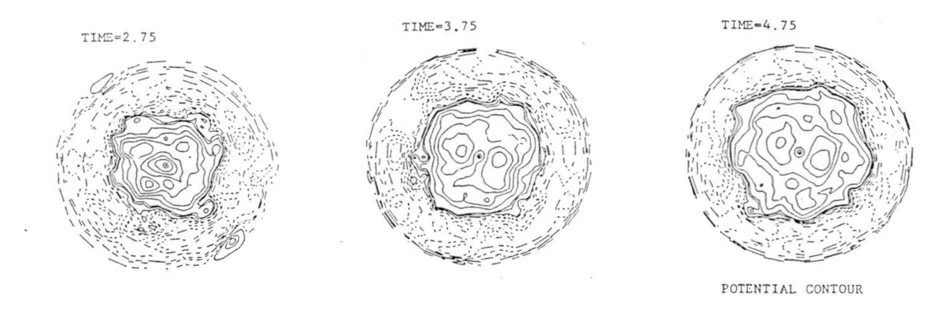

FIGURE 4. Time evolution of electric field potential contours. The dotted lines denote the negative potential region (Ref.20).

Here we assume homogeneous turbulence by assuming many resonant surfaces. We introduce a local coordinates (x, y, z) with relations to the cylindrical coordinates (r, θ, ζ),

$$x = (r - r_0) \left. \frac{dq}{dr} \right|_{r=r_0} \tag{20}$$

$$y = \zeta - q\,\theta \tag{21}$$

$$z = \theta\,. \tag{22}$$

From the reduced MHD equations (8), (9), (10) we obtain normalized equations in the above local coordinates,

$$\frac{d}{d\tau} \Delta_\perp \tilde{\phi} = \frac{1}{q} \frac{\partial}{\partial z} \Delta_\perp \tilde{\Psi} + \frac{\beta^*}{q} \frac{\partial \tilde{n}}{\partial y} \tag{23}$$

$$\frac{1}{q} \frac{\partial \tilde{\phi}}{\partial z} + \eta\, \Delta_\perp \tilde{\Psi} = 0 \tag{24}$$

$$\frac{d\tilde{n}}{d\tau} - \kappa\, q \frac{\partial \tilde{\phi}}{\partial y} = 0\,, \tag{25}$$

where the inductive electric field $\partial\tilde{\Psi}/\partial\tau$ is neglected in Ohm's law, and

$$\frac{d}{d\tau} = \frac{\partial}{\partial \tau} + q^2\, s \left(\frac{\partial \tilde{\phi}}{\partial x} \frac{\partial}{\partial y} - \frac{\partial \tilde{\phi}}{\partial y} \frac{\partial}{\partial x} \right) \tag{26}$$

$$\Delta_\perp = q^2 \, s^2 \, \frac{\partial^2}{\partial x^2} \, . \tag{27}$$

Here it is assumed that $|\partial^2/\partial x^2|$ is much larger than other derivatives in the perpendicular Laplacian. The parameters in eqs.(23), (24), (25) are $\eta, q, \kappa = dn_0/dr$, $s = r(dq/dr)/q$ and $\beta^* = \beta_p r d\Omega/dr$. Here β_p denotes a poloidal beta value. We note that eqs.(23), (24), (25) describe the resistive interchange turbulence under the electrostatic approximation and constant temperature.

One popular way to estimate the turbulent transport is to use the mixing length theory which gives $D = \gamma(\Delta x)^2$, where γ is the linear growth rate and Δx is the radial width of linear mode structure. By substituting γ and Δx obtained from the linear resistive interchange mode theory discussed in the second section,

$$\begin{aligned} D_g &= \eta \, \kappa \, \beta^* \, / \, s^2 \\ &= D_{c\ell} \, \frac{q^2}{\epsilon^2} \left(\frac{r}{q} \, \frac{dq}{dr} \right)^{-2} \left(\frac{-r}{P_0} \, \frac{dP_0}{dr} \right) \left(r \, \frac{d\Omega}{dr} \right) \end{aligned} \tag{28}$$

is obtained, where $D_{c\ell}$ is the classical diffusion coefficient and $\epsilon = r/R_0$ (23).

We also applied the scale invariance developed by Connor and Taylor (24) to eqs.(23), (24), (25). It is interesting that the same diffusivity as eq.(28) is obtained by (23).

It is known that the two point renormalized theory gives turbulent diffusivity with a numerical coefficient (25), which gives 4 D_g (26). Therefore, the three theoretical approaches give the same turbulent diffusivity. There are a few results for comparison between the diffusivity (28) and the diffusivity obtained from the power balance. Generally we need a fairly large numerical coefficient, which is larger than 4, to make D_g comparable to empirical results.

Principally we can apply the scale invariance and the two-point renormalization transport theory to the resistive interchange turbulence with the sheared plasma flow or the resistive drift wave turbulence. For these cases we cannot determine the turbulent diffusivity completely as shown by eq.(28). For example, in the latter case, we did not estimate the phase difference between $\tilde{n}$ and $\tilde{\phi}$, though it is always $\pi/2$ for the resistive interchange turbulence.

SHEARED FLOW GENERATION DUE TO NONLINEAR INTERCHANGE MODES

It seems that the magnetic shear is not essential to generate the sheared flow, when Reynolds stress is finite. Here we consider the most simple slab plasma without the magnetic shear. Thus the magnetic field is constant in the z direction. Also fluctuations induced by the interchange mode are constant in the z direction. Since reconnection of magnetic field line does not occur in the shearless magnetic field, the resistivity does not play a role. For dissipative processes viscosity μ and thermal diffusivity χ are included in the model equations describing the ideal interchange mode,

$$\frac{\partial \nabla_\perp^2 \Phi}{\partial t} + \hat{z} \times \nabla \Phi \cdot \nabla \nabla_\perp^2 \Phi = -g \frac{\partial P}{\partial y} + \mu \nabla_\perp^4 \Phi \tag{29}$$

$$\frac{\partial P}{\partial t} + \hat{z} \times \nabla \Phi \cdot \nabla P = -\kappa \frac{\partial \Phi}{\partial y} + \chi \nabla_\perp^2 P , \tag{30}$$

where $\Phi = \Phi_0(x) + \tilde{\Phi}(x,y,t)$ and $P = P_0(x) + \tilde{P}(x,y,t)$ and $V_y = d\Phi_0(x)/dx$ is a background plasma flow in the y direction (2). Magnetic curvature is denoted by g and background pressure gradient is denoted by $\kappa = dP_0/dx$. The famous parameters for eqs.(29), (30) are Rayleigh number

$$R_a = \frac{g\kappa d^4}{\mu\chi} \tag{31}$$

and Prandtl number

$$P_r = \frac{\nu}{\chi} . \tag{32}$$

The transport is evaluated with Nusselt number

$$N_u = 1 + \frac{\langle \tilde{P} \tilde{v_x} \rangle}{\kappa\mu} , \tag{33}$$

where the second term denotes the turbulent thermal transport while the first term corresponds to the classical thermal transport. Width of the slab plasma is denoted by d. For simplicity fixed boundary conditions are imposed at $x = \pm 1/2$, and periodic boundary conditions in the y direction. In the region with $-1/2 \leq x \leq 1/2$ and $0 \leq y \leq 2$ eqs.(29), (30) are solved numerically. Figure 5 shows numerical results with $R_a = 10^5$ and $P_r = 1.0$. First the nonlinear interchange mode with the dominant mode number $m = 1$ saturates at $T \simeq 40$, where T is normalized with the linear growth time of interchange mode. For $40 \leq T \leq 80$ the saturated state continues; however, at $T \simeq 80$, the $m = 0$ mode grows suddenly. Since the $m = 0$ mode corresponds to the plasma flow in the y direction, the growth of the $m = 0$ mode corresponds to the generation of sheared flow. Once the sheared flow grows,

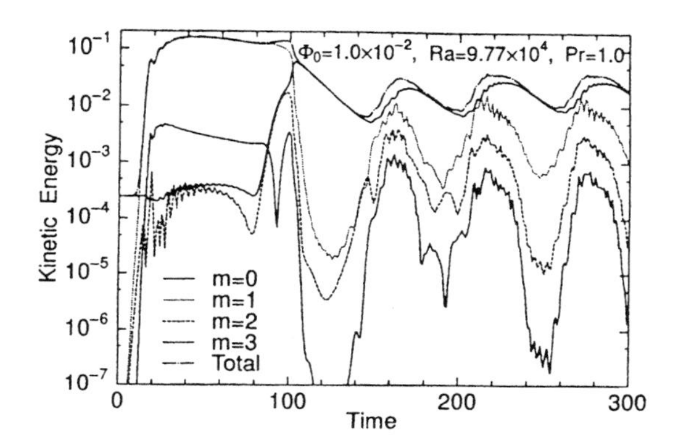

FIGURE 5. Time evolution of the kinetic energy of dominant Fourier modes ($m = 0 - 3$) for $\Phi_0 = 1.0 \times 10^2, R_a = 9.77 \times 10^4$, and $P_r = 1.0$. The total kinetic energy is also shown. Time is normalized by $(g\kappa)^{-1/2}$ (Ref.2).

the $m = 0$ component becomes dominant. The time evolution of the $m = 0$ mode shows a relaxation type oscillation. It is considered that this type oscillation resembles to the ELMs after the L to H transition. The mechanism that the $m = 0$ mode grows in the steady saturated state should be clarified. For this purpose low order dynamical models are useful (27). It is also noted that the $m = 0$ mode suppresses almost all modes with $m = 1, 2, \cdots$. On the contrary, since the $m = 0$ mode is produced by the $m = 1, 2, \cdots$ modes, the $m = 0$ mode decays slowly after they are decreased substantially. This behavior is similar to ELMs; however, it is required to compare numerical results with experiments quantitatively.

ACKNOWLEDGEMENTS

The author acknowledge the Future Energy Research Association for visiting Australian National University at Canberra.

REFERENCES

1. Wagner,F., et al., Phys. Rev. Lett. **49**, 1408 (1982).
2. Takayama,A., Wakatani,M., and Sugama, H., Phys. Plasmas **3**, 3 (1996).
3. Furth,H.P., Killeen,J., and Rosenbluth,M.N., Phys. Fluids **6**, 459 (1963).
4. Rutherford,P.H., Phys. Fluids **16**, 1903 (1973).
5. Terry,P.W., et al., Phys. Fluids 29, 2501 (1986).
6. Mercier,C., Nucl. Fusion **1**, 47 (1960).
7. Suydam,B.R., 2nd Int. Conf. Peaceful Uses At. Energy
(Proc. Conf. Geneva, 1958) **31**, UN, New York, 1959, p.157.
8. Wakatani,M., and Sudo,S., Plasma Phys. Control. Fusion **38**, 937 (1996).
9. Harris,J.H., et al., Phys. Rev. Lett. **53**, 2242 (1984).
10. Strauss,H.R., Plasma Phys. **22**, 733 (1980).

11. Strauss,H.R., Phys. Fluids **20**, 1354 (1977).
12. Wakatani,M., Shirai,H., and Yamagiwa,M., Nucl. Fusion **24**, 1407 (1984).
13. Wakatani,M., et al., Proc. Plasma Phys. Control. Nucl. Fusion Research (Washington, 1990) vol.2, p.567.
14. Sugama,H., and Horton,W., Phys. Plasmas **1**, 345 (1994).
15. Sugama,H., and Wakatani,M., Phys. Fluids B **3**, 1110 (1991).
16. Krall,N.A., Advances in Plasma Physics, vol.1, Interscience Publishers (1968) p.153.
17. Wakatani,M., and Hasegawa,A., Phys. Fluids **27**, 611 (1984).
18. Sugama,H., Wakatani,M., and Hasegawa,A., Phys. Fluids **31**, 1601 (1988).
19. Hasegawa,A., and Mima,K., Phys. Fluids **21**, 87 (1978).
20. Wakatani,M., Watanabe,K., Sugama,H., and Hasegawa,A., Phys. Fluids B **4**, 1754 (1992).
21. Hasegawa,A., and Wakatani,M., Phys. Rev. Lett. **59**, 1581 (1987).
22. Carreras,B.A., Lynch,V.E., and Garcia,L., Phys. Fluids B **3**, 1438 (1991).
23. Yagi,M., Wakatani,M., and Shaing,K.C., J. Phys. Soc. Jpn **57**, 117 (1988).
24. Connor,J.W., and Taylor,J.B., Phys. Fluids **27**, 2676 (1984).
25. Carreras,B.A., and Diamond,P.H., Phys. Fluids B **1**, 1011 (1989).
26. Sugama,H., and Wakatani,M., J. Phys. Soc. Jpn **57**, 2010 (1988)
27. Howard,L.N., and Krishnamurti,R., J. Fluid Mech. **170**, 385 (1986).

Jovian Vortex Dynamics from a Historical Perspective

Jun-Ichi Yano [(1)], Joseph J. Tribbia [(2)], and Fritz Reichmann[(3)]

[(1)] *CRC–SHM, Monash University, Clayton, Victoria, Australia*
[(2)] *National Center for Atmospheric Research, Boulder, CO 80307-3000, USA*
[(3)] *Meteorological Institute, University of Munich, Munich, Germany*

Abstract.
Theories for Jovian vortices are discussed from a historical perspective. Both weakly and strongly nonlinear approaches are reviewed. An observational indication of Jupiter's Great Red Spot as a modon is noted. The historical observations suggest a catastrophic genesis of Jovian vortices, whose basic process appears to be understood by strongly nonlinear geostrophic adjustment.

I INTRODUCTION

The planetary atmospheres often provide ideal natural laboratories for studying geophysical fluid dynamics. The long-lived Jovian vortices such as the Great Red Spot (GRS), and the White Ovals of Jupiter are typical examples. Longevity and stability of those vortices are unprecedented by Earth standards. GRS has been continuously observed from the ground for more than one hundred years and the three White Ovals have remained stable since their formation in the 1930s (Peek, 1958; Rogers, 1995). The close Voyager observations in 1979 revealed that these are very well-defined isolated vortices within shearing flows. The Voyager mission also revealed the existence of a vortex of equal size on the surface of Neptune: the Great Dark Spot (GDS). Unfortunately, the continuous observation by the Hubble Space Telescope showed that the GDS had disappeared by 1994 in a process of gradual drift toward the equator. Rather surprisingly, a mirror image of the original Dark Spot was discovered by the Hubble in the opposite hemisphere a few months later, in November, 1994. A similar dramatic appearance of a spot on the surface of Saturn has been observed from the ground for years (cf., Sanchez–Lavega, 1982, 1994). The Voyager probe as well as ground observations see numerous small short-lived vortices (spots) on the surface of Jupiter. Keeping in mind such a generic nature of the vortices (spots) in the atmo-

CP414, *Two-Dimensional Turbulence in Plasmas and Fluids:* Research Workshop
edited by R. L. Dewar and R. W. Griffiths

spheres of the Major Planets (which we call Jovian atmospheres), this review will be focused on the dynamics of the long-lived vortices like the GRS for the most part. For reviews of Jovian atmospheric dynamics in general, we refer to Ingersoll (1990), Gierasch and Conrath (1993), Yano (1994), and Dowling (1995).

The immense size of the GRS is noted: it extends more than 20,000 km in longitude and 10,000 km in latitude. It was revealed as an anticyclonic vortex by the Voyager wind observation (Mitchell *et al.*, 1981) embedded in one of the anticyclonic shear zones with the wind speed more than 100 m/s (cf., Ingersoll *et al.*, 1981). The White Ovals have about half of the size and are embedded in another anticyclonic shear zone immediately poleward to the GRS. Infrared observations from the Voyager probe revealed some thermal structure of the GRS (Flasar *et al.*, 1981): its uppermost bound is marked by a cold core at 50 mb level above the tropopause (100 mb level). The downward thermal wind inversion from the 20 mb level to the 500 mb level agrees well with the direct wind measurements by cloud drifting, which indicates the vertical scale of the GRS. Although the lower bound of the GRS is not directly observable, it is commonly believed that the GRS is a shallow entity confined to a thin upper part of the deep Jovian atmosphere (*e.g.*, Ingersoll and Cuong, 1981; Yano, 1987a, b). We will also mostly follow this common wisdom in our discussions in the following.

We advocate that the dynamics of the long-lived Jovian vortices can be understood in terms of the free vortex dynamics under almost vanishing forcing and dissipation. This view can be justified by the following scaling argument (Yano and Flierl, 1994): the solar energy (the solar forcing) F_R per unit area absorbed by the atmosphere is estimated in terms of the distance R_s of the planet from the Sun as $F_R \sim R_s^{-2}$. The distances from the Sun to Jupiter and Saturn are 5.2 and 9.5 times, respectively, of the distance to the Earth. Hence, the solar forcing is estimated to be about 4% and 1% of that to the Earth for Jupiter and Saturn, respectively. It is seen that the Major Planets are much more weakly forced by the Sun than is the Earth. Note that the estimated internal energy source is of the same order or less than the solar energy, so that this does not change the estimate for the order of magnitude of forcing. On the other hand, the magnitude of the wind speeds in the Jovian atmospheres is observed to be least 10 times larger than that of the Earth (roughly 100 m/s and 400 m/s for Jupiter and Saturn compared to the typical wind speed of 10 m/s for the Earth). By assuming the same vertical scale for those observed winds, we conclude that the magnitude of the kinetic energy E_K, which is measured by the square of the wind speed, is more than 100 times larger for the Jovian atmospheres than the Earth's atmosphere. Note that this estimate is a conservative one, because the atmospheric motion in the Jovian atmospheres can be deeper than that in the Earth's atmosphere. Finally, the energy decay time-scale τ_D is estimated by a simple dynamical balance:

$$F_r - E_K/\tau_D = 0.$$

This gives an estimate that the energy decay time-scale is more than 1000 times longer for the Jovian atmospheres than for the Earth. Hence, we conclude that the Jovian atmospheres are much more weakly forced and much less dissipative than the Earth's atmosphere.

II WEAKLY NONLINEAR THEORIES

Under what conditions can we obtain an isolated vortex like the GRS in the geophysical flows? A convenient theoretical framework to start from is the shallow water system on the mid-latitude β-plane (cf., Pedlosky, 1987). This system can be further approximated to the one-layer quasi-geostrophic (QG) system in the parameter regime with $\hat{\epsilon} \ll 1$ and $F \sim 1$. Here, $\hat{\epsilon} \equiv U/f_0L$ is the Rossby number, $F \equiv (L/L_R)^2$ the Froude number defined by the typical wind speed U, horizontal scale L, the reference Coriolis parameter $f_0(\equiv 2\Omega \sin\phi$ with the reference latitude ϕ and the planetary rotation rate $\Omega)$ and the deformation radius L_R. The deformation radius is defined by $L_R \equiv (gh_E)^{1/2}/f_0$ with the gravitational acceleration g and the equivalent depth h_E, which measures the vertical scale of the motion.

One possible approach to answer this question is to attempt to reduce this original system into a system known to contain a soliton solution, for example the KdV equation

$$\frac{\partial}{\partial\tau}\psi + \psi\frac{\partial}{\partial\xi}\psi + \frac{\partial^3}{\partial\xi^3}\psi = 0,$$

where ψ the streamfunction, τ a slow time, and ξ a space coordinate, which is going to be in the longitudinal direction in our application.

Redekopp (1977) showed how to derive a KdV system by reduction of the QG system by using an asymptotic expansion method. Assuming a mean zonal state $\bar{\psi}(y)$ only depending on the latitudinal coordinate y and a weak vortex component (eddy) φ of order, say, ϵ, the time t and the longitudinal coordinate x are transformed by

$$\frac{\partial}{\partial t} = -c\frac{\partial}{\partial x} + \epsilon^{2/3}\frac{\partial}{\partial\tau},$$
$$\frac{\partial}{\partial x} = \epsilon^{1/2}\frac{\partial}{\partial\xi}.$$

As a result, a linear equation to define the latitudinal structure of the eddy is obtained to the leading order. To the next order, the KdV equation is obtained by a solvability condition. Note that by transforming the coordinate, a much longer scale is assumed in longitude than in latitude by the stretching rate

$r \equiv \epsilon^{-1/2}$. This coordinate stretching induces a weak wave dispersion in the longitudinal direction to balance with a weak nonlinearity, which leads to the KdV equation.

The Rossby-soliton system shares a nice property with the original KdV system that the two soliton-vortices remain unaffected after interacting with each other in a collision. This feature appeared to guarantee the longevity of Jovian vortices. The Rossby-soliton theory was applied to Jovian vortices by Maxworthy and Redekopp (1976), and Maxworthy *et al.* (1978).

A natural question arises: how general are such Rossby-soliton systems? This question is addressed by analysing the shallow water system by Yano and Tsujimura (1987). By a systematic scale analysis, they defined the upper bound for the vortex amplitude to follow KdV dynamics for a given amplitude of mean flow and the Froude number. Most interestingly, the intermediate-geostrophic (IG) system, which is derived and proposed as a two-dimensional version of the KdV system by Flierl (1979, 1980: see also Charney and Flierl, 1981), Petviashvili (1980), and Yamagata (1982), is naturally obtained in the limit of $r \to 1$ in the phase space by the scaling $1/F \sim \hat{\beta}$ and $\epsilon\hat{\epsilon} \sim \beta^2$ with the nondimensional β-parameter defined by $\hat{\beta} \equiv L\beta/f_0$ and $\beta \equiv df/dy$, where y measures the latitudinal distance. The universality of the KdV dynamics shown by Yano and Tsujimura's analysis may be emphasised: they showed that such a dynamics is possible for any given zonal mean flow and with arbitrary deformation radius (relative to the planetary radius).

The IG dynamics was applied to Jovian vortices by Williams and Yamagata (1984). They show that the IG vortices have a tendency to merge together after a collision unlike the Rossby solitons.

Finally, it must be emphasised that all those KdV-type theories are based on the weakly-nonlinear approximation. Hence, it is formally restricted by the condition

$$|c| \gg |v'|,$$

where c the vortex phase velocity and v' the magnitude of the vortex velocity. For GRS and White Ovals, the phase velocity is known to be only a few meters per second, which is much smaller than the strong wind motion surrounding those vortices. Consequently, the weakly nonlinear theory is at least formally not applicable to those Jovian vortices. This implies the necessity to proceed to the strongly-nonlinear theory for geophysical vortices.

III COMPACTING CONDITION

Oddly enough, one of the possible starting points for constructing a strongly-nonlinear theory for geophysical vortices is the dispersion relation of a linear Rossby wave:

$$c = -\frac{\hat{\beta}}{K^2 + F} \tag{1}$$

under the mid-latitude β-plane approximation, where K is the total wavenumber. The isolation of a vortex requires that the vortex amplitude should decay in the far field, which requires an imaginary wavenumber, *i.e.*,

$$K^2 < 0 \qquad \text{as} \qquad r \rightarrow +\infty. \tag{2}$$

The Jovian vortices are typically embedded in a sinusoidal zonal flow *i.e.*, $u \sim \sin ky$ with wavenumber k. This zonal flow follows the Rossby-wave phase velocity

$$\bar{c} = -\frac{\hat{\beta}}{K^2 + \bar{F}}, \tag{3}$$

where $\bar{F}$ designates the Froude number for the zonal flow. For the steady propagation of the vortex, one requires that those two phase velocities match *i.e.*,

$$c = \bar{c}. \tag{4}$$

The substitution of the dispersion relations (1), (3) into the matching condition (4) and the application of the compacting condition (2) lead to

$$F > k^2 + \bar{F}.$$

This compacting condition can be interpreted in two ways. First, it requires that the Froude number for the zonal flow is smaller than that for the vortex *i.e.*,

$$F > \bar{F}.$$

In other words, the zonal flow ought to have a larger vertical scale than the vortex. In the following, we assume that the zonal flow is purely barotropic (*i.e.*, having an infinitely large vertical scale) and, hence, $\bar{F} = 0$. Consequently, the compacting condition reduces to

$$F > 1$$

by setting $k = 1$. Hence, we expect the radius of deformation is much smaller than the scale of the vortex *i.e.*, $L_R < L$.

The relevance of this compacting condition was tested for the quasi-geostrophic system by Yano and Flierl (1994). We report here the results with the global shallow water system. The global shallow water system is constructed in terms of the normal mode expansion on the rotating sphere. The ellipticity of the Jovian geosurface is taken into account as additional forcing terms. We use the measurements by Limaye (1986) for the zonal winds. The initial condition for GRS is defined in the same way as in Dowling and Ingersoll (1989). The evolution of the height field is plotted every 5 Jovian days in Fig. 1 starting from the 5th Jovian day at the upper left: the frame moves from the left to the right, then moves to the next row. In presenting

1E1_8. t=dt=5 days

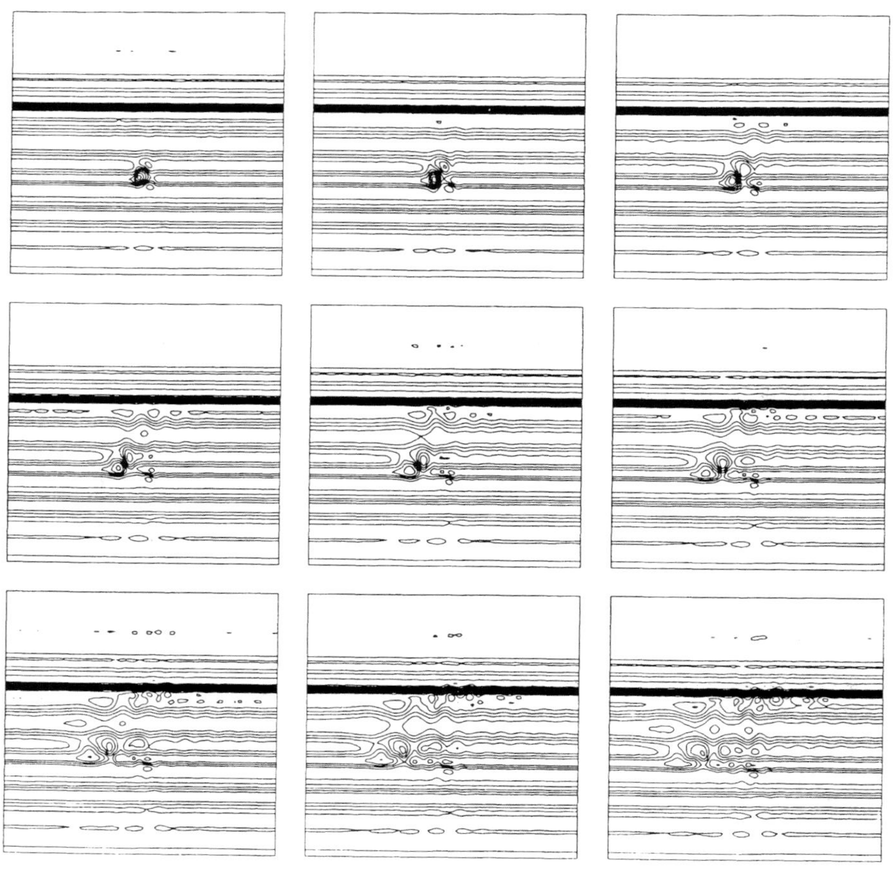

FIGURE 1(a)

5E0_8. t=dt=5 days

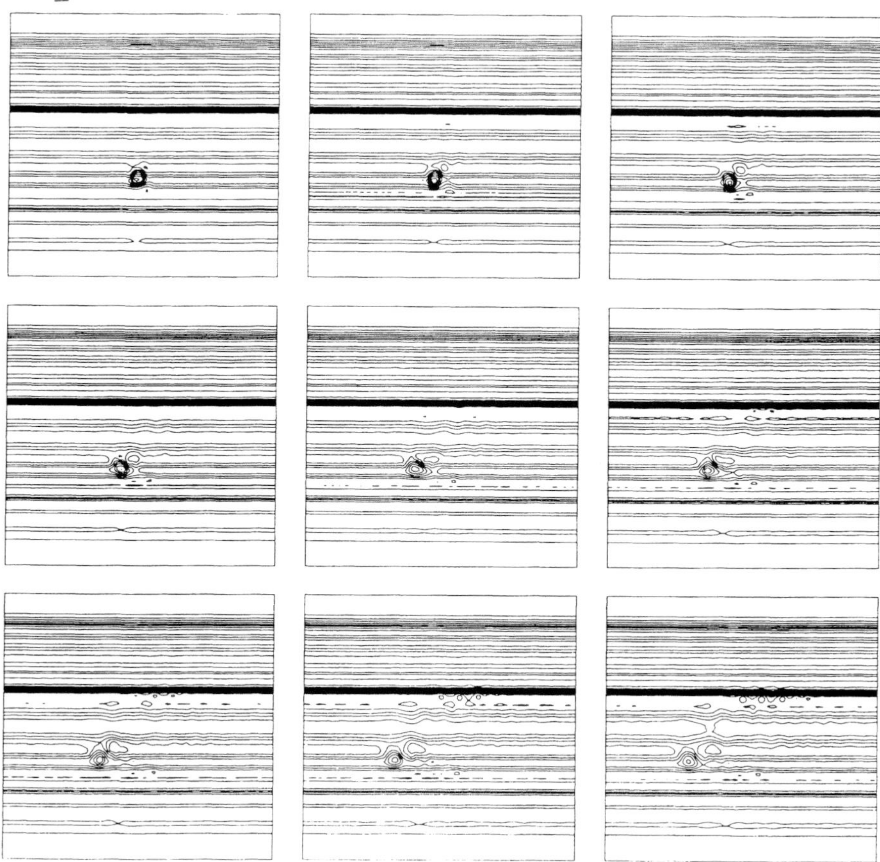

FIGURE 1(b)

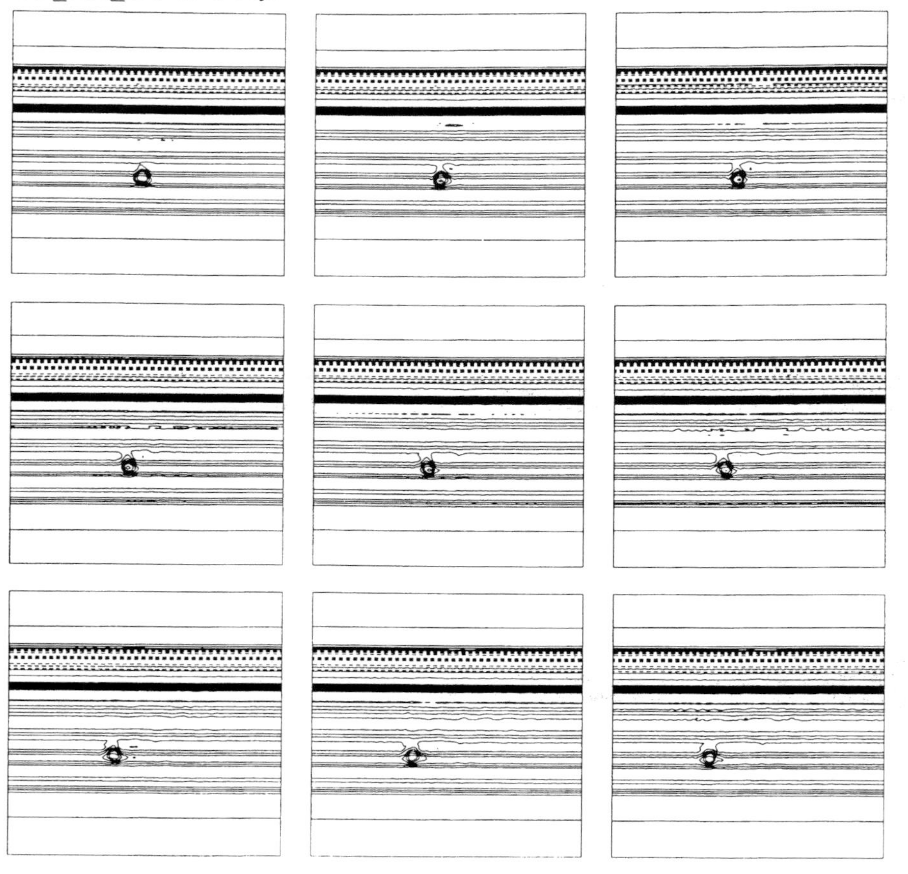

FIGURE 1(c)

FIGURE 1. Test of the compacting condition with a GRS-like vortex. The evolution of the height field is plotted every 5 Jovian days starting from the 5th Jovian day at the upper left. The frame moves from the left to the right, then moves to the next row. The cases are (a) h_E= 10 km, (b) h_E = 5 km, and (c) h_E= 2 km. The contour interval is 0.5 km.

the result, we use the latitude of the GRS ($\phi \simeq 20°$) for the estimate of the deformation radius from the equivalent depth h_E.

The compacting condition is found to be severer than with the quasi-geostrophic system on the mid-latitude β-plane. Even with $L_R = 4000$ km ($h_E = 10$ km: Fig. 1(a)), the initial vortex is completely resolved within 50 days by the Rossby-wave dispersion. Note the initial tendency that the wave is emitted equatorward, presumably due to the smaller Froude number for the lower latitude. This ostensibly makes the compacting condition severer with spherical geometry. With $L_R = 2800$ km ($h_E = 5$ km: Fig. 1(b)), the vortex appears to be marginally stable with a slow decay of the amplitude with time. With sufficiently small deformation radius as $L_R = 1800$ km ($h_E = 2$ km: Fig. 1(c)), the vortex eventually becomes steadily propagating with time.

Strong sensitivity of the model behaviour to the deformation radius L_R is contrasted with the weak sensitivity reported by Dowling and Ingersoll (1989). This presumably reflects the fact that the latter used a channel model, which tends to artificially suppress the Rossby-wave radiation with the help of the side walls.

Nevertheless, this compacting condition does not constitute a sufficient condition for the existence of the long-lived stable GRS-type vortices. Yano and Flierl (1994) showed by the two-layer quasi-geostrophic system that once the active role of the deep lower layer is taken into account, the GRS-like vortex starts to radiate barotropic Rossby waves, and decays in a relatively short time-scale. They showed that this difficulty can be resolved by assuming a deep QG system of the lower layer. This points to the crucial importance for the vertical structure of Jovian vortices in understanding their dynamics.

IV ANOTHER LOOK AT THE COMPACTING CONDITION: MODON THEORY

A slightly different approach can help to see both the weakly-nonlinear and strongly-nonlinear theories in a unified view (cf., Butchart *et al.*, 1989). We take the QG system

$$[\frac{\partial}{\partial t} + J(\psi, \cdot)]Q = 0$$

as the starting point, where Q is the potential vorticity. The weakly-nonlinear theory constitutes in separating the total field into the mean and the eddy components, *i.e.*,

$$\psi = \bar{\psi}(y) + \epsilon\varphi,$$
$$Q = \bar{Q}(y) + \epsilon q.$$

The leading-order problem

$$J(\bar{\psi}, \bar{Q}) = 0$$

is trivially satisfied. To the next order, we obtain

$$[\frac{\partial}{\partial t} + J(\bar{\psi}, \cdot)]q + J(\varphi, \bar{Q}) = 0.$$

In order to further simplify the problem, we assume a steadily-propagating solution by setting $\partial/\partial t = -c\partial/\partial x$ and introduce $\tilde{\bar{\psi}} \equiv \bar{\psi} - cy$. Note that the Jacobian is re-written as $J(\varphi, \bar{Q}) = \Lambda J(\varphi, \tilde{\bar{\psi}})$ with the functional derivative $\Lambda \equiv \delta\bar{Q}/\delta\tilde{\bar{\psi}}$. As a result, the second-order equation reduces to

$$J(\tilde{\bar{\psi}}, q - \Lambda\varphi) = 0,$$

or by using the zonal flow $\bar{u} \equiv -d\bar{\psi}/dy$,

$$(\bar{u} - c)\frac{\partial}{\partial x}(q - \Lambda\varphi) = 0,$$

or

$$q - \Lambda\varphi = 0.$$

By introducing an explicit form for the eddy potential vorticity $q = (\triangle - F)\varphi$, we finally obtain

$$(\triangle + k^2)\varphi = 0,$$

where

$$k^2 = -(F + \delta\bar{Q}/\delta\tilde{\bar{\psi}}) \tag{5}$$

can be interpreted as a square of the local wavenumber. We make the interpretation that an isolated vortex can be formed by having an internal wave structure (*i.e.*, $k^2 > 0$) at the centre (say, $r \leq r_0(\theta)$) of the polar coordinate (r, θ) and an evanescent structure (*i.e.*, , $k^2 < 0$) at the exterior ($r > r_0(\theta)$). Such a structure is obtained by appropriately choosing the zonal mean potential-vorticity to form a potential well with the help of the relation (5).

The extension of this approach to the strongly nonlinear regime is straightforward. A steadily-propagating system under full nonlinearity is given by

$$J(\tilde{\psi}, Q) = 0$$

with $\tilde{\psi} \equiv \psi - cy$, which implies that potential vorticity is solely defined as a functional of $\tilde{\psi}$, *i.e.*, $Q = Q(\tilde{\psi})$. An equivalent expression to the weakly-nonlinear case is obtained by Taylor expanding the functional as,

$$Q(\tilde{\psi}) = (\delta Q/\delta\tilde{\psi})\tilde{\psi} + \frac{1}{2}(\delta^2 Q/\delta\tilde{\psi}^2)\tilde{\psi}^2 \cdots.$$

Since $Q = (\triangle - F)\psi + \hat{\beta}y$ on the mid-latitude β-plane, we obtain

$$(\triangle + k^2)\tilde{\psi} + (\hat{\beta} - Fc)y - \frac{1}{2}(\delta^2 Q/\delta\tilde{\psi}^2)\tilde{\psi}^2 + \cdots = 0$$

with

$$k^2 = -(F + \delta Q/\delta\tilde{\psi}).$$

Note that the definition of the local wavenumber (refractive index) k^2 remains the same as in the case of the weakly-nonlinear theory apart that it is now defined in terms of the total field in stead of the mean zonal state. So long as $|\delta^2 Q/\delta\tilde{\psi}^2| \ll |\delta Q/\delta\tilde{\psi}|$ *etc*, the same theoretical framework applies, if we assume $c = \hat{\beta}/F$, *i.e.*, the long-wave limit for the Rossby wave.

The major difference from the weakly-nonlinear case is that the vortex field now forms a potential well by itself. In particular, the modern formulation (cf., Flierl, 1987; Flierl, *et al.*, 1980) assumes a piecewise constant distribution for the refractive index, *i.e.*,

$$k^2 = \begin{cases} \alpha & \text{for } r < r_0 \\ -\mu & \text{for } r > r_0 \end{cases}$$

as the simplest configuration to construct an isolated vortex. Here, α, μ, r_0 are positive constants.

Dowling and Ingersoll (1989) attempted, by a diagnosis of Voyager wind observations around the GRS and the White Oval BC, a retrieval of the functional relationship between the potential vorticity Q and the Bernoulli functional B, which can be considered as a nonlinear equivalent to the stream-function. They determined this functional relationship by dividing the domain namely into the three regions: the interior (I, the central part of the vortex), the core (a ring region surrounding I), and the exterior, which is further divided into the North and South branches. Remarkably, their retrieval (their Fig. 3) indicates that the functional slope around the both Jovian vortices is almost piecewise constant, which is in accordance with the modern formulation.

However, unlike the standard formulation presented above, their diagnosis indicates that it is external (*i.e.*, $k^2 < 0$) in the interior region (I) and internal (*i.e.*, $k^2 > 0$) in the exterior with an almost neutral (*i.e.*, $k^2 \simeq 0$) core (ring) region (see Fig. 1 of Yano, 1994). Although the whole implication of this diagnosis is not yet clear, this is in accordance with the view by Markus and Lee (1994) that the scale of the core (ring) region defines the Jovian vortex dynamics.

V GENESIS?

How are these long-lived Jovian vortices generated? Once a Jovian vortex is considered as a free vortex, it becomes a semi-permanent entity by assumption.

Hence, no question of the genesis arises. Nevertheless, this question is asked, presumably by the analogy with the tropical cyclones. Here the vortices are maintained against dissipation is also often asked in the same context.

The commonly accepted view is that these long-lived Jovian vortices are generated by a series of coalescences of smaller vortices to form a larger one. This picture is consistent with the merging character of geophysical vortices (Ingersoll and Cuong, 1981; Williams and Yamagata, 1984). Some numerical experiments are performed with the shallow water system, adding a weak perturbation to a Jupiter-like zonal basic flow, to show that a GRS-like vortex can be generated by this process (Williams and Wilson, 1988; Dowling and Ingersoll, 1989). The process can furthermore be interpreted as a consequence of the inverse cascade of two-dimensional turbulence (cf., McWilliams, 1984). An accompanying popular view is that these long-lived vortices are kept maintained against dissipation by feeding on the smaller vortices as the energy source. This hypothesis became particularly appealing when the Voyager probe showed a particular event in which a small vortex was fed into the GRS as it circulated around the latter (cf., Fig. 2 of Ingersoll, 1990).

However, a survey of historical observations of Major Planets indicates otherwise: the genesis of the Jovian vortices is more a sudden catastrophic event than a gradual process as presupposed in the above commonly accepted picture.

The genesis of Saturn's Great White Spot may be the best example to demonstrate the point. It was observed several times during the last one hundred years. Among them, the event in 1990 is most well documented (Sanchez-Lavega *et al.*, 1991; Barnet *et al.*, 1992; Beebe *et al.*, 1992; Westphal *et al.*, 1992: see also Sanchez-Lavega, 1994). The event was literally characterised by a sudden appearance of a huge spot of an equivalent size to the GRS. It grew twice in size in one week, and then further stretched in longitude to form a planetary-scale disturbance. A similar sudden appearance of a well-defined dark feature and its growth into a planetary-scale disturbance is also often observed on Jupiter with lesser extents (cf., Peek, 1958: Rogers, 1995). Lastly the genesis of a new Dark Spot (or re-genesis?) in Neptune can be classified as a similar, more dramatic event, although its relation to the original Dark Spot is not clear. Those cases make the point that the sudden genesis of vortices is a very common event in the Major Planets, although most of them are not stable and not long-enduring.

Among the long-lived Jovian vortices, the genesis of the White Ovals is better documented: it was initially observed as a single bright feature very elongated in longitude. As it shrunk, it separated into three bright features (it appeared as if like a creation of three darks spots in-between initially) and gradually settled into well-defined three white spots after a decade.

The history of the GRS is much less clear: the first observation of the GRS is commonly attributed to Hook and Cassini in the late 17th century, and, hence, the GRS is supposed to be more than 300 years old. However, no

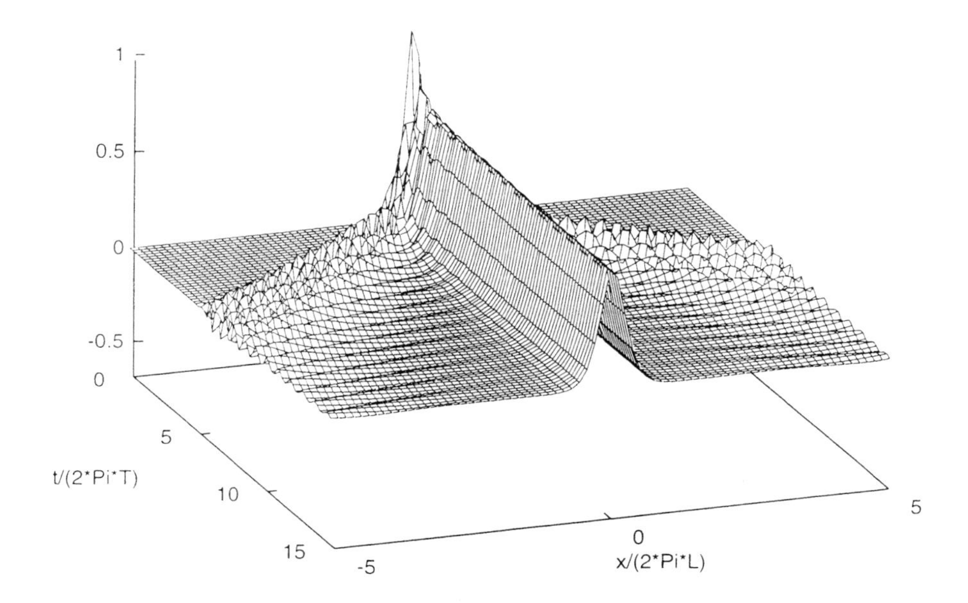

FIGURE 2(a)

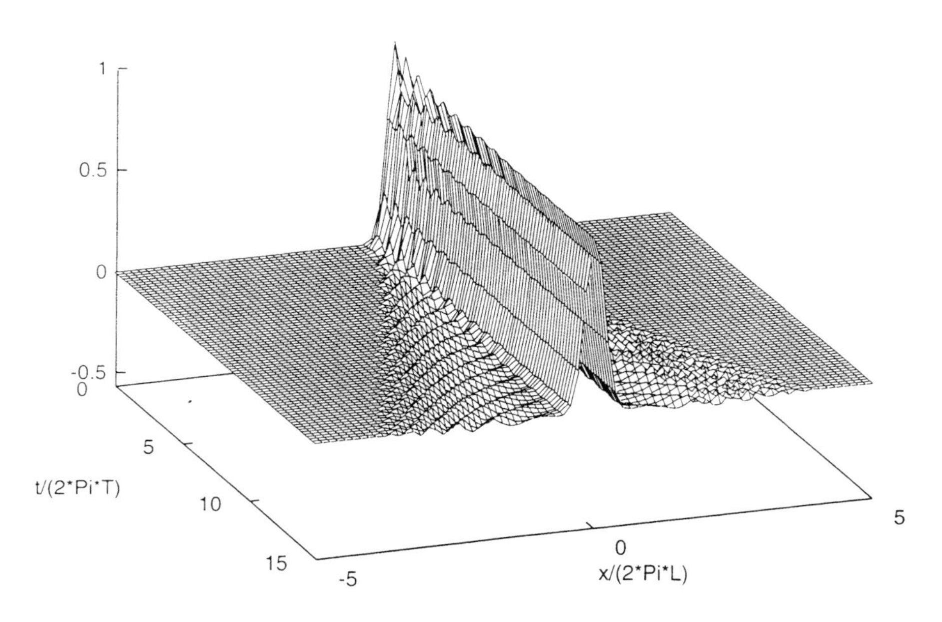

FIGURE 2(b)

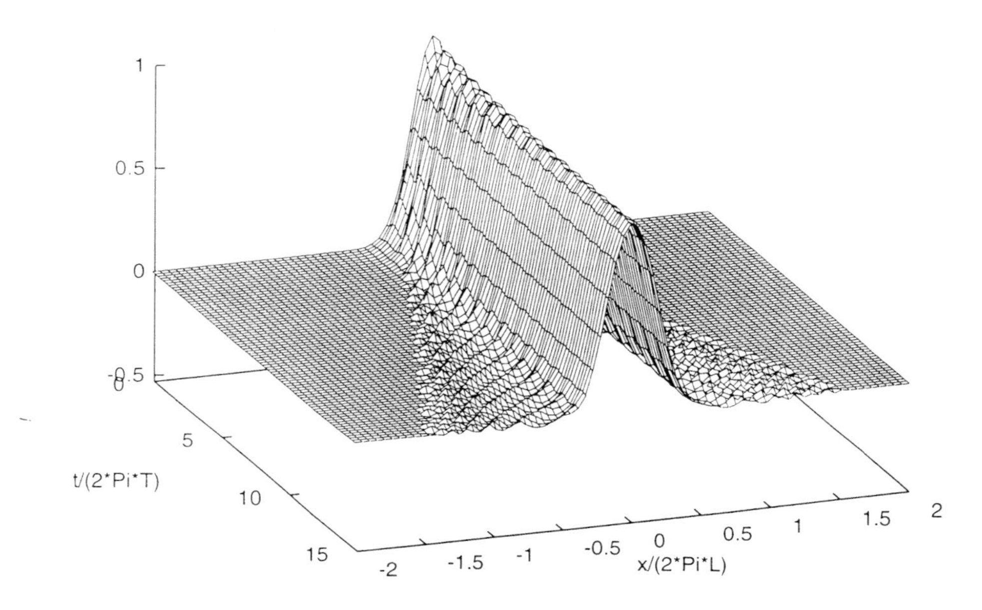

FIGURE 2(c)

FIGURE 2. Geostrophic adjustment experiments as an idealisation of a Jovian vortex genesis with the initial condition $h = \exp(-x^2/2)$. The evolution of the height field h is shown three-dimensionally for (a) $F = 0.1$, (b) $F = 1$, and (c) $F = 10$. The computation domain is taken wide enough to avoid the boundary problem.

follow up observation of this feature is available. The continuous observation of the current GRS is only available from mid-19th century. The analysis of the historical observation from this period suggests that the GRS has also gradually shrunk with time. Rogers (1980, 1995) proposes that the GRS has followed a similar history as the three White Ovals: it was generated as a very long elongated disturbance (*i.e.*, a South Tropical Disturbance) in the 1870s and gradually took a smaller, more coherent structure.

These historical observations are not consistent with the two-dimensional turbulence inverse cascade view of the Jovian vortices, which predicts the growth of a vortex with time, or least it should remain quasi-steady under the balance of inverse cascade and dissipation. The observations, on the other hand, indicate that the Jovian vortices gradually shrink in their size after the initial formation.

Here, we propose that the genesis of the Jovian vortices are rather understood simply in terms of the adjustment of an initial large-scale thermal anomaly to a steady state. We speculate that the sudden appearance of a large-scale feature is provided by an intrusion of a thermal plume generated deep in the interior by thermal convective instability into the upper atmospheric layer. Such a sudden intrusion will likely create a cold anomaly by an adiabatic cooling of the feature to the upper level. The adjustment of the atmosphere to such a sudden appearance of a cold anomaly (accompanied by a warm counterpart to the lower level) is most easily understood by the shallow water system. Such an adjustment problem is called the geostrophic adjustment.

The geostrophic adjustment for the parameter $F \leq 1$ has been intensively investigated since the classical study by Rossby (1938: cf. Blumen, 1972). However, the Jovian atmospheres are likely to be in the opposite regime, *i.e.*, $F \gg 1$, because they are close to the convective neutrality due to the internal heat sources. As a result, the adjustment is strongly nonlinear in the sense that the depth fluctuation (temperature anomaly) is much larger than the mean depth (mean stratification).

For a simple demonstration, we use the Gaussian shape

$$h = \exp(-x^2/2)$$

for the initial condition of the depth perturbation with no initial motion on the one-dimensional f-plane. The results with various Froude numbers are shown in Fig. 2. When the initial anomaly is smaller than the deformation radius ($F = 0.1$, a), it is mostly radiated away by gravity-wave radiation. When the initial anomaly is of the same scale as the deformation radius ($F = 1$, b), still about one third of the mass is removed from the initial anomaly by the gravity-wave radiation. However, once the initial anomaly substantially exceeds the deformation radius ($F = 10$, c), most of the initial mass anomaly is retained after the adjustment and a huge vortex is formed under the geostrophic balance. This ostensibly simplified experiment appears to explain why a sudden intrusion of a large-scale anomaly can generate a stable long-lived vortex in the atmosphere. Against intuition, in order for a vortex to be stable and long-lived, its size must be substantially larger than the deformation radius. Conversely, it also explains the shorter life for smaller vortices in the Jovian atmosphere.

On the other hand, when the initial thermal plume fails to reach the stably-stratified layer, a different evolution is expected. A weak, negative stratification of the layer, where the plume ceased the uplifting by achieving neutrality with the environment, makes the Rossby-waves strongly dispersive, and hence the initial plume structure consequently disintegrates rapidly (Yano and Sommeria, 1997). We make the interpretation that such a process is typically observed with the Great White Spots in Saturn.

VI ACKNOWLEDGEMENTS

J.I.Y. is supported by the Australian Government Cooperative Research Centres Program. The National Center for Atmospheric Research is sponsored by the National Science Foundation.

References

Butchart, N., K. Haines, and J.C. Marshall, 1989: A theoretical and diagnostic study of solitary waves and atmospheric blocking, *J. Atmos. Sci.*, **46**, 2063-2078.

Barnet, C.D., J.A. Westphal, R.F. Beebe, 1992: Hubble Space Telescope Observations of the 1990 Equatorial Disturbance on Saturn: Zonal Winds and Central Meridian Albedos. *Icarus*, **100**, 499-511.

Beebe, R.F., C. Barnet, P.V. Sada, and A.S. Murell, 1992: The Onset and Growth of the 1990 Equatorial Disturbance on Saturn. *Icarus*, **95**, 163-172.

Blumen, W., 1972: Geostrophic adjustment. *Rev. Geophy.*, **10**, 485-528.

Butchart, N., K. Haines, and J.C. Marshall, 1989: A theoretical and diagnostic study of solitary waves and atmospheric blocking. *J. Atmos. Sci.*, **46**, 2063-2078.

Charney, J.G., and G.R. Flierl, 1981: Oceanic analogues in large-scale atmospheric motions. in *Evolution of Physical Oceanography*, ed. by B. Warren and C. Wunsch, MIT Press, pp. 502-546.

Dowling, T.E., 1995: Dynamics of Jovian atmospheres. *Ann. Rev. Fluid Mech.*, **27**, 293-334.

Dowling, T.E., and A.P. Ingersoll, 1989: Jupiter's Great Red Spot as a shallow water system. *J. Atmos. Sci.*, **46**, 3256-3278.

Flasar, F.M., B.J. Conrath, J.A. Pirraglia, P.C. Clark, R.G. French, and P.J. Gierasch, 1981: Thermal structure and dynamics of the Jovian atmosphere 1: The Great Red Spot. *J. Geophys. Res.*, **86 A10**, 8759-8767.

Flierl, G.R., 1979: Planetary solitary waves. *POLYMODE News*, **62**, 7-14.

Flierl, G.R., 1980: Introduction to Coherent Features. *1980 Summer Study Program in Geophysical Fluid Dynamics*, Woods Hole Oceanographic Institution.

Flierl, G.R., 1987: Isolated eddy models in geophysics. *Ann. Rev. Fluid Mech.*, **19**, 493-530.

Flierl, G.R., V.D. Larichev, J.C. McWilliams, and G.M. Reznik, 1980: The dynamics of baroclinic and barotropic solitary eddies. *Dyn. Atmos. Ocean*, **5**, 1-41.

Gierasch, P.J., and B.J. Conrath, 1993: Dynamics of the atmosphere of the outer planets: Post-Voyager measurement objective. *J. Geophys. Res.*, **98**, 5459-5459.

Ingersoll, I.P., 1990: Atmospheric dynamics of the outer planets. *Science*, **248**, 308-315.

Ingersoll, A.P., R.F. Beebe, J.L. Mitchell, G.W. Garneau, G.M. Yagi, and J.-P. Müller, 1981: Interaction of eddies and mean zonal flow on Jupiter as inferred from Voyager 1 and 2 images. *J. Geophys. Res.*, **86**, 8733-8743.

Ingersoll, A.P., and P.G. Cuong, 1981: Numerical model of long-lived Jovian vortices. *J. Atmos. Sci.*, **38**, 2067-2076.

Limaye, S.S., 1986: Jupiter: New estimates of the mean zonal flow at the cloud level. *Icarus*, **65**, 335-352.

Marcus, P.S., and C. Lee, 1994: Jupiter's Great Red Spot and zonal winds as a self-consistent, one-layer quasigeostrophic flow. *Chaos*, **4**, 269-286.

Maxworthy, T., and L.G. Redekopp, 1976: A solitary wave theory of the Great Red Spot and other observed features in the Jovian atmosphere. *Icarus*, **29**, 261-271.

Maxworthy, T., L.G. Redekopp, and P.D. Weidman, 1978: On the production and interaction of planetary solitary waves. *Icarus*, **33**, 388-409, 1978.

Pedlosky, J., 1987: *Geophysical Fluid Dynamics*, 2nd ed., Springer, 710pp.

Peek, B.M., 1958: *The Planet Jupiter*, Faber and Faber, London.

Petviashvili, V.I., 1980: Red Spot of Jupiter and drift soliton in a plasma. *JETP Lett.*, **32**, 619-622.

Redekopp, L.G.,1977: On the theory of solitary Rossby waves. *J. Fluid Mech.*, **87**, 725-745.

Rogers, J.H., 1980: Disturbances and dislocations on Jupiter. *J. of British Astronomical Association*, **90**, 132-147.

Rogers, J.H., 1995: *The Giant Planet Jupiter*, Cambridge University Press, 418pp.

Rossby, C.G., 1938: On the mutual adjustment of pressure and velocity distributions in certain simple current system, II. *J. Mar. Res.*, **1**, 239-263.

Sanchez-Lavega, A., 1982: Motions in Saturn's atmosphere: observations before Voyager encounters *Icarus*, **49**,1-16.

Sanchez-Lavega, A., 1994: Saturn's Great White Spots. *Chaos*, **4**, 341-353.

Sanchez-Lavega, A., F. Colas, and J. Lecacheux, 1991: The Great White Spot and disturbances in Saturn's equatorial atmosphere during 1990. *Nature*, **353**, 397-401.

Sanchez-Lavega, A., J. Lecacheux, J.M. Gomez, F. Colas, P. Laques, K. Noll, D. Gilmore, I. Miyazaki, and D. Parker, 1996: Large-scale storms in Saturn's atmosphere during 1994, *Science*, **271**, 631-634.

Westphal, J.A., W.A. Baum, A.P. Ingersoll, C.D. Barnet, E.M. DeJong, G.E. Danielson, and J. Caldwell, 1992: Hubble space telescope observations of the 1990 equatorial disturbance on Saturn; Images, albedos, and limb darkening. *Icarus*, **100**, 485-498.

Williams, G.P., and R.J. Wilson, 1988: The stability and genesis of Rossby vortices. *J. Atmos. Sci.*, **45**, 207-241.

Williams, G.P., and T. Yamagata, 1984: Geostrophic regimes, intermediate solitary vortices and Jovian eddies. *J. Atmos. Sci.*, **41**, 453-478.

Yamagata, T., 1982: On nonlinear planetary waves: A class of solutions missed by the quasi-geostrophic approximation. *J. Oceanogr. Soc. Japan*, **38**, 236-244.

Yano, J.-I., 1987a: Rudimentary considerations of the dynamics of Jovian atmospheres. Part I: The depth of motions and energetics. *J. Meteor. Soc. Japan*, **65**, 313-327.

Yano, J.-I., 1987b: Rudimentary considerations of the dynamics of Jovian atmospheres. Part II: Dynamics of the atmospheric layer. *J. Meteor. Soc. Japan*, **65**, 329-340.

Yano, J.-I., 1994: A critical review on the dynamics of Jovian atmospheres. *Chaos*, **4**, 287-297.

Yano, J.-I., and G.R. Flierl, 1994: Jupiter's Great Red Spot: compactness condition and stability. *Ann. Geophysicae*, **12**, 1-18.

Yano, J.-I., and J. Sommeria, 1997: Unstably-stratified geophysical fluid dynamics, *Dyn. Atmos. Ocean*, **25**, 233-272.

Yano, J.-I., and Y.N. Tsujimura, The domain of validity of the KdV-type solitary Rossby waves in the shallow water β-plane model. *Dyn. Atmos. Ocean*, **11**, 101-129, 1987.

Research Papers

Low-Dimensional Model of Resistive Interchange Convection in Magnetized Plasma

Sergey Bazdenkov and Tetsuya Sato

Theory and Computer Simulation Center, National Institute for Fusion Science 322-6 Oroshi-cho Toki-shi Gifu-ken 509-52, Japan

Abstract. Self-organization and generation of a large shear flow component in turbulent resistive interchange convection in magnetized plasma is considered. The effect of plasma density-electrostatic potential coupling via the inertialess electron dynamics along the magnetic field is shown to play a significant role in the onset of the shear component. The results of large-scale numerical simulation and low-dimensional (reduced) model are presented and compared.

I INTRODUCTION

One of the serious problems in thermonuclear plasma confinement is anomalous particle and energy transport caused by plasma convection across the magnetic field. This convection is driven by the release of free energy during the growth of perturbations in generically unstable thermonuclear plasma.

The most dangerous consequences for plasma confinement might arise from the development of a magnetic perturbation which can cause rearrangement of the magnetic configuration and trigger its complete disruption. For this reason, the magnetic configuration and plasma parameters are usually chosen in such a way as to suppress the perturbation as much as possible. There are many experimental evidences that, indeed, in large tokamaks and stellarators the magnetic perturbations are basically absent, especially in the peripheral plasma near magnetic separatrix and scrape-off-layer (SOL).

In the absence of magnetic perturbation, plasma convection across strong confining magnetic field is mainly controlled by electrostatic field and pressure gradient forces which cause drift vortex flow perpendicular to both the magnetic field and the force. This flow, in turn, leads to self-consistent spatial redistribution of plasma density and electrostatic potential. Within one turn-over time scale, such a nonlinear vortex flow becomes complicated and is usually treated in terms of drift turbulence. The corresponding turbulent

CP414, *Two-Dimensional Turbulence in Plasmas and Fluids:* Research Workshop
edited by R. L. Dewar and R. W. Griffiths

convection is widely believed to be the cause of anomalous transport in magnetized plasma. In such a scenario, stronger fluctuations of plasma density and electrostatic potential correspond to higher convective transport, i.e., exactly what is observed in the low-confinement (L) mode in tokamaks and other machines.

In contrast to L-mode, there also exists the high-confinement (H) mode of operation when fairly intensive turbulent convection near the separatrix spontaneously undergoes drastic self-reorganization with consequent improvement in plasma confinement. Depending on plasma parameters, the transition from L- to H-regime may occur in reversible way in the form of edge-localized modes (ELMs). In this case, the configuration spontaneously, but repeatedly, returns to L-regime with further transition to H-mode and so on.

A key and universal feature of the improved confinement mode observed in various plasmas in tokamaks [1–6], stellarators [7–10] and linear machines [11] is the generation of a radial electric field and poloidal shear flow (here the radial direction corresponds to the minor radius of the toroidal plasma, i.e., the direction of confinement, while the poloidal direction corresponds to minor circuit around the magnetic axis). Generation of shear flow is usually accompanied by considerable suppression of plasma density and potential fluctuations, so that the convective turbulent transport is essentially reduced as well. Consequently, a mechanism for self-consistent shear flow generation is a necessary component in the understanding of the physics of the observed L-H transition.

Concerning this problem, two different scenarios are usually considered: either i) generation of radial electric field via the process of charged particle orbit loss (i.e., the kinetic effect), or ii) shear flow self-generation via nonlinear interaction and reconnection of the instability-driven convective cells (see [12–14]). The latter seems to be strong candidate because it is based on a robust effect which could be easily treated in terms of drift plasma fluid dynamics. The details of nonlinear interaction of driving instability and self-consistently generated shear flow depend, of course, on the peculiarities of the considered plasma dynamics.

In the case of edge localized modes (ELMs) in the scrape-off layer (SOL) plasma, which is our main concern in the present paper, experimentally observed edge plasma perturbations are usually of a flute-type character, i.e., look like a "dense" or "hot" plasma filament strongly elongated along the magnetic field (see, e.g., [15,16]). They are localized at the outside of the torus and exist even in the case when pressure gradient near the edge plasma is well below the ideal ballooning stability threshold (hence, no considerble magnetic perturbation is developed). Such a features indicate that driving instability in the SOL is, probably, a kind of pressure-driven resistive interchange (RI) mode akin to a flute-like mode in open systems (this is because of a lack of closed magnetic surfaces in the SOL). Interchange instability is also akin to Rayleigh-Taylor instability in stratified fluid, so that one can expect the onset of Bénard-like convection at the nonlinear stage of interchange instability. Indeed, proposed in [17] and also considered in [18–22], the model of self-consistent generation of Bénard-like convective cells with their periodic rearrangement into the shear flow at the nonlinear stage of RI-instability reproduces qualitatively well many important features of the ELMs.

In the present paper we consider a further development of this model mainly with regard to the effect of plasma density-electrostatic potential coupling, i.e., the tendency to establish Boltzmann distribution, which arises as a result of dynamical force balance for inertialess electrons in the presence of pressure and potential inhomogeneity along the field lines. This is a well-known effect which plays an important role, for example, in drift plasma dynamics based on the Hasegawa-Mima equation. The point, however, is that the Boltzmann coupling is usually considered for poloidally periodic perturbations only. I.e, only the deviations of plasma density and potential from their poloidally averaged background profiles are assumed to be inhomogeneous along the magnetic field and, hence, coupled via electron longitudinal dynamics, while poloidally averaged background profiles are supposed to be longitudinally homogeneous. In the bulk plasma with closed magnetic surfaces this assumption is quite reasonable because the poloidal averaging, in this case, corresponds simultaneously to the averaging along a magnetic field line. But in the case of SOL plasma with open magnetic field lines the procedure of poloidal averaging does not necessarily correspond to the averaging along the field line, and the background pressure and potential profiles are not necessarily homogeneous along the magnetic field. In the toroidal SOL plasma such a longitudinal inhomogeneity of the poloidally averaged profiles seems to be quite natural because of a certain asymmetry of the outer and inner parts of the torus. Indeed, let us assume for a while that there exists a longitudinally homogeneous poloidally averaged pressure profile in the SOL plasma, and that this profile is unstable against, say, RI-mode, otherwise there is no driving force for turbulent convection. This convection, in turn, causes deformation of the background pressure profile, namely, its flattening, within a time scale of about one-turn-over time of the conective cell, $\tau_\perp \sim l_\perp / v_*$ (here $l_\perp$ is the characteristic scale length of convective cell, and v_* is drift flow velocity in the cell; in the case of highly nonlinear flow this velocity is about a sound speed c_s). Then, in the toroidal geometry, the unfavourable curvature of the magnetic field lines, which is a driving force of the RI instability, is localized on the outside of the torus. Respectively, the growth of instability, accompanied by generation of convective cells, as well as the corresponding flattening of pressure profile, takes place outside of the torus only, while in the remaining part of the SOL no driving force exists, and both the plasma convection and pressure profile flattening are considerably suppressed. Hence, the flattened background, i.e., poloidally averaged, pressure profile might be longitudinally inhomogeneous with the characteristic scale length of inhomogeneouty $l_\parallel$ of the order of the torus length. Usually $l_\parallel \gg l_\perp$, and the corresponding characteristic time scale of heavy ion longitudinal dynamics, $\tau_{\parallel}^{(i)} \sim l_{\parallel}/c_s$, is much larger than the convection time scale $\tau_{\perp}$, i.e, the longitudinal ion dynamics does not play an essential role and can be neglected. As for inertialess electrons, they react immediately to any force imbalance along the field line and redistribute their density in such a way that the longitudinal force balance is recovered, mostly through the appearance of a longitudinal electrostatic field. This process inevitably imposes changes in the poloidally averaged profile of the electrostatic potential which, in turn, determines the poloidally averaged shear flow structure.

In the present paper, the mechanism of shear flow generation described above is considered in the simplest case of a slab SOL plasma geometry in the plane perpendicular to the strong magnetic field, while the longitudinal plasma inhomogeneity and corresponding differential operators are treated in a "finite-difference" or "single-mode" approximation. This allows us to consider still a two-dimensional problem while taking into account some important features of three-dimensional plasma dynamics.

The paper is organized as follows. In section 2 basic equations are introduced and the model is described with some details. In section 3 the results of large scale numerical simulation are presented which clearly demonstrate the onset of ELM-like activity and shear flow generation. Two different regimes controlled by either interchange instability or Boltzmann coupling are found and discussed. In section 4 a low-dimensional model of the phenomenon is considered. The conclusions are summarized in section 5.

II BASIC EQUATIONS

In many respects, the dynamics of scrape-off-layer plasma is described by two-fluid magnetohydrodynamic equations which are the continuity and momentum equations for the electron (e) and ion (i) components:

$$\frac{\partial n_\alpha}{\partial t} + \nabla \cdot (n_\alpha \mathbf{v}^\alpha) = 0, \tag{1}$$

$$\frac{d\mathbf{v}^\alpha}{dt} = \frac{e_\alpha}{m_\alpha}(\mathbf{E} + \frac{1}{c}\mathbf{v}^\alpha \times \mathbf{B}) - \frac{1}{n_\alpha m_\alpha}\nabla P_\alpha + \frac{1}{\tau_{ei}}(\mathbf{v}^\beta - \mathbf{v}^\alpha) + \nu_\alpha \Delta \mathbf{v}^\alpha. \tag{2}$$

Here $\frac{d\mathbf{v}^\alpha}{dt} \equiv \frac{\partial \mathbf{v}^\alpha}{\partial t} + (\mathbf{v}^\alpha \cdot \nabla)\mathbf{v}^\alpha$; $\alpha = e, i$, $\beta = i, e$; n_α , $\mathbf{v}^\alpha$ and P_α are density, velocity and pressure of the α-th component, respectively; τ_{ei} is the characteristic momentum exchange time between ions and electrons, closely related to the plasma electric resistivity; ν_α is the kinematic viscosity coefficient; other notations are as usual.

Drift convection, which we are interested in, is a relatively slow process in comparison with plasma oscillations, and charge separation effects are averaged out on the convective time scale. In this case, with high accuracy, plasma is assumed to be quasineutral, $n_e = n_i = n$, although an electrostatic potential of the plasma, φ, is not necessarily equal to zero. Slow evolution of the electrostatic potential is described by the equation for electric charge density, $(n_i - n_e)$, with the fast time derivative term averaged out,

$$\nabla \cdot \mathbf{J} = 0, \tag{3}$$

where $\mathbf{J} \equiv e(n_i \mathbf{v}_i - n_e \mathbf{v}_e)$ is the electric current density. In explicit form, the relationship between equation (3) and evolution of the electrostatic potential becomes clear when the plasma dynamics, i.e., the velocity $\mathbf{v}^{e,i}$, is substituted explicitly [see equation (8)]. Then, the electron continuity equation (1) can be considered as the equation for the quasineutral plasma density n ,

$$\frac{\partial n}{\partial t} + \nabla_{\perp} \cdot (n \mathbf{v}_{\perp}) - \frac{1}{e} \nabla_{\parallel} J_{\parallel} = 0, \tag{4}$$

where "$\parallel$" and "$\perp$" stand for the directions along and across the magnetic field $\mathbf{B}$, respectively, i.e., for example, $\nabla_{\parallel} \equiv (\frac{\mathbf{B}}{B} \cdot \nabla)$, $\nabla_{\perp} \equiv \nabla - \frac{\mathbf{B}}{B} \nabla_{\parallel}$ and so on. The longitudinal part of the divergence $\nabla \cdot (n\mathbf{v})$ in equation (4) is expressed in terms of the electron longitudinal velocity only, i.e., in terms of the longitudinal electric current, $\nabla_{\parallel}(n v_{\parallel}) \approx -(1/e)(\nabla_{\parallel} J_{\parallel})$, while the longitudinal dynamics of heavy ions is neglected.

In order to close the set (3) and (4) we have to express the perpendicular flow velocity $\mathbf{v}_{\perp}$ and longitudinal current density $J_{\parallel}$ in terms of the principal variables n and φ .

Hereafter, we assume that the magnetic field perturbation is absent and, hence, only the potential electric field $\mathbf{E} = -\nabla_{\perp} \varphi$ should be taken into account in equation (2). Then, the steady-state magnetic field in the SOL is as strong as in the bulk plasma. So, in the main order of B^{-1} expansion, the perpendicular flow velocity $\mathbf{v}_{\perp}$ corresponds to the $\mathbf{E} \times \mathbf{B}$-drift,

$$\mathbf{v}_{\perp} = \frac{c}{B^2} \mathbf{E} \times \mathbf{B} = \frac{c}{B^2} \mathbf{B} \times \nabla_{\perp} \varphi \tag{5}$$

Expression (5) follows from equation (2) with neglected pressure and inertia terms which, however, should be retained in equation (3) for the electrostatic potential [see below, equation (8)].

In the considered approximation, the longitudinal component of the electric current density is determined by longitudinal electron motion only and can be found from the momentum equation (2). For inertialess electrons one obtains:

$$\frac{1}{\sigma} J_{\parallel} = -\nabla_{\parallel} \varphi + \frac{1}{en} \nabla_{\parallel} P_e, \tag{6}$$

where $\sigma^{-1} = \dfrac{m_e}{ne^2\tau_{ei}} = \dfrac{4\pi}{\tau_{ei}\omega_{pe}^2}$ is plasma resistivity. The momentum exchange time scale τ_{ei} is not specified in the present analysis, and it is sufficient to treat σ as a phenomenological plasma parameter (σ = constant). For the same reason, it is sufficient to assume that the plasma temperature T is a constant as well. Then the longitudinal part of the electric current divergence can be written as

$$\nabla_{\parallel} J_{\parallel} \approx \sigma \nabla_{\parallel}^2 \left[-\varphi + \frac{T}{e} \ln n \right]. \tag{7}$$

Thus, in weakly-resistive $(\sigma^{-1} \to 0)$ longitudinally inhomogeneous $(\nabla_{\parallel} \neq 0)$ plasma the longitudinal electron dynamics tends to establish Boltzman distribution with strongly coupled plasma density and electrostatic potential, $n \sim \exp(e\varphi/T)$. Note that equation (7) and, hence, the Boltzmann coupling effect are valid for the poloidally averaged n and φ as well.

Substituting expression (7) into equation (3) and also using the general equality $\nabla_{\perp} \cdot \mathbf{J}_{\perp} \equiv \nabla_{\perp}\{\frac{1}{B^2}[\mathbf{B} \times \mathbf{J}_{\perp} \times \mathbf{B}]\}$ with the magnetic ponderomotive force $\mathbf{J}_{\perp} \times \mathbf{B}$ taken from the momentum equation (2), one obtains equation for plasma flow vorticity, $(\nabla \times \mathbf{v}_{\perp})_{\parallel} \approx (c/B)\nabla_{\perp}^2\varphi$ (or, equivalently, for the electrostatic potential). Namely,

$$\frac{\partial \nabla_{\perp}^2 \varphi}{\partial t} + \frac{c}{B}[\nabla_{\perp}\varphi \times \nabla_{\perp}\nabla_{\perp}^2\varphi]_{\parallel} - (\nu_{\perp}\nabla_{\perp}^2 + \nu_{\parallel}\nabla_{\parallel}^2)\nabla_{\perp}^2\varphi =$$

$$= -\frac{TB^3}{m_i nc}[\nabla_{\perp}(\frac{1}{B^2}) \times \nabla_{\perp} n]_{\parallel} + \frac{\sigma B^2}{m_i nc^2}\nabla_{\parallel}^2(-\varphi + \frac{T}{e}\ln n). \tag{8}$$

Here we neglect all plasma inhomogeneity and inertia effects, retaining only the most important effects: i) convective and diffusive vorticity transport [see the left hand side of equation (8)], ii) an effective "gravitational" driving force in a curved magnetic field, $-\frac{TB^3}{m_i nc}[\nabla_{\perp}(\frac{1}{B^2}) \times \nabla_{\perp} n]_{\parallel}$, and iii) the Boltzmann coupling term in the r.h.s., respectively.

There exists an analogy between interchange instability in magnetized plasma and Rayleigh-Taylor instability in stratified fluid. It is based on the analogy between the magnetic curvature effect and the action of gravity. Indeed, in the case of Rayleigh-Taylor instability, flow evolution is described by the Euler equation, $d\mathbf{v}/dt = -g\mathbf{e}_x - 1/\rho\nabla P$, where g is the gravitational acceleration along, say, the x-direction, and ρ is the spatially inhomogeneous mass density. Respectively, the flow vorticity, $\nabla \times \mathbf{v}$, is driven by the forcing term $\frac{\partial}{\partial t}\nabla \times \mathbf{v} \sim -[\nabla(\frac{1}{\rho}) \times \nabla P] \sim \frac{g}{\rho}[\mathbf{e}_x \times \nabla\rho]$ where it is assumed that, for slow fluid dynamics, the pressure gradient is mainly determined by its hydrostatic quasi-equilibrium value, $\nabla P \approx -g\rho\mathbf{e}_x$. This gravitational driving force is obviously analogous to the magnetic curvature driving force in equation (8) with an effective "gravitational acceleration" $g \sim \frac{T}{m_i}\frac{|\nabla_\perp B|}{B} \sim \frac{c_s^2}{R}$ (here $R \sim l_\parallel$ is the tokamak major radius).

So far, we considered three-dimensional plasma dynamics. Let us now simplify the problem and reduce its dimensionality by representing the longitudinal differential operator $\nabla_\parallel^2$ in its "finite-difference" form: $\nabla_\parallel^2 \approx -l_\parallel^{-2}$. This, actually, corresponds to a single-mode approach with the only characteristic harmonic in the direction along the magnetic field. Such an approach is quite reasonable unless the details of nonlinear flow (or wave) propagation along the field lines become important. In the present qualitative analysis we neglect any longitudinal plasma dynamics, besides the Boltzmann coupling effect, and consider only the "robust" longitudinal inhomogeneity in the form of a bump-like localization of the instability-driven convection mainly in the inner part of an open magnetic field line in the SOL, while outside the convection region, along the field line, there exist unperturbed (e.g., not flattened) background plasma density and potential profiles, n_b and φ_b , respectively. In the simplest case, we assume that no externally driven plasma flow exists in the SOL, i.e., $\varphi_b = 0$. Then, without loss of generality, we also assume that $| n - n_b | \ll n_b$ and, hence, $\nabla_\parallel^2 \ln(n) \sim -l_\parallel^{-2}[(n-n_b)/n_b]$. With such a simplifications, system (4) and (8) can be written in the following dimensionless form (see [17–26]):

$$\frac{\partial N}{\partial \tau} + [\nabla_\perp \Phi \times \nabla_\perp N]_Z + \frac{\partial \Phi}{\partial Y} - \sigma_2(\Phi - N) = D_\perp \nabla_\perp^2 N, \tag{9}$$

$$\frac{\partial \nabla_\perp^2 \Phi}{\partial \tau} + [\nabla_\perp \Phi \times \nabla_\perp \nabla_\perp^2 \Phi]_Z + g_B \frac{\partial N}{\partial Y} - \sigma_1(\Phi - N) = \mu_\perp \nabla_\perp^4 \Phi. \tag{10}$$

Here $\Phi \equiv e\varphi/T$, $N \equiv (n - n_b)/n_{b0}$, n_{b0} and $(dn_b/dx)_0$ are the characteristic density and radial density gradient of the background SOL plasma, $n_b \approx n_{b0}[1 + x/n_{b0}(dn_b/dx)_0]$. Then, x, y and z are the Cartesian coordinates which correspond locally to the radial, poloidal and toroidal directions, respectively. Dimensionless coordinates in the perpendicular ($\perp$) plane are $X \equiv x/x_0$ and $Y \equiv y/x_0$, where $x_0 \approx n_{b0} \mid (dn_b/dx)_0 \mid^{-1}$ is the SOL width. Dimensionless time $\tau \equiv t/t_0$ is normalized by an effective Bohm diffusion time scale $t_0 = x_0^2(cT/eB)^{-1}$. In these notations, the shear flow component corresponds to the poloidally averaged poloidal velocity, $< v_Y > \equiv \partial < \Phi > /\partial X$, where $< \cdot > \equiv (1/Y_0)\int_0^{Y_0}(\cdot)dY$ represents poloidally averaged quantity.

In system (9) and (10), there are five dimensionless parameters though only three of them, actually, control the regime of flow evolution. Namely, the parameter $g_B \equiv 2(x_0/R)(x_0/\rho_s)^2$, where $\rho_s \equiv c_s(m_i c/eB)$ is the ion Larmor radius, represents the magnetic curvature effect and determines the growth rate of an ideal interchange, or Rayleigh-Taylor, instability, $\gamma_{\rm RT} \approx \sqrt{g_B}$. Parameter $\sigma_1 \equiv \sqrt{m_i/m_e}\lambda_{ei}x_0^4/\rho_s^3 l_\parallel^2$ controls the coupling between interchange and drift waves ($\lambda_{ei} = \tau_{ei}\sqrt{T/m_e}$ is the electron mean free path which determines plasma resistivity along the magnetic field). For a perturbation with sufficiently large wave number, $k \gg (\sigma_1/4)^{\frac{1}{3}}$, parameter σ_1 determines the characteristic time scale of a particular type of drift instability, $\tau_\sigma \approx \sqrt{2k^2/k_Y\sigma_1}$. This follows from the expression for instability growth rate obtained from the linearized equations (9) and (10),

$$\gamma \approx -k^2 D - \frac{\sigma_1}{2k^2} + \mathrm{Re}\sqrt{\frac{\sigma_1^2}{4\mathrm{k}^2} + \mathrm{g_B}\frac{\mathrm{k_Y^2}}{\mathrm{k}^2} - \mathrm{i}\frac{\sigma_1 \mathrm{k_Y}}{\mathrm{k}^2}}, \qquad (11)$$

where we neglect σ_2, for simplicity. The third control parameter, which is a rather phenomenological one, is the diffusivity $D \approx D_\perp \approx \mu_\perp$. We assume that it is small but finite, in order to suppress small scale length perturbations with $k > k_*$, where k_* is a characteristic "cut-off" wave number. In the case when Boltzmann coupling effect dominates, i.e., for $\tau_\sigma\gamma_{\rm RT} \ll 1$, this wave number is $k_* \sim (\sigma_1/2D^2 k_Y/k)^{\frac{1}{5}}$, while in the opposite case of $\tau_\sigma\gamma_{\rm RT} \gg 1$ it is $k_* \approx \sqrt{\gamma_{RT}/D}$. As is also shown below, the diffusivity D controls slow relaxation of stable shear flow structure towards the instability threshold and, thus, determines the period of ELM-like flow evolution. The other coupling parameter, $\sigma_2 \equiv \sigma_1(\rho_s/x_0)^2$, is relatively small (this is because the ion Larmor radius is small) and does not play an essential role until finite Larmor radius effects, i.e., the transition to Hasegawa-Mima equation, is considered.

In a hidden form, system (9) and (10) is also characterized by an important geometrical parameter, namely, the aspect ratio of the SOL, which is equal to the dimensionless length of the SOL in the poloidal direction, Y_0. Usually, the SOL width x_0 is about 1 cm while tokamak minor radius is about 0.5 m, so that the aspect ratio is expected to be more than one hundred. However, for qualitative analysis, it is sufficient to consider its rather moderate value, say, $Y_0 = 5$, in order to model self-consistent nonlinear dynamics of a long chain of instability-driven vortices with about ten convective cells per the period Y_0.

Thus, in the present paper, we consider the following range of the parameter values: $g_B = 0 - 10^2$, $\sigma_1 = 0 - 10^3$, $D = 10^{-4} - 10^{-3}$ and $\sigma_2 = (10^{-4} - 10^{-2})\sigma_1$, which, hence, includes both the limiting cases of strong and weak Boltzmann coupling effect. A relatively small diffusivity value is chosen in such a way that it allows the development of about ten Bénard-like convective cells per one poloidal period, Y_0.

With such parameters, governing equations (9) and (10) have been solved numerically. The simulation domain $(0 \leq X \leq 1;\ 0 \leq Y \leq 5)$ was implemented on a 101×501 grid point, respectively. As the boundary condition, we assumed periodicity along the poloidal Y-direction with the period $Y_0 = 5$, and the free-slip ($\nabla_{\perp}^2 \Phi \mid_{X=0;1} = 0$) condition at the rigid wall, where it was also assumed that $\Phi \mid_{X=0;1} = N \mid_{X=0;1} = 0$. The initial small amplitude (less than 10^{-6}) random perturbation of plasma density was considered, while the initial electrostatic potential was not perturbed at all, $\Phi \mid_{\tau=0} \equiv 0$.

III SIMULATION RESULTS

Numerical simulation clearly demonstrates that, in all the cases within the considered parameter range, a large shear flow component with the poloidal velocity $< v_Y > \sim 1$ is always generated, but its temporal evolution strongly depends on the coupling parameter value.

Let us consider, for the beginning, a typical flow evolution in the case when the Boltzmann coupling effect dominates, i.e., when $\tau_\sigma \gamma_{\mathrm{RT}} \leq 1$ (namely, we consider the case of $\sigma_1 = 6$, $g_B = 1$ and $D = 7 \times 10^{-4}$).

In Figure 1, few successive snapshots showing plasma density (a) and potential (b) contour lines, as well as radial profiles of the poloidally averaged plasma density (c) and poloidal velocity (d), are presented for the very first cycle of shear flow generation. For $\tau < 15$, linear growth of the interchange instability is accompanied by the development of the most unstable mode with characteristic wave number $k \approx 2/3k_* \approx 14$. Then, at the nonlinear stage, $15 < \tau < 20$, the instability-driven convection causes significant spatial redistribution of the poloidally averaged plasma density profile (namely, its flattening) with corresponding rearrangement of the electrostatic potential and, hence, flow velocity profiles. The flattening of density profile and generation of a large shear flow component, which stirs and suppresses the convective cells, also arrest the growth of the instability. As a result, in the absence of free energy release, both the shear-flow component and the perturbation of the radial density profile slowly decay within the relatively long period, $\delta\tau$, of quiet evolution determined by plasma diffusivity, $\delta\tau \approx 1/D(2\pi)^2 \approx 35$, (see time interval $20 < \tau < 60$ in Fig. 1). However, when the poloidal shear flow velocity becomes sufficiently small and the spatial distribution of the plasma density approaches its initial linear profile, the flow suffers another cycle of instability with the consequent generation of a shear-flow component (see Figure 2 which represents the second cycle of shear-flow generation immediately successive to the first one given in Figure 1). Besides the very initial period, $\tau < 15$, flow evolution during all the other successive cycles goes on in a similar periodic way and follows the same scenario described above. The only difference concerns the shape of the unstable plasma density perturbation.

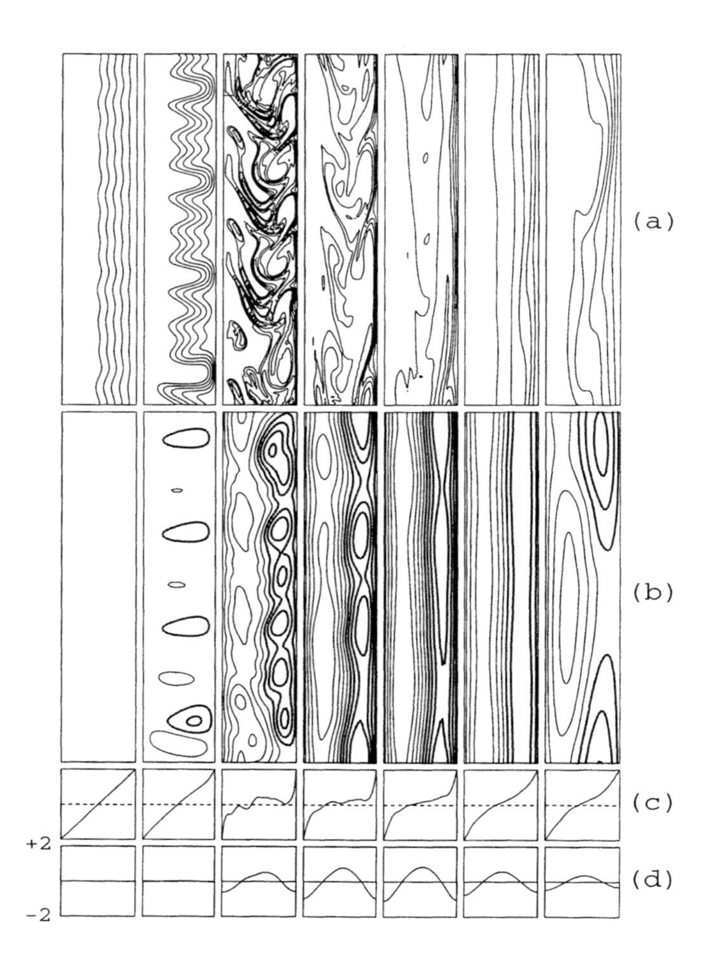

FIGURE 1. Snapshots of the contour lines for plasma density (*a*) and electrostatic potential (*b*) at the time moments $\tau = 11; 14; 17; 20; 24; 40; 60$ (from left to right) corresponding to the very first cycle of shear flow generation. Poloidally averaged profiles of plasma density (*c*) and poloidal flow velocity (*d*) are also shown (the case of $g_B = 1,\ \sigma_1 = 6$).

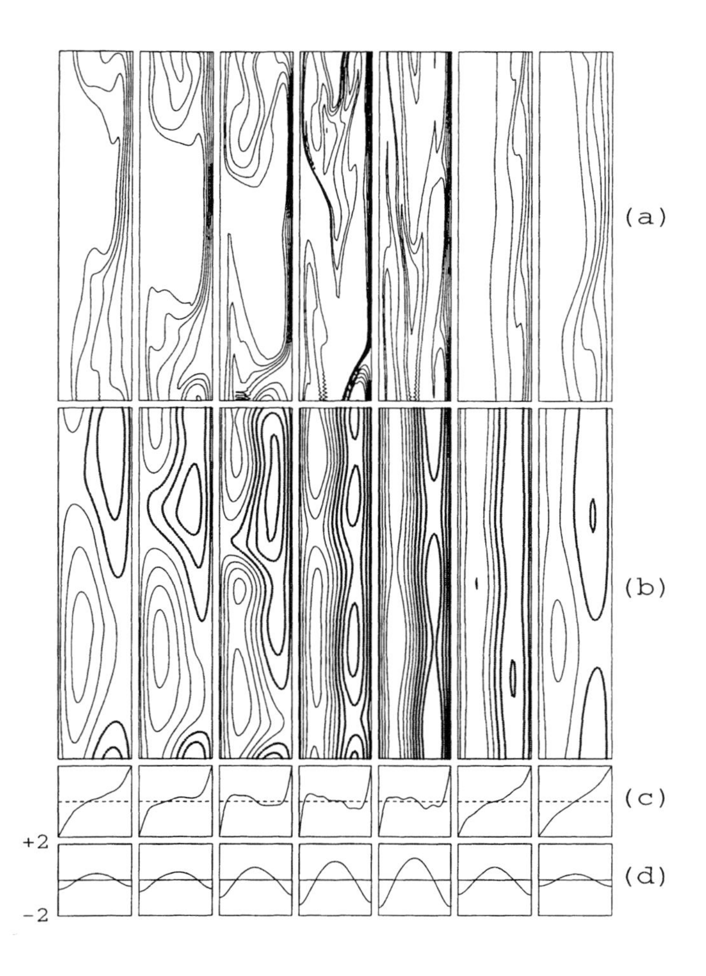

FIGURE 2. The same as in Figure 1 but for the second cycle of shear flow generation ($\tau = 62; 64; 66; 68; 70; 90; 116$, from left to right)

For $\tau < 20$, Rayleigh-Taylor instability grows in the absence of any considerable shear flow component and is characterized by the development of narrow, radially elongated spikes of dense plasma. In contrast, plasma density perturbation during all the successive cycles of shear flow generation is usually of the form of a poloidally elongated "bubble". When such a "bubble" approaches the wall (see, e.g., the time moment $\tau = 66$ in Fig. 2), the plasma density gradient and, hence, the particle flux, $q = D\int_0^{Y_0}(1-(\partial N/\partial X)_{|X=0;1})\,dY$, grow there significantly. However, efficient contact of the "bubble" with the wall takes place during a short period of time only, $\delta\tau < 10$, while during all the rest of a cycle the "bubble" is separated from the wall by the shear flow separatrix, and the corresponding particle flux is small. Consequently, the time-averaged particle flux, $\bar{q} = (1/\tau_c)\int_\tau^{\tau+\tau_c} q\,d\tau$, is reduced and does not greatly exceed its value in the absence of turbulent convection. This is clearly seen in Figure 3.a where the time evolution of both the fluxes q and $\bar{q}$ is shown together with the temporal behaviour of the shear flow kinetic energy per unit length in the poloidal direction, $W_s \equiv 1/2\int_0^1(\partial <\Phi> /\partial X)^2 dX$. In order to clarify the role played by the Boltzmann coupling effect in the observed strong correlation between the processes of plasma density flattening and shear flow generation, let us consider the temporal evolution of the shear flow kinetic energy, $dW_s/d\tau = P_R + P_B - Q$, and compare the driving terms which are the rate of Reynold stress,

$$P_{\mathrm{R}} = -\int_0^1 \frac{\partial^2 <\Phi>}{\partial X^2} < \frac{\partial\Phi}{\partial X}\frac{\partial\Phi}{\partial Y} > dX\ ,$$

the rate of Boltzmann coupling, $P_{\mathrm{B}} = -\sigma_1\int_0^1 <\Phi> (<\Phi> - <N>)\,dX$, and dissipation rate, $Q = D\int_0^1(\partial^2 <\Phi> /\partial X^2)^2 dX$. Temporal evolution of these quantities is shown in Figure 3.b. One can easily see that the Boltzmann coupling effect always dominates over Reynolds stresses and, hence, plays important role in flow evolution. Then, during the phase of dissipative relaxation, the coupling rate P_{B} goes down much faster than the dissipative decay process itself. This can be satisfactorily explained only by the establishment of the Boltzmann distribution, $<\Phi>\approx<N>$. Hence, the Boltzmann coupling effect controls flow evolution during slow dissipative relaxation as well, even though, in this case, the dissipation rate Q formally exceeds the coupling rate P_B.

In the opposite case of $\tau_\sigma\gamma_{\mathrm{RT}} \geq 1$ (namely, for $g_B = 7$ with other parameters unchanged), the magnetic curvature effect dominates, and the flow evolution is changed considerably. First of all, this concerns the generation of a saturated poloidally averaged shear flow component. In contrast to the previous case, the shear flow component, being initially generated within the period of two inverse ideal growth rates γ_{RT}, is then maintained at the same saturated level $< v_Y >\approx 1$ (see Fig. 4, Fig. 5 and Fig. 6, which are the analogies

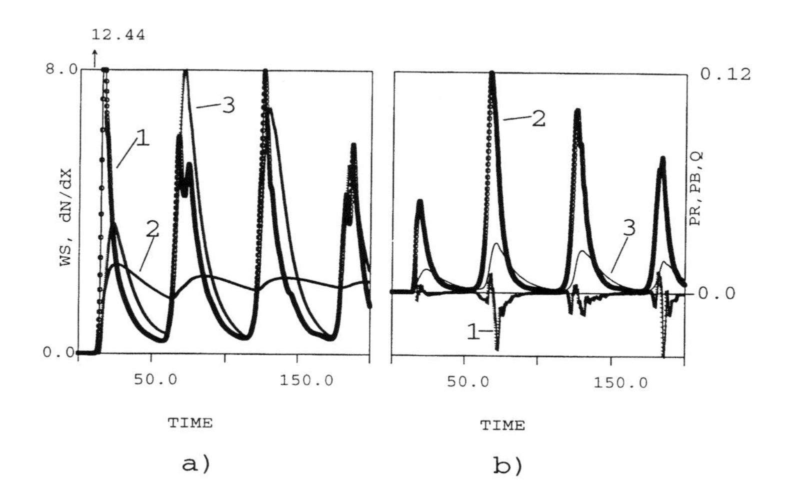

FIGURE 3. (*a*) ELM-like temporal evolution of the poloidally- (curve 1) and time-averaged (curve 2) density gradient at the wall, as well as the kinetic energy of the shear-flow component, W_s (curve 3); (*b*) shear-flow energy balance: the rates of Reynolds stresses (curve 1), Boltzmann coupling (curve 2) and dissipation (curve 3). The case $g_B = 1$, $\sigma_1 = 6$.

of Fig. 1, Fig. 2 and Fig. 3, respectively, but correspond to higher g_B value). The corresponding poloidally averaged density profile is flattened permanently as well. Hence, no great difference between the particle fluxes q and $\bar{q}$ appears (see Fig. 6.a). Both the fluxes exceed at least twice their values in the case $\tau_\sigma \gamma_{\mathrm{RT}} \leq 1$, what reminiscent of the degradation of plasma confinement in L-regime. However, an interesting feature of such a "strange" L-mode is the generation of a steady-state shear flow component which, however, does not suppress interchange instability (this is because g_B is high enough) and which co-exists with instability-driven Bénard-like convective cells (see Fig. 6.b). As the vortices are strong, neither "spikes" nor "bubbles" are effectively separated from the walls by the shear flow separatrix, and this explains the degradation of plasma confinement even in the presence of shear flow.

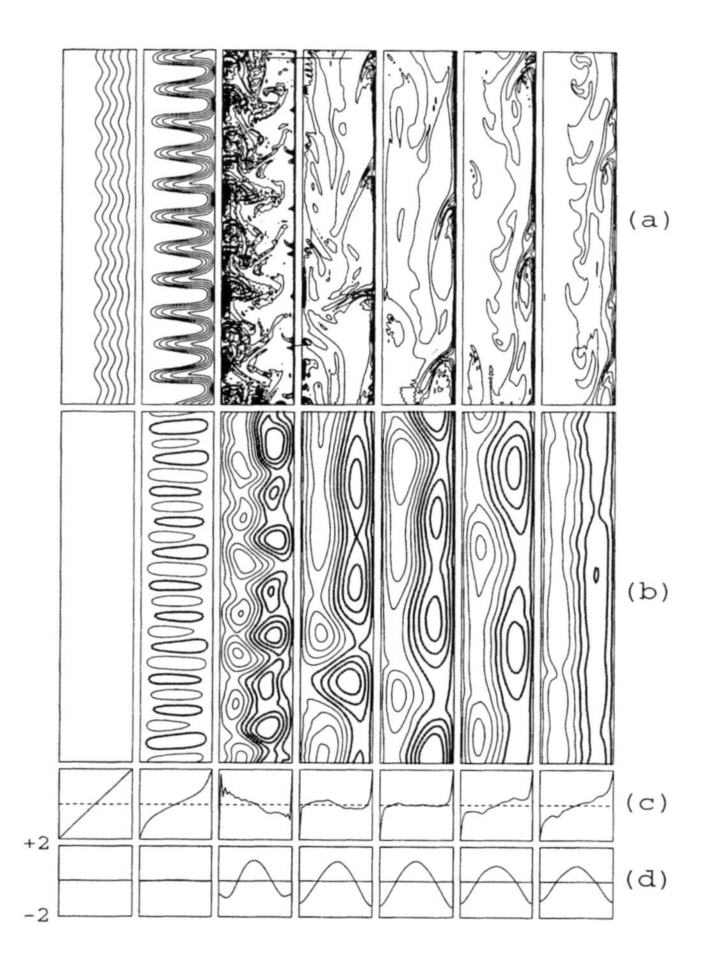

FIGURE 4. Snapshots of the contour lines for plasma density (a) and electrostatic potential (b) at the time moments $\tau = 4; 5; 6; 8; 10; 15; 20$ (from left to right) corresponding to the very first cycle of shear flow generation. Poloidally averaged profiles of plasma density (c) and poloidal flow velocity (d) are also shown (the case of $g_B = 7,\ \sigma_1 = 6$).

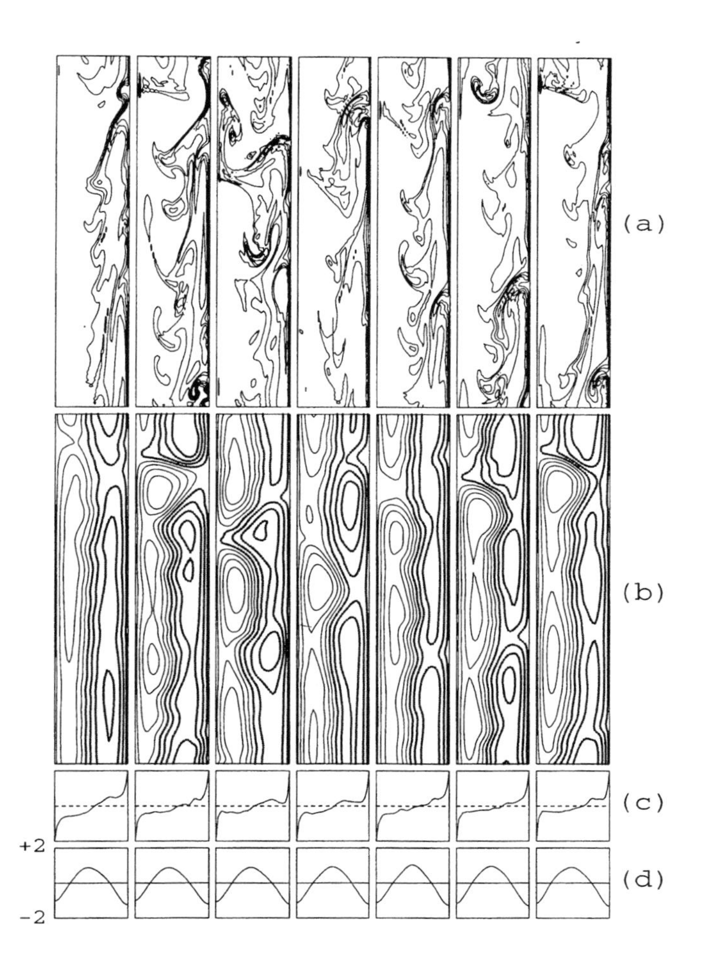

FIGURE 5. The same as in Figure 4 but for the time moments $\tau = 40; 42; 44; 46; 50; 56; 60$ (from left to right)

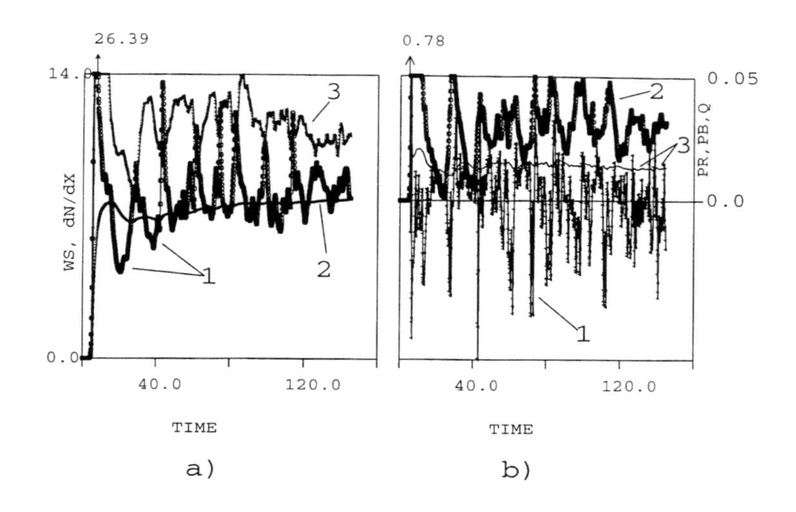

FIGURE 6. The same as in Figure 3 but for $g_B = 7$, $\sigma_1 = 6$.

In the ($g_B - \sigma_1$)-plane, the two different regimes of plasma flow evolution described above, namely, ELMs and "strange" L-mode, are separated by the curve $\tau_\sigma \gamma_{\mathrm{RT}} \sim 1$ or, equivalently, $g_B/\sigma_1 \sim 1/2k$ with the observed effective wave number of the "bubble" structure $k \sim 1$. A ather sharp transition from one regime to another takes place while moving across the separatrix. For example, the case $g_B = 5$, $\sigma_1 = 4$ corresponds to a typical "strange" L-mode, while the case $g_B = 5$, $\sigma_1 = 6$ already corresponds to the ELM-like evolution, although with relatively short ELMs period, about 50 % of its typical value observed in the case of $g_B = 5$, $\sigma_1 = 10$.

IV LOW-DIMENSIONAL MODEL

The observed process of shear flow generation is, roughly speaking, a combination of four "robust" processes: i) density profile flattening via nonlinear convection; generation of shear component via ii) Boltzmann coupling effect (when density profile is flattened) and/or iii) Reynolds stresses; and iv) slow dissipative relaxation of a stable configuration towards the instability threshold. These processes, according to the simulation results, seem to be insensitive to the particular details of plasma density and potential spatial structure and, hence, their description in terms of a low-dimensional model is quite reasonable. Such a model, if it represents the process qualitatively well, could help in scanning a wider range of plasma parameters and elucidate the physics of the phenomenon.

There exists a well-known and successful example of such a low-dimensional model, namely, the Lorentz set for Bénard convection in unstably stratified fluid. In the Lorentz model (but in our notations), the only flow structure taken into account is a regular chain of vortices, $\Phi^{(L)} = b_{\Phi}^{(L)} \cos(k_Y Y) \sin(\pi X)$. As for the density structure, it includes both profile flattening (the amplitude $A_N^{(L)}$) and a spike-like convective deformation (the amplitude $a_N^{(L)}$), so that $N^{(L)} = A_N^{(L)} \sin(2\pi X) + a_N^{(L)} \sin(k_Y Y) \sin(\pi X)$. This model describes well many features of Bénard convection, so it is meaningful to use the Lorentz set as a "kernel" of an extended low-dimensional model which takes into account the effect of shear flow generation (this effect is not included into the original Lorentz set).

One example of such an extended Lorentz model was considered in (18). However, the Boltzmann coupling effect did not play an essential role in that simulation, and shear flow generation was mainly controlled by Reynolds stress forcing. Respectively, the density and potential structures considered in (18) did not represent intrinsically the coupling effect.

In the present paper we consider another type of an extended Lorentz set which is mainly oriented towards taking into account Boltzmann coupling effect. In the original Lorentz set, density representation already includes both the important effects of profile flattening and spike- (or bubble)-like convective deformation, so it is quite reasonable to use the same density representation in our model as well. As for the representation of plasma potential, we have to explore the following observations (see the Figures presented above): i) poloidally averaged potential profile, $< \Phi >$, is always akin to density perturbation profile, and ii) there exists a tendency to establish Boltzmann distribution, i.e., $(\Phi - N)$-coupling. This means that it is necessary (and, actually, sufficient) to include the density counterpart into the potential representation. With such a minimal correction, we consider the following density and potential structures:

$$N = A_N^{(L)} \sin(2\pi X) + a_N^{(L)} \sin(k_Y Y) \sin(\pi X),$$

$$\Phi = A_{\Phi} \sin(2\pi X) + [b_{\Phi}^{(L)} \cos(k_Y Y) + a_{\Phi} \sin(k_Y Y)] \sin(\pi X). \qquad (12)$$

Here (L) denotes an amplitude from the original Lorentz set. In the present analysis, the characteristic poloidal wave number, $k_Y = 2\pi m / Y_0$, where m is the corresponding poloidal mode number, is not specified self-consistently and should be considered as a parameter. According to the simulation results described above, the poloidal mode number is about 10 at the very first cycle of shear flow generation, and it is reduced to $m = 1$ or $m = 2$ during all the successive cycles.

Substituting expressions (12) into the governing equations (9) and (10), we obtain the following extended Lorentz set:

$$\frac{d\, A_N^{(L)}}{d\tau} = -\lambda_1 A_N^{(L)} + \sigma_2 A_\Phi - \frac{m\pi^2}{Y_0}\, a_N^{(L)}\, b_\Phi^{(L)}$$

$$\frac{d\, a_N^{(L)}}{d\tau} = -\lambda_2 a_N^{(L)} + \sigma_2 a_\Phi + \frac{2\pi m}{Y_0} b_\Phi^{(L)} + \pi\, A_N^{(L)}\, b_\Phi^{(L)}$$

$$\frac{d\, b_\Phi^{(L)}}{d\tau} = -\lambda_3 b_\Phi^{(L)} + \gamma_B a_N^{(L)} + \xi\, A_\Phi\, a_\Phi \tag{13}$$

$$\frac{d a_\Phi}{d\tau} = -\lambda_4 a_\Phi + \gamma_\sigma a_N^{(L)} - \xi\, A_\Phi\, b_\Phi^{(L)}$$

$$\frac{d\, A_\Phi}{d\tau} = -\lambda_5 A_\Phi + \frac{\sigma_1}{4\pi^2}\, A_N^{(L)}$$

Here

$$\lambda_1 = 4\pi^2 D + \sigma_2,\ \lambda_2 = k^2 D + \sigma_2,\ \lambda_3 = \lambda_4 = k^2 D + \frac{\sigma_1}{k^2},\ \lambda_5 = 4\pi^2 D + \frac{\sigma_1}{4\pi^2},$$

$$\xi = \frac{\pi^2 m}{k^2 Y_0}\left(\left(\frac{2\pi m}{Y_0}\right)^2 - 3\pi^2\right),\ \gamma_B = \frac{g_B}{k^2}\frac{2\pi m}{Y_0},\ \gamma_\sigma = \frac{\sigma_1}{k^2},\ k^2 = \pi^2 + \left(\frac{2\pi m}{Y_0}\right)^2.$$

The last equation in the set (13) describes dissipative relaxation towards the Boltzmann distribution for the poloidally averaged potential and density profiles. This is, actually, the process of shear flow generation in the presence of the flattened density profile (see Introduction). As this follows from the first and the last equations (13), both the shear flow component and the flattening perturbation of plasma density profile tend to zero because of dissipative relaxation with characteristic dissipative time scale λ_5^{-1}. However, there exists a "source"-term in the right-hand-side of the equation for $A_N^{(L)}$, namely, the term with $a_N^{(L)} b_\Phi^{(L)}$, which generates nonzero amplitudes $A_N^{(L)}$ and, hence, A_Φ. This "source"-term describes nonlinear convective flattening of the plasma density profile because of the development of a chain of vortices (amplitude $b_\Phi^{(L)}$) with consequent spike-like deformation of density profile (amplitude $a_N^{(L)}$). In turn, convection is driven by the development of instability which can be saturated by two nonlinearities arising from deformations of the poloidally averaged density and potential profiles: i) change in effective density gradient, i.e., $b_\Phi^{(L)} A_N^{(L)}$ term in the second equation (13), and ii) stirring of the vortices by the shear flow component, i.e., the nonlinear terms in the third and fourth equations (13). Complicated interaction of these effects leads to ELM-like temporal behaviour of all the amplitudes.

Numerical solution of the set (13) is in a good agreement with the solution of the original set (9) and (10). In principle, the described low-dimensional model could be useful in the analysis of ELM activity in a wide range of plasma parameters. However, such an analysis is beyond the scope of the present paper.

V CONCLUSIONS

We have studied, by two-dimensional large-scale numerical simulation, the process of shear flow generation in a plasma that is unstable against the growth of flute-like interchange perturbations. Particular attention was drawn to the effect of density and electrostatic potential coupling in a longitudinally inhomogeneous plasma. This effect is shown to play a significant role in the onset of the shear flow component.

Two different limiting cases of shear flow generation and evolution are found, depending on the importance of the Boltzmann coupling effect and driving force of the interchange instability (magnetic curvature effect). When the coupling effect dominates, flow evolution is akin to ELM-activity in the SOL plasma with essential no degradation of plasma confinement. In the opposite case, plasma transport is enhanced considerably, even though a shear flow component still exists. Between such a "strange" L-mode of plasma confinement, on the one hand, and an ELM-like regime, on the other hand, there are various "transitional" regimes characterized by different values of the frequency of ELM-like events. In the $(g_B - \sigma_1)$-plane, however, there exists a rather sharp boundary between these two regimes, qualitatively corresponding to the balance between the Boltzmann coupling effect and the interchange instability driving force.

A low-dimensional model of the process is proposed. The model is based on the well-known Lorentz set for Bénard convection with new elements which take into account shear flow generation and the Boltzmann coupling effect.

ACKNOWLEDGEMENT

This work was performed by using the Advanced Computing System for Complexity Simulation at NIFS under the support of Grants-in-Aid of the Ministry of Education, Science, Sports and Culture in Japan (No. 08044109).

One of the authors (S.B.) expresses his gratitude to Professor R.L. Dewar and Dr R.W. Griffiths (the Australian National University) for their kind hospitality, and the excellent working conditions and fruitful discussions during the Research Workshop on 2D Turbulence in Plasmas and Fluids.

REFERENCES

1. Wagner F. et al., *Phys. Rev. Lett.* **49**, 1408 (1982)
2. Groebner R.J. et al., in *Proceeings of the 16th European Conference on Controlled Fusion and Plasma Physics*, Venice, 1989 (European Physical Society, Petit-Lancy, Switzerland, 1989), p.245
3. Burrell K.H. et al., *Plasma Phys. Controlled Fusion* **31**, 1649 (1989)
4. Doyle E.J. et al., in *Proceedings of the 18th European Conference on Controlled Fusion and Plasma Physics*, Berlin, 1991 (Vienna: IAEA), **1**, p.285
5. Zohm H. et al., *Nuclear Fusion* **32**, 489 (1992)
6. Kerner W. et al., *Bull. Am. Phys. Soc.* **36**, 2R13 3210 (1991)
7. Erckmann V. et al., *Phys. Rev. Lett.* **70**, 2086 (1993)
8. Wagner F. et al., *Plasma Phys. Controlled Fusion* **36**, A61 (1994)
9. Toi K. et al., *Plasma Phys. Controlled Fusion* **36**, A117 (1994)
10. Shats M. et al., *Phys. Rev. Lett.* **77**, 4190 (1996)
11. Sakai O., Yasaka Y., and Itatani R., *Phys. Rev. Lett.* **70**, 4071 (1993)
12. Drake J.F. et al., *Phys. Fluids* **B 4**, 488 (1992)
13. Finn J.M., Drake J.F., and Guzdar P.N., *Phys. Fluids* **B 4**, 2758 (1992)
14. Finn J.M., *Phys. Fluids* **B 5**, 415 (1993)
15. Niedermeyer H. et al., in *Proceedings of the 18th European Conference on Controlled Fusion and Plasma Physics*, Berlin 1991 (Vienna: IAEA) **1**, p.301
16. Endler E. et al., in *Proceedings of the 20th International Conference on Controlled Fusion and Plasma Physics*, Lisboa 1993 (Vienna: IAEA) **2**, p.583
17. Kukharkin N.N., Osipenko M.V., Pogutse O.P., and Gribkov V.M., in *Proceedings of the 14th IAEA Conference on Plasma Physics and Controlled Nuclear Fusion Research* Würzberg, 1992 (Vienna: IAEA) **2**, p.293
18. Pogutse O., Kerner W., Gribkov V., Bazdenkov S., and Osipenko M., *Plasma Phys. Control. Fusion* **36**, 1963 (1994)
19. Bazdenkov S., and Pogutse O., *JETP Lett.* **57**, 426 (1993)
20. Sugama H., and Horton W., *Phys. Plasmas* **1**, 345 (1994)
21. Takayama A., Wakatani M., and Sugama H., *Phys. Plasmas* **3**, 3 (1996)
22. Takayama A., and Wakatani M., *Plasma Phys. Control. Fusion* **38**, 1411 (1996)
23. Yagi M., Wakatani M., and Hasegawa A., *J. Phys. Soc. Japan* **56**, 973 (1987)
24. Carreras B.A., Garcia L., and Diamond P.H., *Phys. Fluids* **30**, 1388 (1987)
25. Sugama H., and Wakatani M., *J. Phys. Soc. Japan* **57**, 2010 (1988)
26. R.B. White, *Theory of Tokamaks* Amsterdam: North-Holland, 1989, sec.6.

2D and 3D convection motions from a local source in rotating fluids

B.M. Boubnov

A.M. Obouhov Institute of Atmospheric Physics, 109017, Moscow, Russia

I INTRODUCTION

Spatial inhomogeneity of heating of fluids in gravity fields is the cause of all motions in nature: in the atmosphere and oceans on the Earth, in astrophysical and planetary objects. All astronomical objects rotate and so convective motions in rotating fluids are of interest in many geophysical and astrophysical phenomena. Two simplest cases of the convective motions in rotating fluids are usually considered: an infinite plane layer with a temperature difference between the horizontal boundaries [1-3] (vertical temperature gradient) and rotating annuli with temperature differences between the vertical walls [7] (horizontal temperature gradient).

Convection motions from a local (but not small) source of density is the more usual case in geophysics, but the most complicated, because of the inhomogeneous density distribution in the horizontal, and vertical directions. In addition to the density distributions, the local source makes the geometry more complicated. Here we will consider different regimes of convective motions from the local source of density with and without rotation and the influence of the geometrical configuration on these motions.

According to the Proudman-Taylor theorem, in a rotating homogeneous inviscid fluid for stationary slow motions when non-linear effects are small, the vertical velocity component should not change in the direction of the rotation vector and if it is zero at some surface it should be absent throughout the fluid. The slow motions in an inviscid fluid should be two-dimensional. The Proudman-Taylor theorem does not hold in the presence of viscosityor density stratification. This means that in a rotating fluid layer vertical motions can exist only due to the viscosity and should be observed only in the regions with strong shear where the viscous forces become important. Depending on the external parameters, in some parts of the convective motions became two dimensional (laminar or turbulent) and some attention is paid to these flow regimes.

CP414, *Two-Dimensional Turbulence in Plasmas and Fluids:* Research Workshop
edited by R. L. Dewar and R. W. Griffiths

II NON-DIMENSIONAL PARAMETERS AND SOME APPROXIMATIONS

The main nondimensional parameters which describe convective motions in rotating fluids [1] are the buoyancy parameter B^* , Reynolds number Re and Rossby number Ro

$$B^* = \frac{U^2}{Lg(\delta\rho/\rho)}$$

$$Re = \frac{UL}{\nu}$$

$$Ro = \frac{U}{2\Omega L} .$$

Here U and L are scales of velocity and length, $\delta\rho$, ρ and ν are the density variation, mean density and kinematic viscosity of the fluid and Ω is a constant rotation rate. All of these nondimensional parameters B^*, Re, Ro are the measures of the ratio of relevant forces (buoyancy, viscosity or Coriolis) to inertia forces.

When we use the nondimensional parameters, we should define the main scales of motion (in our case U and L), and aim to find the connection of these scales with the main external parameters. Let us consider some simplifications, which can be carried out by using various approximations that take into account the specifics of the problem.

A Rayleigh approximation

When we study convective motions in a plane horizontal layer of depth H, with or without rotation, there are no systematic motions and the initial velocity is zero when we consider the stability of layer heated from below. Following Lord Rayleigh we can use a scale of velocity in terms of viscosity ν: $U = \frac{k}{L}$. If we will consider that the buoyancy is only caused by the thermal expansion of the fluid with an equation of state $\rho = \rho_0(1 + \alpha(T - T_0))$, then

$$B^* = \frac{U^2}{\alpha g \Delta T L} = \frac{k^2}{\alpha g \Delta T L^3} = (Ra Pr)^{-1} .$$

Here $Pr = \frac{\nu}{k}$ is a Prandtl number (k is thermodiffusivity of the fluid) and the buoyancy parameter B^* is replaced by the well known Rayleigh number

$$Ra = \frac{\alpha g \Delta T L^4}{k\nu} = (B^* Pr)^{-1} .$$

For this velocity scale

$$Re = Pr^{-1} \text{ and } Ro = Ta^{-1/2} Pr^{-1} ,$$

where $Ta = \frac{4\Omega^2 L^4}{\nu^2}$ is a Taylor number. Under the Rayleigh approximation, the Taylor and Rayleigh numbers are analogous to the buoyancy parameter and Rossby number. This approximations holds not far from the critical curve for the onset of convection motions in a plain layer of rotating fluid.

For developed turbulent convective motions, or for horizontal temperature gradients, this velocity scale does not work well. 2D convective flows in the frame of this approximation can be not far from the critical curve of the onset of the convection for very large Taylor numbers ($Ta > 10^7$). In this case a regular laminar vortex grid with vertical vortices is the main convective regime [1-3].

B Local approximation

Large height convection in a plane infinite layer becomes turbulent, but if we consider a case where the source of density is local (for example a heated disk with diameter D in plane layer [4]) then horizontal density (or temperature) differences will create entrainment flows from the side. If the buoyancy flux $B = (\frac{\delta\rho}{\rho})gQ$ and Q is a flow rate of dense fluid per unit area the velocity U near the local region depends only on the buoyancy flux and a length scale and we can construct the velocity scale

$$U = (BD)^{1/3} .$$

The nondimensional buoyancy parameter and Rossby number will transform to

$$B^* = (\tfrac{\delta\rho}{\rho})gD(BD)^{-2/3} = (\tfrac{\delta\rho}{\rho})B^{-2/3}D^{1/3}$$

$$Ro = B^{1/3}D^{-2/3}\Omega^{-1} .$$

Note that for both the above approximations the buoyancy parameter B^* does not depend on the rotation rate Ω and both are valid for rotating and for nonrotating fluids. Rotation will transform the structure, but will not drastically change the velocity scales. As a consequence, the Rossby number will be proportional to Ω^{-1}.

C Geostrophic sloping approximation

When $Ro << 1$ the so-called geostrophic approximation is used: the Coriolis force is balanced by buoyancy, where the buoyancy force under the Boussinesq approximation is the part of the pressure not balanced by the hydrostatic component in the rest state. Then $B^* = Ro << 1$ and

$$\frac{U}{2\Omega L} = \frac{U^2}{Lg(\delta\rho/\rho)} \text{ or } U = \frac{g(\delta\rho/\rho)}{2\Omega} .$$

Under the geostrophic approximation, the velocity in a fluid with a horizontal temperature gradient is also a function of the aspect ratio $\delta_1 = \frac{L}{H}$, where H is a height of the fluid layer and L is the horizontal distance for which the horizontal temperature difference is prescribed. For this case a velocity scale, called thermal wind velocity is

$$U_{tw} = \frac{g(\delta\rho/\rho)}{2\Omega}\frac{L}{H}$$

and in this case

$$Ro = B^* = \frac{gL(\delta\rho/\rho)}{4\Omega^2 H^2} = Ro_T$$

where Ro_T is called the thermal Rossby number because the thermal wind enters its definition as a velocity scale. Under this approximation the Rossby number $Ro = B^*$, and both are proportional to Ω^{-2}. The Taylor number Ta - is often used as a second nondimensional parameter.

In the geostrophic approximation 2D flows are the basic motions. The geostrophic approximation describes the experiments in differentially-heated rotating annuli [7,8], in which two coaxial cylinders with a different temperatures are rotating around the axis of symmetry. The basic flow for this experiment is a convective cell with the up flow near the warm outer cylinder and down flow near a cold inner cylinder. Due to this cell a stable temperature gradient in the vertical direction is established, and the isopycnals are at an angle to the horizontal. Sometimes this type of convection is termed "sloping convection". Vertical motions are very close to the cylinders walls and in the fluid interior the flow is quasi-twodimensional. Depending on the external parameters Ro_T and Ta, there are four main regimes. The simplest is an axisymmetric zonal laminar flow (Hadley circulation, when fluid moves from one cylinder to the other in a spiral motion). For larger rotation rates this circulation became unstable and laminar regimes of regular or oscillating waves are established. For very large rotation rates a non-regular geostrophic turbulence regime existed.

Two-dimensional turbulent vortex interaction is the main characteristic of this regime when the mean vertical temperature gradient in the fluid is stable.

D Geostrophic turbulence approximation

When the vertical density gradient is unstable, then vertical mixing becomes important. For convection in a rotating plane fluid layer with a vertical temperature gradient a Rayleigh approximation is suitable only for the onset of motion. For the developed turbulent flow, the geostrophic condition can be used. In this case we assumed that the main velocity gradients are in the Ekman layers. Therefore, most of the dissipation ϵ takes place on the scale $l_E = (\frac{\nu}{2\Omega})^{1/2} = HTa^{-1/4}$ and is of order

$$\epsilon = \frac{\alpha g f}{\rho c_P} \approx \nu(\frac{U}{l_E})^2 ,$$

where f is a heat flux, and c_P is the specific heat capacity of the fluid. For $Pe = RePr >> 1$, $f = \rho c_P w T$ is primarily advective, w is a vertical component of velocity. For this approximation a velocity scale is

$$U = (\frac{\epsilon}{2\Omega})^{1/2} \text{ and } Ro = (\frac{\epsilon}{8\Omega^3 H^2})^{1/2} .$$

Note that for this approximation the Rossby number is proportional to $\Omega^{-3/2}$.

III EXPERIMENTAL APPARATUS AND PROCEDURE

A Geometry

Let us consider the usual experimental set up for studying convective motions from a local source of density in rotating fluids (Fig. 1).

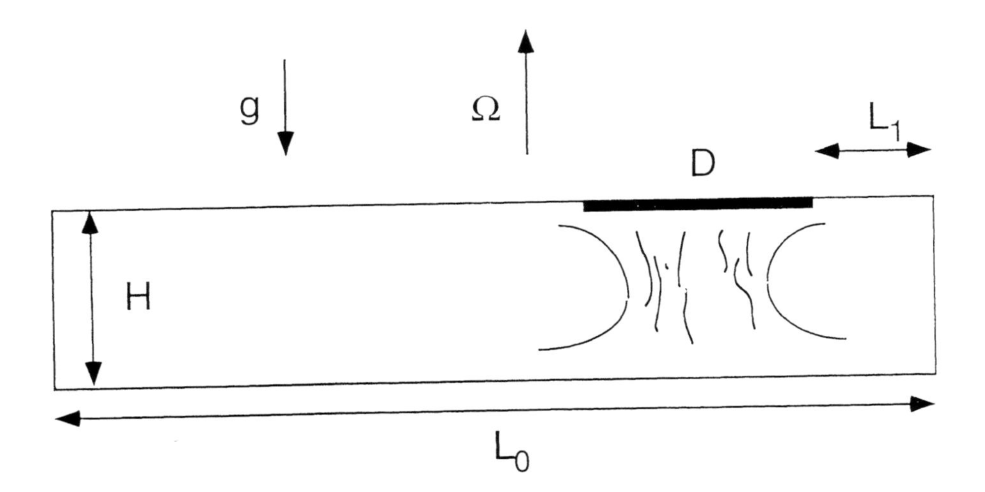

FIGURE 1.

A tank with horizontal scale L_0 is filled with fluid to a depth H. A disk with diameter D, is mounted at the upper surface at distance L_1 from the side boundary. The tank is rotated about its axis of symmetry with a constant angular velocity Ω, aligned in the same vertical direction as the gravitational acceleration g.

For this geometry we will have three aspect ratios which define the flow structure near the disk:

$$\delta_1 = \frac{D}{H}, \delta_2 = \frac{D}{L_0}, \delta_3 = \frac{H}{L_1}.$$

We will consider the case when the diameter of disk is rather large (usually of the order of 20 cm), and below this disk, the convection is turbulent [4]. For non-rotating fluids when $\delta_1 << 1$, $\delta_2 << 1$ and $\delta_1\delta_3 = \frac{H}{L_1} << 1$, then the flow below the disk at the beginning of convection depends mostly on the diameter D and a local approximation can be applied to describe the flow. If $\delta_1 >> 1$, $\delta_2 = 1$ (or $D = L_0$) and the height H is small, the Rayleigh approximation can be applied. In a rotating fluid, on the other hand, the most appropriate approximation depends on the rotation rate or Rossby number.

B Density sources

Two types of buoyancy sources were used in our experiments: "thermal" and "saline". In the "thermal" experiments cooled or heated disks were used, while in the "saline" method a source of salt water was added to ambient water with a lesser density. The configuration in Fig.1 is for a cooled disk or a source of salt water. When the heated disk was used, it was put on the bottom of the tank. Usually, the temperature of the disk T_0 was fixed and remained constant during the experiments. The temperature difference $\Delta T = T - T_0$, which determined the density as $\rho = \rho_0(1 + \alpha(T - T_0))$, depended on the temperature distribution T. The second fixed temperature T_1 is the temperature of the ambient fluid (for a local source in a large tank $\delta_1 << 1$, $\delta_2 << 1$), the temperature at the bottom (for a plane layer with a temperature difference between the horizontal boundaries, $\delta_1 << 1$, $\delta_2 >> 1$, $\delta_3 >> 1$), or the temperature difference between the side walls (for geostrophic sloping convection). Sometimes cooling due to evaporation at the upper surface wass used as a buoyancy source and in this case the heat flux f supplied was fixed and constant. A constant flux was also in some experiments by heating at the bottom. As the heat flux or Nusselt number were not measured in all experiments, a comparison between the experiments driven by a constant temperature difference and a constant heat flux is difficult.

In "saline" source experiments the behaviour is determined by the density difference between the fluid in the tank (fresh or salted water) and the salty water introduced at the surface (or over part of the surface), and the flow rate Q from the source. The main external parameter is the buoyancy flow B per unit area $B = (\frac{\delta\rho}{\rho})gQ$, which corresponds to the buoyancy flux per unit area $g\alpha f/\rho_0 c_P$ (c_P the spcific heat) in "thermal" experiments. For "saline" source experiments it is possible to achive a much greater flux B compared with that due to heat. This is because the value of α is small and the range of ΔT is limited (order $50K$). In air it is possible to generate a very large temperature difference ($\Delta T > 500K$) and therefore, the heat flux f can be relatively large.

C Time development

There are many time scales in these experiments [5-6,9]. One scale is the rotation period $\tau_1=\Omega^{-1}$ which is usually constant during experiments (spin-up and spin-down cases are more complicated). The second scale τ_2 is the time between the start of convection and a steady flow to the source. A third scale τ_3 is the time the convection flow takes to reach the bottom of the tank. A fourth scale τ_4 is the time taken for the flow to propagate to the side boundaries. A fifth scale τ_5 is the time required to establish quasi-stationary flow in the tank.

It was impossible to reach stationary conditions in "saline" source experiments, but the "thermal" source experiments often reached stationary conditions. After some time τ_6 a stable stratification will be established in the tank.

IV FLOW REGIMES

First let us consider convection from a local source in a very large tank without rotation. For this case $H >> D$ and $L_0 >> D$, or $\delta_1 << 1$, $\delta_2 << 1$ and the value of δ_3 is not important. A picture of convective motions for this case is presented in Fig. 2.

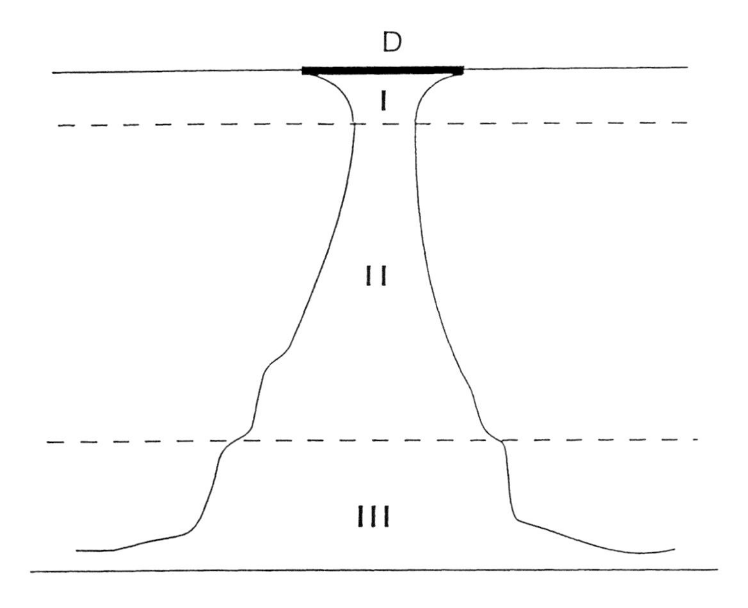

FIGURE 2.

Below the disk there are three regions of flow. In region I ($\delta_z = \frac{D}{z} >> 1$), the horizontal component of velocity is large (due to entrainment from the side), and can be defined as velocity scale from the local approximation $U = (BD)^{1/3}$ [8]. The diameter R of the turbulent plume in region I decreases with height to a minimum value $R \approx 0.4D$. In region II, δ_z is of O(1) and the vertical component of velocity is much larger than the horizontal velocity and $U = C_{II}z^{1/3}$ [8]. At very large distances from the disk $\delta_z << 1$. This is region III, where the flow from the disk is similar to the flow from a point source, with the velocity dependence $U = C_{III}z^{-1/3}$.

Rotation of the syste, will change the flows in region I (and consequently in regions II and III) , because the Coriolis force causes the flow to be horizontal. For $Ro << 1$ the Proudman-Taylor theorem suggests that horizontal turbulent entrainment flows can not exist, and convective motions below the disk are similar to the motions in a plane layer of fluid. A system of a vertical vorticies are arranged as 2D turbulent motions.

For $Ro \approx 1$ the Coriolis force changes the direction of the entrainment in-flow and creates a cyclonic circulation below the disk. Depending on the buoyancy flux B (or the heat flux f) a single vortex with horizontal scale R can originate below (in the case of heating [10]) or above the disk (cooling). For a large flux, there are three types of single vortex (Fig. 3) depending on the value of R/H. When $R/H << 1$ there is a quasi-laminar cylindrical vortex, because of the weak influence of the upper (in the case of heating) boundary. When $R/H \approx 1$ the upper boundary changes the local flow and a conical vortex forms. For $R/H >> 1$ there exists a turbulent toroidal vortex.

For a small buoyancy flux there is not enough energy to create an intensive single vortex. In this case there are three main regimes of flow in the space between disk and boundaries: a symmetric circulation (low rotation rate), baroclinic vortices with a size comparable to the gap between the disk and the side wall, and small unsteady vortices. The last two regimes are examples of 2D turbulent motions with complicated vortex interactions.

V SOME GEOPHYSICAL APPLICATIONS

Local heating or cooling are the main sources of motions in geophysics. In the atmosphere, local heating is the cause of all intensive vortices, and the structure of these vortices is similar to its experimental analog (Fig. 3): when the scale of a vortex is much less than the height of the atmosphere, they have a cylindricaldust devil form with a horizontal scale of the order of a meter. When $R/H \approx 1$ a conical vortex (tornado) with similar horizontal and vertical scales forms. In the case of a hurricane the horizontal scale is much greater that vertical scale, and it is a toroidal vortex. In all cases rotation (local or global) is very important.

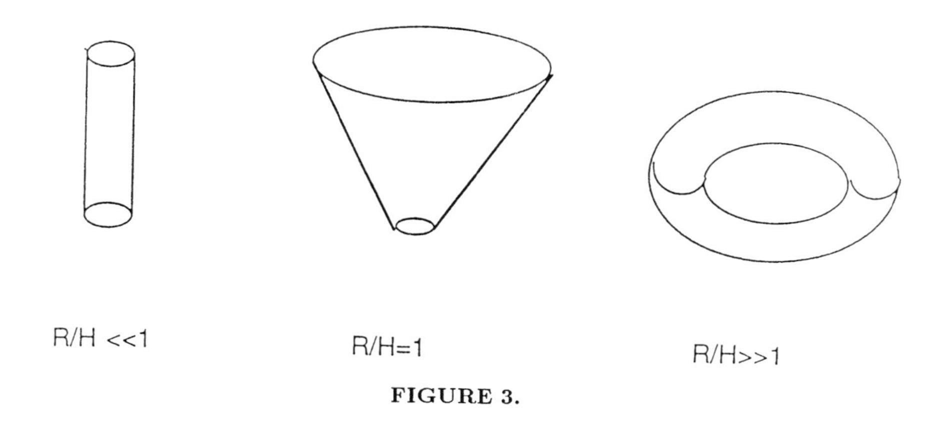

FIGURE 3.

For large scale circulations in the atmosphere convective cells from the equator to the poles are subjected to the Coriolis force, and a geostrophic approximation can be successfully applied. In the ocean, there are two main types of local convection: deep convection from cooling surfaces, and heating at the bottom due to geothermal waters. For deep convection, the horizontal scale D is of order 50 km, $D/H >> 1$ and the Earth rotation is important. In some regions of the ocean, quasi-2D turbulent flows can exist. For thermal water, the scale of heating is about 100-400 meters and $D/H << 1$ so a local approximation can be used.

ACKNOWLEDGEMENTS

This work was supported in part by the National Center for Theoretical Physics (Australian National University) and grants INCO-COPERNICUS and INTAS. Thanks to Graham Hughes for critically reading the paper.

REFERENCES

1. Boubnov B.M. & Golitsyn G.S. 1995 Convection in Rotating Fluids. Kluwer, Dordrecht, Boston, London, 224p.
2. Boubnov B.M. & Golitsyn G.S. 1986 Experimental study of convective structures in rotating fluids. J. Fluid Mech. V.167, p.p.503-531.
3. Boubnov B.M. & Golitsyn G.S. 1990 Temperature and velocity field regimes of convective motions in a rotating plane fluid layer. J. Fluid Mech. V.219, p.p. 215-239.
4. Boubnov B.M. & Heijst van G.J.F. 1994 Experiments on convection from horizontal plate with and without rotation. Experiments in Fluid. (to be published).

5. Maxworthy T. & Narimosa S. 1991 Vortex generation by convection in a rotating fluid. Ocean Moddeling, V. 92, p.p.1037-1040.
6. Maxworthy T. 1992 Convection and shear turbulence with rotation. In: Rotating Fluids in Geophysical and Industrial Applications, (Hopfinger E.J., ed.), Springer-Verlag, Wien -New York, p.p.339-357.
7. Hide R. & Mason P.J. 1975 Sloping convection in a rotating liquid. Adv. Physics, V.24, No1, p.p.47-100.
8. Boubnov B.M.& Fernando H.J. 1997. Private communication.
9. Hignett P., Ibberson A. & Kilworth P. 1981 On rotating thermal convection driven by non-uniform heating from below. J. Fluid Mech., V. 109, p.p.161-187.
10. Bengtsson L. & Lighthill J. (Ed) 1982 Intensive Atmospheric Vortices. Springer-Verlag, New York, 368p.

Two-dimensional magnetohydrodynamics and turbulent coronal heating

P. Dmitruk [1], D. Gómez [1,2]

[1] *Departamento de Física, Universidad de Buenos Aires, Pabellon I, Ciudad Universitaria, Buenos Aires (1428), Argentina*
[2] *Instituto de Astronomía y Física del Espacio, CC 67 Suc 28, Buenos Aires (1428), Argentina*

Abstract. We consider the heating of solar coronal active regions within a turbulent scenario. A direct numerical simulation of the equations governing the dynamics of a coronal magnetic loop is performed, assuming that the essential features can be described by an externally driven two-dimensional magnetohydrodynamic system. A stationary and large-scale magnetic forcing was imposed, to model the photospheric motions at the magnetic loop footpoints. A turbulent stationary regime is reached with an energy dissipation rate consistent with the heating requirements of coronal loops.

The energy dissipation rate time series shows an intermittent behavior, in the form of impulsive events, superimposed on the stationary component. We associate the impulsive events of magnetic energy dissipation with the so-called nanoflares. A statistical analysis of these events yields a power law distribution as a function of their energies with a slope consistent with those obtained for flare energy distributions reported from X-ray observations.

We also show the development of small scales in the spatial distribution of electric currents.

I INTRODUCTION

Soft X-ray observations of the solar corona, reveal a highly structured brightness distribution, which is the consequence of magnetic fields confining the X-ray emitting plasma and governing its dynamics. Observations made with high spatial resolution [1] show coronal magnetic loops with a highly filamentary internal structure.

The source of energy for coronal heating lies in photospheric convective motions, the question is how the energy is transferred and which is the dissipation mechanism. The natural candidate is Joule dissipation, but the typical timescale to dissipate coronal magnetic stresses on the length scale of the

CP414, *Two-Dimensional Turbulence in Plasmas and Fluids:* Research Workshop
edited by R. L. Dewar and R. W. Griffiths

driving photospheric motions is exceedingly long. This timescale can be estimated as $l^2/\eta \sim 10^6$ years ($l = 10^3$ km : length scale of phostospheric motions, $\eta = 10^3$ cm^2/s : resistivity). Therefore, most of the current theories of coronal heating deal with different mechanisms to speed up Joule dissipation [2–4]. One of the promising scenarios is the assumption that the magnetic and velocity fields of the coronal plasma are in a turbulent state [5,6]. In a turbulent regime, energy is transferred from photospheric motions to the magnetic field, this energy cascades toward small scales due to nonlinear interactions, highly structured electric currents are formed, and the subsequent Joule dissipation of these structures provide for the heating of the confined plasma.

II DESCRIPTION OF THE EQUATIONS

We consider a simplified model of a coronal magnetic loop with length L and cross section $2\pi l \times 2\pi l$, where l is the lengthscale of typical photospheric motions. We discard toroidal effects because $2\pi l << L$. The main magnetic field $\mathbf{B}_0$ is assumed to be uniform and parallel to the axis of the loop (the z axis). The planes at $z = 0$ and $z = L$ correspond to the loop footpoints at the photosphere. Under these assumptions, we are able to use the reduced MHD approximation [7], in which the plasma moves incompressibly in planes perpendicular to the axial field $\mathbf{B}_0$, and the transverse component of the magnetic field is small compared to $\mathbf{B}_0$. The very high electric conductivity allows photospheric motions to easily drive magnetic stresses in the corona (frozen field) [8], the field lines twist and bend due to these motions and this generates transverse components of velocity $\mathbf{u}$ and magnetic field $\mathbf{b}$. Therefore,

$$\mathbf{B} = B_0\mathbf{z} + \mathbf{b}(x, y, z, t) \ , \quad \mathbf{b} \cdot \mathbf{z} = 0 \tag{1}$$

$$\mathbf{u} = \mathbf{u}(x, y, z, t), \quad \mathbf{u} \cdot \mathbf{z} = 0 \ . \tag{2}$$

Since both $\mathbf{b}$ and $\mathbf{u}$ are two-dimensional and divergence-free fields, they can be represented by scalar potentials:

$$\mathbf{b} = \nabla \times (a\mathbf{z}) = \nabla a(x, y, z, t) \times \mathbf{z} \tag{3}$$

$$\mathbf{u} = \nabla \times (\psi\mathbf{z}) = \nabla\psi(x, y, z, t) \times \mathbf{z} \tag{4}$$

where ∇ indicates derivatives in the x, y plane.

The reduced MHD equations for the potentials ψ and a are [7]:

$$\partial_t a = v_A \partial_z \psi + [\psi, a] + \eta\nabla^2 a \tag{5}$$

$$\partial_t w = v_A \partial_z j + [\psi, w] - [a, j] + \nu\nabla^2 w \tag{6}$$

where vorticity w and current density j relate to the potentials through:

$$w = -\nabla^2 \psi \ , \quad j = -\nabla^2 a \ . \tag{7}$$

The brackets $[A, B] = \partial_x A \partial_y B - \partial_y A \partial_x B$ are the standard Poisson brackets and $v_A = B_0/\sqrt{4\pi\rho}$ is the Alfvén velocity (ρ is the plasma density).

Equation (5) describes the advection of the potential a and equation (6) corresponds to the evolution of vorticity w. The terms $v_A \partial_z$ represent the coupling between neighboring z =constant planes. The role of these terms is to transfer energy from the footpoints into the coronal part of the loop. The ∇^2 terms represent dissipative effects, the constants η and ν being the resistivity and viscosity coefficients. The nonlinear terms are represented by the Poisson brackets. Their role is to couple normal modes in such a way that energy, and other ideal invariants, can be redistributed among them. The fields $\mathbf{u}$ and $\mathbf{b}$ depend on the axial z-coordinate, though their nonlinear interaction proceeds independently on different constant z planes across the loop.

Since the kinetic (R) and magnetic (S) macroscale Reynolds numbers in coronal active regions are extremely large ($R \sim S \sim 10^{10-12}$) we expect the system to be in a strongly turbulent regime.

To perform extended time simulations we focus on the dynamics of a given transverse plane, that is, we study the evolution of a generic two-dimensional slice of a loop to which end we model the $v_A \partial_z$ terms as external forces (see [9] [10] for similar approaches). We assume the vector potential to be independent of z and the stream function to interpolate linearly between $\Psi(z = 0)$ and $\Psi(z = L) = \Psi(x, y, t)$ where $\Psi(x, y, t)$ is the stream function for the photospheric velocity field. These simplifying assumptions yield $v_A \partial_z \psi = v_A \Psi / L$ in equation (5) and $v_A \partial_z j = 0$ in equation (6). Then, the 2D equations for a generic transverse slice of a loop are:

$$\partial_t a = f + [\psi, a] + \eta \nabla^2 a \ , \quad f = (v_A/L)\Psi \tag{8}$$

$$\partial_t w = [\psi, w] - [a, j] + \nu \nabla^2 w \tag{9}$$

We performed numerical simulations of equations (8) and (9) on a square box of sides 2π, with periodic boundary conditions. The vector potential and the stream function are expanded in Fourier series. The forcing term f is constant in time and non zero only in a narrow band in k-space, $3 \leq k \leq 4$. We worked with a resolution of 192×192, the code is of the pseudo-spectral type, with 2/3 de-aliasing [11]. The temporal integration scheme is a fifth order predictor-corrector, to achieve almost exact energy balance over our extended time simulations.

III RESULTS

In spite of the narrow forcing and even though velocity and magnetic fields are initially zero, nonlinear terms quickly populate all the modes across the spectrum and a turbulent state develops. Fig. 1 shows the energy dissipation rate as a function of time for a long simulation. After about 50 photospheric turnover times (which is of the order of 10^3 s) a statistically steady state is reached. The behaviour is strongly intermittent, despite the fact that the forcing is constant and coherent. This intermittency is an ubiquitous characteristic of turbulent systems.

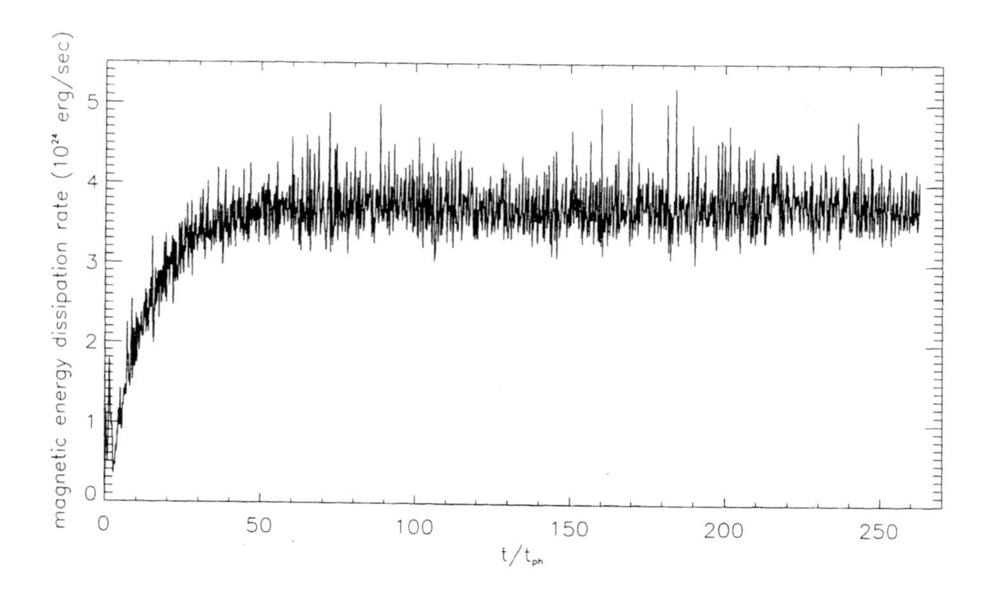

FIGURE 1. Simulation of the energy dissipation rate in a coronal loop driven by photospheric turbulence, vs. time measured in units $t_{\text{ph}} \equiv l/u$.

The energy flux is the dissipation rate ϵ by unit area $F = \epsilon/(2(2\pi l)^2)$. The quantitative value of the energy flux obtained with the simulation as well as the explicit dependence with the relevant parameters of the problem is:

$$F = 4 \times 10^6 \frac{\text{erg}}{\text{cm}^2\ \text{s}} \left(\frac{n}{5 \times 10^9\ \text{cm}^{-3}}\right)^{\frac{1}{4}} \left(\frac{B_0}{100\ \text{G}}\right)^{\frac{3}{2}} \left(\frac{u_{\text{ph}}}{10^5\ \text{cm s}^{-1}}\right)^{\frac{3}{2}} \frac{\left(\frac{l}{10^8\ \text{cm}}\right)^{\frac{1}{2}}}{\left(\frac{L}{5\times 10^9\ \text{cm}}\right)^{\frac{1}{2}}} \tag{10}$$

where B_0, l and L have been defined before, n is the number density of the coronal plasma and u_{ph} is a typical photospheric velocity.

This energy flux compares quite well with the heating of active regions reported from observations [12] which span the range $3 \times 10^5 - 10^7$ erg/cm^{2}s

As can be seen in the plot of dissipation rate vs time, there are peaks or impulsive events of energy dissipation. We have performed a statistical analysis of these peaks. This technique has been applied in several observational analysis related to solar flares [14,15], which are spectacular dissipation events involving typical energy releases of 10^{30-32} erg in very short timescales (10^{2-3}s). As hypothesized by Parker [13] when dealing with coronal heating, the dissipation events, called nanoflares, involve smaller energy releases, of the order of 10^{24-26} erg. This is still a theoretical prediction, because there is not enough resolution in present day detectors to observe these smaller events. Here, in our simulation of a coronal loop, we associate the peaks of energy dissipation with the nanoflares predicted by Parker. We computed the energy released in each event and then estimated the occurrence rate for the nanoflares, that is, the number of events per unit energy and time. This distribution is a power law with a negative slope of 1.5 ± 0.2 over an energy range from 10^{24} to 10^{26} erg. This value is remarkably similar to the slopes obtained from the analysis of data from solar flares such as those by Shimizu [14] (from Yohkoh soft X-ray data he obtained a slope between $1.5 - 1.6$) and those by Crosby et al. [15] (from SMM hard X-ray data they obtained a slope of 1.53). The correspondence between these numbers is indicative of the presence of a common physical process behind the dissipation events, ranging from 10^{24} erg to 10^{33} erg for the largest flares. This is another example of a system in a state of self-organized criticality, a concept introduced by Bak [16] in his study of sandpile avalanches with cellular automata simulations. In fact, there are cellular automata simulations of flares performed by Lu [17] in which the same power law behaviour for the occurrence rate was obtained.

We have mentioned before that the Joule dissipation of the system is enhanced by the presence of small-scale electric currents. These structures are a manifestation of spatial intermittency in the system. Fig. 2 shows some of these structures. The arrows in this figure correspond to the magnetic field overlayed on a halftone of the current density. The zones of intense dissipation are those indicated in black or white. There are also big structures, which are formed by the inverse cascade of squared vector potential ($\int a^2 \, d^2x$) in MHD 2D systems. Between these magnetic islands there are small structures which can be interpreted as current sheets [18] with a thickness of the order of the dissipative scale (the grid resolution) and arbitrary width.

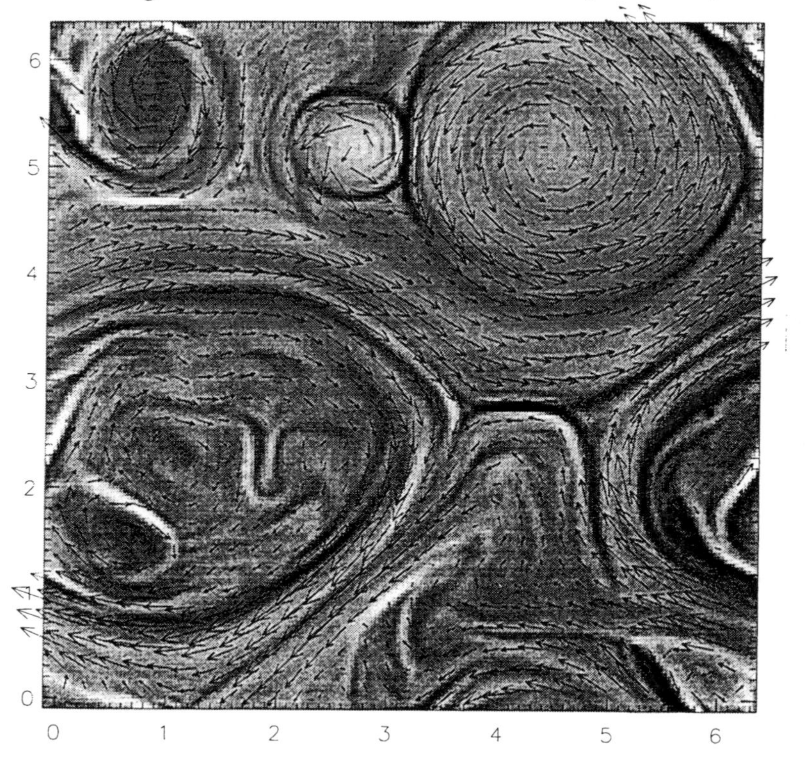

FIGURE 2. Transverse magnetic field and parallel current density in a coronal loop, showing small-scall structures.

IV CONCLUSIONS

Our simulation of a slice of coronal loop show that a stationary turbulent regime is reached even though the external forcing is constant and narrow-band in wavenumber. The mean value of energy dissipation rate obtained for typical footpoint velocities is consistent with the heating rate of active regions reported from observations. Furthermore, the system displays a highly intermittent behaviour, both in time and space. By performing a statistical analysis of the energy dissipation rate time series we computed the occurrence rate of energy dissipation events and obtained a power law distribution with a slope of 1.5 which is fully consistent with the slopes obtained for flare distributions from observations [10].

In summary, we believe that the assumption of a turbulent scenario provides a better understanding of the heating and dynamics of coronal active regions.

REFERENCES

1. Golub, L., Herant, M., Kalata, K., Louvas, I., Nystrom, G., Pardo, F., Spiller, E. & Wilczynski, J.S. 1990, Nature, **344**, 842
2. Parker, E.N. 1983, ApJ, **264**, 642
3. van Ballegooijen, A.A. 1986, ApJ , **311**, 1001
4. Mikic, Z., Schnack, D.D. & van Hoven, G. 1989, ApJ, **338**, 1148
5. Heyvaerts, J., & Priest, E.R. 1992, ApJ, **390**, 297
6. Gómez, D.O., & Ferro Fontán, C. 1992, ApJ, **394**, 662
7. Strauss, H. 1976, Phys. Fluids, **19**, 134
8. Parker, E.N. 1972, ApJ, **174**, 499
9. Einaudi, G., Velli, M., Politano, H., & Pouquet, A. 1996, ApJ, **457**, L113
10. Dmitruk, P., & Gómez, D.O. 1997, ApJLett, in press
11. Canuto, C., Hussaini, M.Y., Quarteroni, A., & Zang, T.A. 1988, Spectral Methods in Fluid Dynamics (Springer, New York).
12. Withbroe, G.L., & Noyes, R.W. 1977, ARAA, **15**, 363
13. Parker, E.N. 1988, ApJ, **330**, 474
14. Shimizu, T. 1995, PASJ, **47**, 251
15. Crosby, N.B., Aschwanden, M.J., and Dennis, B.R. 1992, Solar Phys., **143**, 275
16. Bak, P., Tang, C. & Wiesenfeld, K. 1988, Phys. Rev. A, **38**, 364
17. Lu, E.T., Hamilton, R.J. 1991, ApJ, **380**, L89
18. Biskamp, D. & Welter, H. 1989, Phys. Fluids, B1, 1964

Turbulent transport of passive tracers and the onset of diffusivity

D. Gurarie[1], V. Klyatskin[2,3]

[1]*Dep. of Mathematics, CWRU, Cleveland, Ohio 44106*
[2]*Institute of Atmospheric Physics Russian Academy of Sciences, 109017, Moscow, Russia*
[3]*Pacific Oceanological Institute, Russian Academy of Sciences, Vladivostok, Russia*

Abstract. The paper reviews some basic results of turbulent transport of passive tracers by a random Gaussian velocity field with short temporal correlation. Our emphasis is on the early "transport" phase of the process and the onset of diffusivity. We study the tracer mean-field and fluctuations via the reduced Fokker-Planck equation. The early (transport) phase exhibits steepening of gradients, fractalization of isocontours and exponential growth of statistical moments (fluctuations). We estimate the effective parameters of the process and give several applications, including randomly perturbed shear flows and the "sedimentation problem".

I INTRODUCTION

Passive tracers are abundant in Nature (temperature, salinity, pollutants) and used extensively in physical experiments and measurement (dyes and tracers in flow visualization, buoys and air balloons in air flows and ocean currents, etc.). When transported by a velocity field $\vec{u}(\boldsymbol{r},t)$ their evolution could be viewed from two perspectives:

- as Lagrangian particles advected by $\vec{u}$ together with small scale Brownian (molecular, subgrid) noise of strength ν, described by a stochastic differential system

$$\frac{d\boldsymbol{r}}{dt} = \vec{u}(\boldsymbol{r},t) + \nu\frac{dB}{dt} \qquad (1)$$
$$\boldsymbol{r}(t_0) = \boldsymbol{r}_0$$

CP414, *Two-Dimensional Turbulence in Plasmas and Fluids:* Research Workshop
edited by R. L. Dewar and R. W. Griffiths

- as Eulerian densities $q(\boldsymbol{r}, t)$ evolving according to a partial differential equation

$$\begin{aligned} &\partial_t q + \vec{u} \cdot \partial_r q = \nu \nabla^2 q \\ &q|_{t=t_0} = q_0 \ . \end{aligned} \tag{2}$$

The driving velocities could be deterministic (mean flows) or could involve significant random (turbulent) component, that makes equation (2) stochastic.

The subject of turbulent transport is to study the combined effect of the advective and (molecular) diffusive fluxes on the tracer evolution and compute the "effective rates". Two components of the total density flux in (2) have opposing tendencies: the advection component $\vec{u} \cdot \partial_r$ tends to stretch q-isocontours and steepen its gradients (fig. 1), while the dissipative term $\nu \nabla^2 q$ tends to smooth out q. The molecular diffusion rates (and temporal scales) however, are often several orders of magnitude smaller than the advective fluxes. For instance, regular salt at room temperature has $\nu \simeq 10^{-5}\mathrm{m}^2/\mathrm{sec}$, whereas a small amount of mixing (like a tea cup) could bring it up to $10^{-1}\mathrm{m}^2/\mathrm{sec}$. Low molecular diffusivities of temperature and salinity (along with Coriolis force) are largely responsible for the stable density stratification in the ocean.

Small value of ν allows one to neglect it, take $\nu = 0$, at early stages of the turbulent mixing, and concentrate on the effective role of $\vec{u}$. In this paper we shall briefly review some basic results and principles of turbulent transport. Then we shall outline an approach to the early stages for the "short-time correlated" or "δ-correlated" velocities.

We use some standard statistical tools (average, moments, p.d.f. etc.) to describe the process. Our approach, however does not involve arbitrary "moment closure" assumptions that are often used in "stationary" or "decaying" models of turbulence and transport (cf. [4,17,18,21] etc.). We adopt instead the Furutsu-Novikov formalism — a combination of statistical and variational techniques for "functionals" of random Gaussian fields, that lends itself to the perturbation analysis.

The goal is to derive the evolution of the mean-field $\langle q \rangle$ as well its random fluctuations via "Fokker-Planck equation". The reduced FP-equation of section V A) acts in the $(\boldsymbol{r}, q, \boldsymbol{p})$ phase-space, that encodes the joint statistics of q and its gradient $\boldsymbol{p}= \nabla q$. Its analysis exhibits some essential features of the early turbulent transport, like explosive growth of gradients and fractalization of isocontours.

We develop the general theory to determine the corresponding effective equations, rates and (Péclet-type) parameters, outline a few perturbation schemes and give applications to random shear flows and randomly advected sedimentation.

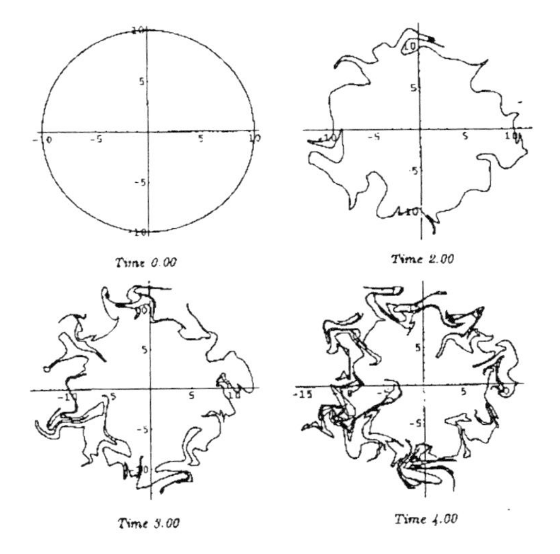

FIGURE 1. Numeric simulations of time snapshots of isocontour $q(r,t) = \text{const}$ advected by random Gaussian velocity field $\vec{u}$ (courtesy of Dr. Y. Hu, at CWRU).

II LAGRANGIAN AND EULERIAN DIFFUSIVITY

A Taylor and Richardson laws

We start by observing that in the absence of velocity drift $\vec{u} = 0$ Lagrangian particle (1) undergoes Brownian motion with the mean and variance

$$\langle \boldsymbol{r}(t) \rangle = 0; \left\langle |\boldsymbol{r}(t)|^2 \right\rangle = 2\nu t$$

increasing linearly in t. The coefficient of the mean-square-distance measures the molecular diffusion rate.

Taylor (1921) proposed that random velocity should produce similar diffusive behavior with the effective coefficient (tensor) D^*_{ij}

$$\langle [\boldsymbol{r}_i(t) - \boldsymbol{r}_i(0)] [\boldsymbol{r}_j(t) - \boldsymbol{r}_j(0)] \rangle = 2D^*_{ij} t \; .$$

Furthermore he gave an expression of D^* as cumulative velocity correlations along the Lagrangian path[1] (1)

$$D^*_{ij} = \int_0^\infty \langle u_i(\boldsymbol{r}(t), t)\, u_j(\boldsymbol{r}(0), 0) \rangle \, dt \; . \tag{3}$$

1) This form of D^* is not very practical due to presence of undetermined Lagrangian path.

So the corresponding Eulerian tracer density, properly averaged, should asymptotically obey the evolution

$$\begin{aligned} &\langle q\rangle_t + U\cdot\partial_r\langle q\rangle = \textstyle\sum_{ij}\partial_i D^*_{ij}\partial_j\langle q\rangle \\ &\langle q\rangle|_{t=t_0} = q_0 \end{aligned} \tag{4}$$

with the mean drift velocity $U = \langle\vec{u}\rangle$.

Richardson (1926) analyzed experimental data on the atmospheric diffusion over the large range of scales $\{\lambda\}$ and proposed a scale-dependent diffusion law

$$D^*_\lambda \propto \lambda^{4/3}\,. \tag{5}$$

Another way to interpret (5) is via mean-square separation of two Lagrangian path

$$Z(t) = \left\langle |\boldsymbol{r}_1(t) - \boldsymbol{r}_2(t)|^2\right\rangle$$

that obeys a DE

$$\frac{d}{dt}Z \propto Z^{2/3} \Rightarrow Z(t) \propto t^3\,.$$

Later Batchelor (1952) et al have derived the Richardson law from the Kolmogorov-Obukhov turbulent velocity spectrum $u_\lambda \approx \lambda^{1/3}$.

One could interpret (5) in terms of the anomalous "super-diffusive" behavior of the system. Lately anomalous scaling laws and exponents drew considerable attention, both in relation to intermittency of the turbulent hydrodynamic flows and the tracer transport problems ([8,5,7,12,6,26,27,29]). We won't go into the detailed discussion of the anomalous diffusion, but make a few brief comments.

Anomalous diffusion laws could be described in terms of the so called Levy α-stable processes (in place of the standard Brownian motion), generated by fractional powers of the Laplacian $(-\Delta)^{\alpha/2}$. So the differential evolution equation (2) is replaced by a "non-local" integro-differential (convolution) one

$$\partial_t\langle q\rangle = D^*(-\Delta)^{\alpha/2}[\langle q\rangle]\,. \tag{6}$$

The Richardson law (5) corresponds to a "$\frac{1}{3}$-fractional evolution"

$$\partial_t\langle q\rangle = (-\Delta)^{1/3}[\langle q\rangle]\,. \tag{7}$$

Zaslavsky (1992), Bouchard at al (1990) advocate more general form of anomalous diffusion with both space and time represented by fractional derivatives,

$$\partial_t^\beta\langle q\rangle = \partial_x^\alpha\langle q\rangle \tag{8}$$

- the fractional time-derivative ∂_t^β accounts for the relaxation effects (time-memory) in the system, as opposed to the Markovian dynamics (2). The basic mechanism underlying fractional diffusion laws has to do with long trapping periods of tracer particles inside vortex cores followed by rapid transitions (flights) through conducting layers (cf. [8,9]).

B Large-scale limit: homogenization and local averaging

One general result of diffusive behavior in random velocity fields goes under the name "homogenization". It claims that at large space-time scales transport dynamics becomes asymptotically diffusive. Precise results (McLaughlin et al 1985, see also [2]) are often stated for time-independent, stationary, zero-mean random field $\vec{u}$, whose velocity correlator $U_{ij}(\boldsymbol{r}) = \langle u_i(\boldsymbol{r}+y)\, u_j(y)\rangle$ decays sufficiently fast at ∞. Then taking ensemble-average solution $\langle q\rangle$ of (2) with initial input $q_0(\varepsilon \boldsymbol{r})$ and letting $\varepsilon \to 0$ we get function $Q(\boldsymbol{r}/\varepsilon, t/\varepsilon^2)$ that obeys the effective diffusion law

$$Q_t = \sum_{ij} \partial_i D^*_{ij} \partial_j Q \tag{9}$$

Furthermore, the effective diffusion tensor always exceeds the molecular rate[2] $D^*_{ij} \geq \nu$.

The mean-field $\langle q\rangle$ in (9) is taken over the entire statistical ensemble $\{\vec{u}\}$. An alternative way to produce the effective diffusion is via local space-time averaging. The latter could be relevant both in the physical setup and in numeric simulation.

One generic (sufficient) condition for the onset of effective diffusivity was found in [24] (see also [8,10]). It requires stream-potential ψ of $\vec{u}$ to be bounded: $|\psi(\boldsymbol{r},t)| \leq \psi_0$. One could show that (9) holds approximately after suitable "local averaging" $\ll q \gg$ over appropriate space-time *mixing scales* $l_{\mathrm{m}}; \tau_{\mathrm{m}}$ discussed below (section II D). Furthermore, the effective diffusion coefficient satisfies: $\nu < D^* < \nu + \frac{\psi_0^2}{\nu}$.

C Péclet number and effective diffusion rate

An important parameter that measures the relative strength of the advection term $q\vec{u}$ versus molecular $\nu \nabla q$ (hence their relative contributions to the q-flux) goes under the name *Péclet number* P. Depending on the context it can be defined either as

$$P = \frac{UL}{\nu} \tag{10}$$

-in terms of characteristic velocity and length scales U, L, or

$$P = \frac{\langle |\vec{u}|^2 \rangle \tau}{\nu} \tag{11}$$

---

[2] This fact was first observed by Zeldovich (1937) in pure deterministic context. He asked to maximize the mixing rate $\mu = [\text{flux}(q)]/[\text{grad}(q)]$, and showed that any nontrivial flow would bring about the increase of μ from its least possible value ν.

-in terms of ensemble-average velocity and characteristic time τ, or

$$P = \frac{\langle |\psi| \rangle}{\nu} \tag{12}$$

-in terms of stream-function ψ.

One of the basic problems in turbulent transport is to find the effective diffusion rate D^* as function of P and ν. Here we shall mention a few typical results, further details and references could be found in Isichenko (1992).

- Taylor (1953) followed by Aris (1956) studied diffusion for the classical Poisseuille flow $u_0 \left(1 - \frac{\boldsymbol{r}^2}{a^2}\right)$ in a pipe of radius a and derived the effective (horizontal) rate

$$D_x^* = \nu + K \frac{(a u_0)^2}{\nu} = \nu \left(1 + C\, P^2\right)$$

with some constant C.

- Avellaneda-Majda (1990) and Isichenko-Kalda (1991a) took random 2D shear flow of zero mean, $\langle \psi (y) \rangle = 0$ and produced the effective diffusion coefficient

$$D_x^* = \nu + \frac{\langle |\psi|^2 \rangle}{\nu} = \nu \left(1 + \, P^2\right)$$

- Moffat (1983), Solomon-Golub (1988) among others, studied the diffusive regime in a periodic array of vortex rolls $\psi = \psi_0 \sin mx \sin ky$ and Rayleigh-Bénard convection. These results produced a different law for the effective coefficient in terms of ν, P

$$D^* = C \sqrt{\nu \psi_0} = C\, \nu \sqrt{P}\,.$$

A variety of other examples and models were investigated theoretically, numerically and experimentally (see [8]). The expected form of D^* that transpired from these studies can be surmised as

$$D^* (\nu, P) \simeq \begin{cases} \nu \left(1 + a_1 P^2 + ...\right) & \text{for small } P \ll 1 \\ \nu P^{\alpha}\, ; (\alpha \leq 2) & \text{for large } P \gg 1\,. \end{cases} \tag{13a}$$

All results discussed so far are related to the fully developed (asymptotic) normal diffusive regime. To study the onset of diffusivity we need to examine the proper space-time scales.

D Mixing scales

The mixing process could be schematically viewed as the balance between two opposing trends. Pure transport ($\nu \approx 0$) in (2) leads to stretching and striation of isocontours q, while preserving iso-areas (incompressibility), hence a drive to smaller scales and steepening gradients (fig. 1) (cf. [6,8,9]). Steep gradients in turn create strong molecular diffusive fluxes $\nu \nabla q$, the latter "smooth" out small scales and flatten gradients.

The mixing scales l_{m} and τ_{m} express the correlation properties of such transport process. One expects the correlation-function of the flux density $J = q\vec{u} - \nu \nabla q$

$$J_{ij}(\boldsymbol{r}, t) = \langle J_i(\boldsymbol{r}, t) J_j(0, 0) \rangle$$

to fall off exponentially for $\boldsymbol{r} > l_{\mathrm{m}}; t > \tau_{\mathrm{m}}$, and have a power law at smaller scale. We shall give more precise definition following Isichenko at al (1991).

1 *Length scale* l_{m}

The density flux $J(q)$ in the Euler equation $q_t = \nabla \cdot J$ (2) tends to mix iso-levels $q(\boldsymbol{r}) = q_1$ and $q(\boldsymbol{r}) = q_2$. We pick a pair of isocontours of q separated at a distance a (fig. 2). Assuming the "horizontal" (contour-wise) scale of q far exceeding a we estimate the total advective-diffusive flux from q_2 to q_1 as

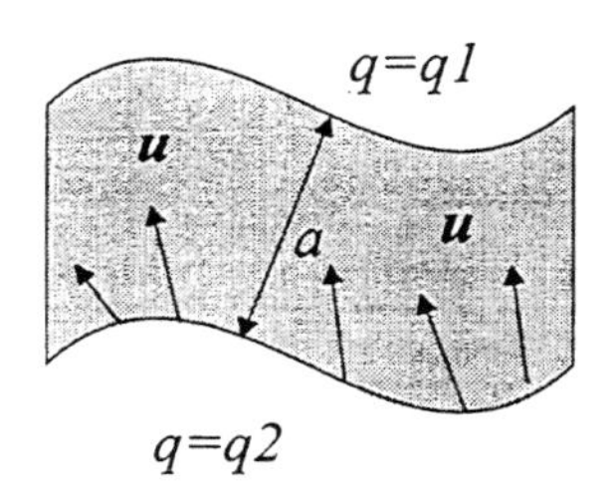

FIGURE 2. Isocontours of density, q, as discussed in the text.

$$J_{21} \approx [u]_a (q_1 - q_2) + \frac{\nu}{a}(q_1 - q_2) .$$

Here $[u]_a$ in the first term means the "horizontally" averaged normal component of $\vec{u}$ along the isocontour, while the second term approximates $\nu \nabla q$. Hence we get the "a-th approximate mixing rate" = [flux]/[grad]

$$D(a) \approx a [u]_a + \nu .$$

Function $D(a)$ is shown to increase with a, and could saturate at the value D^* for some l_{m}, which defines the *mixing length* of the process (fig. 3-top).

2 *Time scale* τ_{m}

This time we take the initial delta-density $q_0 = \delta(\boldsymbol{r} - \boldsymbol{r}_0)$, let it evolve according to (2) and look at the expansion rate of its second (inertia) moment

$$D(t) = \tfrac{1}{4t} \int (\boldsymbol{r} - \boldsymbol{r}_0)^2 \, q(\boldsymbol{r}, t) \, d\boldsymbol{r} \; .$$

Once again function $D(t)$ increases with time and could saturate at the level D^* at a certain time τ_{m} (fig. 3-bottom).

One could expect the two functions $D(a), D(t)$ to have the same limit

$$\lim_{a \to \infty} D(a) = \lim_{t \to \infty} D(t) = D^*$$

but there is no formal proof of such result.

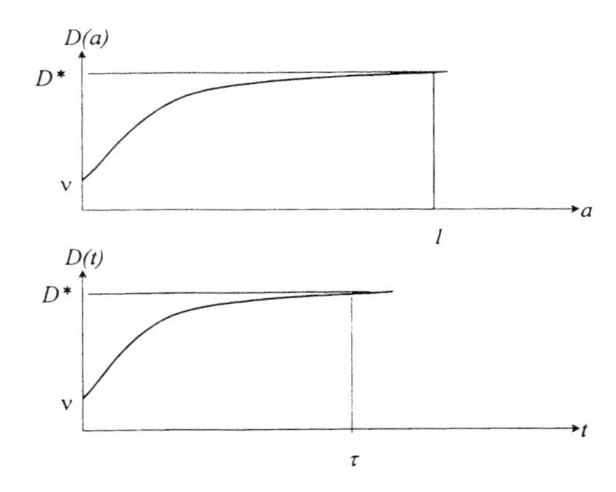

FIGURE 3. Saturation of effective "diffusion coefficients", D, as discussed in the text.

The mixing scales mark the onset of the turbulent diffusion. The diffusivity is expected to set at times scales $t \gg \tau_{\mathrm{m}}$, whereas $t \approx \tau_{\mathrm{m}}$ (or $\tau_{\mathrm{m}} = \infty$) should result in the anomalous law (7-8). In the following sections we shall focus our attention on the early (pre-diffusive) phase, where the transport process dominates the mixing.

III SHORT-CORRELATED FIELDS AND EFFECTIVE PARAMETERS

We take statistically stationary and homogeneous Gaussian velocity field $\vec{u}(\boldsymbol{r}, t)$, or more general $U + \vec{u}(\boldsymbol{r}, t)$ (mean-flow plus random fluctuation) and assume fluctuating component $\vec{u}$ to be δ-correlated or have short-range correlation τ_0, compared to the mixing time-scale: $\tau_0 \ll \tau_{\mathrm{m}}$.

The statistics of a Gaussian field $\vec{u}$ are completely determined by the auto-correlation function

$$b_{ij}\left(\boldsymbol{r}-\boldsymbol{r}', t-t'\right) = \left\langle u_i\left(\boldsymbol{r}, t\right) u_j\left(\boldsymbol{r}', t'\right)\right\rangle \ . \tag{14}$$

Our basic assumption on b_{ij} read

$$b_{ij}\left(\boldsymbol{r}, t\right) \simeq B_{ij}\left(\boldsymbol{r}\right)\left\{\begin{array}{c}\delta\left(t\right) \\ \frac{1}{\tau_0} e^{-t/\tau_0}\end{array}\right. .$$

The Fourier transform of the velocity correlations gives the energy spectral tensor

$$E_{ij}\left(\boldsymbol{k}, t\right) = \int e^{i\boldsymbol{k}\cdot\boldsymbol{r}} b_{ij}\left(\boldsymbol{r}, t\right) d\boldsymbol{r} \ .$$

In case of homogeneous and isotropic random incompressible $\vec{u}$ it depends on a single scalar spectral density $E\left(\boldsymbol{k}, t\right)$,

$$E_{ij}\left(\boldsymbol{k}, t\right) = E\left(\boldsymbol{k}, t\right)\left(1 - \frac{k_i k_j}{\boldsymbol{k}^2}\right) \tag{15}$$

We furthermore assume the time-integrated correlation tensor

$$B_{ij}\left(\boldsymbol{r}\right) = \int_0^\infty b_{ij}\left(\boldsymbol{r}, t\right) dt \tag{16}$$

to be regular at $\boldsymbol{r} = 0$ (fig. 4). So the energy spectrum $E\left(\boldsymbol{k}\right)$ should fall off sufficiently fast at ∞, or be truncated at some finite (dissipation) value k_d.

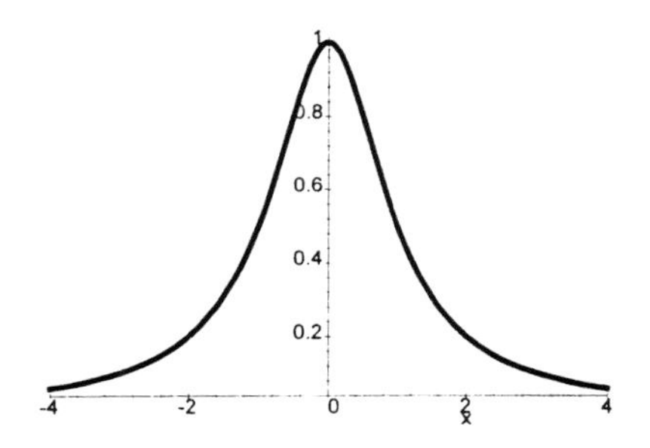

FIGURE 4. Typical shape of time-integrated correlation function $B\left(r\right)$.

Two parameters of the integrated correlation tensor $B_{ij}\left(\boldsymbol{r}\right)$ will play an important role here: its value at $\boldsymbol{r} = 0$, $B_{ij}\left(0\right)$, and its second derivatives tensor

$\partial^2_{ml} B_{ij}(0)$. For incompressible, homogeneous, isotropic field $\vec{u}$ in $\mathbb{R}^N$ $(N = 2, 3)$ two tensors are essentially determined by a pair of scalar coefficients: D_1 and D_2. Precisely,

$$B_{ij}(0) = \int \left(\delta_{ij} - \frac{k_i k_j}{\boldsymbol{k}^2}\right) E(\boldsymbol{k})\, d^N\boldsymbol{k} = D_1 \delta_{ij} \tag{17}$$

and

$$\begin{aligned} \partial^2_{ml} B_{ij}(0) &= \tfrac{1}{N(N+2)} D_2 \int \{(N+1)\,\delta_{ml}\delta_{ij} - \delta_{mi}\delta_{lj} - \delta_{mj}\delta_{il}\}\, \boldsymbol{k}^2 E(\boldsymbol{k})\, d^N\boldsymbol{k} \\ &= \tfrac{1}{N(N+2)} D_2 \left\{ \begin{array}{lll} N-1 & (i,i) = (m,m) & \\ N+1 & (i,i) \neq (m,m) & \\ -1 & (i,j) = (m,l) \text{ or } (l,m) & \end{array} \right\} \int \boldsymbol{k}^2 E(\boldsymbol{k})\, d^N\boldsymbol{k}\,. \end{aligned} \tag{18}$$

Both definitions involve the energy-spectrum $E(\boldsymbol{k})$. We shall show coefficient D_1 to give the effective diffusion rate for the tracer-density q, whereas D_2 will play a similar role for the q-gradient.

Let us also observe that coefficient D_1 is proportional to the time-integrated (over $[0;\tau_0]$) total energy of $\vec{u}$, while D_2 gives time-integrated total enstrophy.

IV FURUTSU-NOVIKOV FORMALISM

Our goal is to study solutions of the stochastic diffusive-transport equation

$$\begin{array}{l} q_t + \vec{u} \cdot \nabla q = \nu \Delta q \\ q|_{t_0} = q_0(\boldsymbol{r}) \end{array} \tag{19}$$

with random Gaussian $\vec{u}$ via statistical tools: mean-fields, correlation, moments, p.d.f. etc. In general q has complicated, nonlinear, implicit dependence on $\vec{u}$, accumulating in time. The idea of the Furutsu-Novikov method is to view q as a "functional" of random Gaussian field $\vec{u}$, to employ variational derivatives $\frac{\delta q}{\delta u_i}$ (their evolution and statistics) and use correlation splitting techniques for such functionals. We shall follow the exposition of Klyatskin (1980).

Let us start with the mean field $\langle q \rangle$. It can be shown to satisfy the integro-differential equation ([13,15])

$$\begin{aligned} (\partial_t + U \cdot \partial_r)\langle q \rangle + \int_0^t dt' \int d\boldsymbol{r}' \sum_{ij} b_{ij}(\boldsymbol{r} - \boldsymbol{r}', t - t')\, \partial_j \left\langle \frac{\delta q(\boldsymbol{r}, t)}{\delta u_i(\boldsymbol{r}', t')} \right\rangle \\ = \nu \nabla^2 \langle q \rangle\,. \end{aligned} \tag{20}$$

The latter is clearly, not closed, as the mean variational derivative $\frac{\delta q}{\delta u_i'}$ enters it (for the sake of notation "primed functions" refer to functions of primed variables, e.g. $u_i' = u_i(\boldsymbol{r}', t')$, etc.). The derivative in turn obeys a stochastic differential equation with the δ-source initial value

$$
\begin{aligned}
(\partial_t + \vec{u}\cdot\partial_r)\left(\tfrac{\delta q}{\delta u_i'}\right) &= \nu\Delta_r\left(\tfrac{\delta q}{\delta u_i'}\right) \\
\left.\tfrac{\delta q}{\delta u_i'}\right|_{t=t'} &= (\partial_i q')\,\delta(\boldsymbol{r}-\boldsymbol{r}')\ .
\end{aligned}
\tag{21}
$$

When we continue averaging procedure with (21), it would involve yet higher variational derivatives of q for $\left\langle \frac{\delta q}{\delta u_i'} \right\rangle$, etc. So eventually one ends up with an infinite chain of such equations, similar to a typical moment chain (cf. [17,18,4]). Here we shall work at the first level approximation (20), using its exact or approximate (perturbed) form depending of specific assumption about $\vec{u}$.

As a first illustration we let velocity $\vec{u}$ to be δ-correlated. Then equation (20) "closes", its integral term turning into the differential one, and yields the standard effective diffusion law

$$
(\partial_t + U\cdot\partial_r)\langle q\rangle = \sum_{ij}\partial_i\left(\nu\delta_{ij} + \underbrace{B_{ij}(0)}_{D_1}\right)\partial_j\langle q\rangle
$$

with coefficient $\nu + D_1$.

The higher q-moments, however would involve mixed correlators of q and the gradient $\boldsymbol{p} = \nabla q$

$$
(\partial_t + U\cdot\partial_r)\langle q^n\rangle = (\nu + D_1)\,\Delta\langle q^n\rangle - \nu n(n-1)\left\langle q^{n-2}\boldsymbol{p}^2\right\rangle \tag{22}
$$

and require more detail analysis to be developed in the next section.

V DELTA-CORRELATED VELOCITIES

Here we shall elaborate the case of *delta-correlated* velocities, focusing on p.d.f. $\rho(\boldsymbol{r}, q, \boldsymbol{p})$ that describes random fluctuation of q and its evolution via the Fokker-Planck equation (cf. [12,15]). Another basic hypothesis will be *zero molecular diffusivity* $\nu = 0$, appropriate at the initial transport-dominated phase. Precisely, we assume spatial scale of the initial q_0-variations $l_0 = \frac{q_0}{|\nabla q_0|}$ to obey

$$
l_0\langle|\vec{u}|\rangle \gg \nu
$$

- far exceeding the molecular rate. As no small scale spatial structures are present in q_0 the molecular flux is negligible and could be dropped from (19).

A Reduced Fokker-Planck equation

In the standard context of stochastic dynamics we are given a vector field $V(X,t)$ on the phase-space $\mathcal{P}$ and a differential system with initial-value problem

$$\begin{cases} \frac{d}{dt}X = V(X,t) \\ X|_{t_0} = Y \end{cases} \Rightarrow X(t) = X(t|t_0;Y) \,. \tag{23}$$

The corresponding evolution of densities on $\mathcal{P}$ is described by the 1-st order pde, for which (23) form a set of characteristics. We are particularly interested in the δ-concentrated densities, i.e.[3]

$$\begin{cases} \partial_t \Phi + V \cdot \nabla \Phi = 0 \\ \Phi|_{t=t'} = \delta(X-Y) \end{cases} \Rightarrow \Phi = \Phi(X|Y,t|t') = \delta(X(t)-X) \tag{24}$$

whose evolution follows the characteristic path. Here $X|Y,t|t'$ indicate initial conditioning (location of δ-source).

If vector field V is random, δ-correlated in time, and one sums sources Φ of (24) over all realizations (solutions) of (23) that connect Y to X in time $t' < t$, the resulting density-function

$$\rho(X|Y,t|t') = \langle \Phi(X|Y,t|t') \rangle = \langle \delta(X(t|t';Y)-X) \rangle \tag{25}$$

on $\mathcal{P}$ would give the (Markovian) transition probability to jump from state Y to X in time $t-t'$. It obeys the appropriate Fokker-Planck equation determined by field V.

In our context $\mathcal{P}$ is an infinite space of all tracer configurations $\{q(\boldsymbol{r}) : \boldsymbol{r} \in \mathbb{R}^{\mathrm{N}}\}$, and the analog of the "complete F-P equation" should be defined on "functionals $F[q]$" on $\mathcal{P}$. To make it a useful (computational) tool, however, one would like to reduce such infinite-D evolution to a finite-dimensional one.

It turns out the transport equation (2) does allow such reduction from $\mathcal{P}$ to a finite-D "one-particle" phase-space $\{(\boldsymbol{r},q,\boldsymbol{p})\}$. The latter is often associated to a classical mechanical system, whose configuration space $\mathbb{R}^{\mathrm{N}} = \{\boldsymbol{r}\}$, and is used extensively in the method of characteristics for linear and nonlinear pde's, like Euconal in optics.

The role of "delta-point" functionals on $\mathcal{P}$, $\delta(q(\boldsymbol{r}) - q_0(\boldsymbol{r}'))$ will be played by the reduced *level-delta*,

$$\phi(\boldsymbol{r},q) = \delta(q(\boldsymbol{r}) - q) \tag{26}$$

---

[3] The reader should not confuse V with transport velocity and X, Φ with the Lagrangian coordinate and tracer concentration. Here X means a generalized state and V -its evolutionary law. Of course, tracer transport could be viewed as a special case of (23)-(24).

More generally, for a given configuration $q(\boldsymbol{r})$ we fix its level $\{q(\boldsymbol{r})=q\}$ and the value of its gradient $\boldsymbol{p}=\nabla q(\boldsymbol{r})$, and introduce joint *delta*

$$\Phi(\boldsymbol{r},q,\boldsymbol{p})=\delta(q(\boldsymbol{r})-q)\,\delta(\nabla q(\boldsymbol{r})-\boldsymbol{p})\ . \tag{27}$$

If density $q(\boldsymbol{r},t)$ evolves according to (24) any flow-map (23) preserves isocontours of q and maps $\boldsymbol{p}'=\nabla q'(\boldsymbol{r}')$ into $\boldsymbol{p}=\nabla q(\boldsymbol{r})=(\frac{\partial \boldsymbol{r}}{\partial \boldsymbol{r}'})\boldsymbol{p}'$. So each "contour-delta" Φ (27) is mapped into another "contour-delta".

$$\Phi_0(\boldsymbol{r},q,\boldsymbol{p})=\delta(q_0(\boldsymbol{r})-q)\,\delta(\nabla q_0(\boldsymbol{r})-\boldsymbol{p})\to\Phi(t,\boldsymbol{r},q,\boldsymbol{p}|q_0)\ .$$

In fact, both deltas (26) and (27) obey the transport equations

$$\begin{aligned}(\partial_t+\vec{u}\cdot\partial_r)\,\phi&=0\\ (\partial_t+\vec{u}\cdot\partial_r)\,\Phi&=\sum_{ij}\frac{\partial u_i}{\partial p_j}\partial_j(p_i\Phi)\end{aligned} \tag{28}$$

with the proper initial conditions. In other words, transport-evolution of q in the configuration space $\mathcal{P}$, and the induced evolution on "functionals $\mathcal{F}(\mathcal{P})$" admit a finite-D invariant manifold of $\mathcal{F}(\mathcal{P})$ made of iso-level deltas (27).

By analogy with (25) the reduced transitional probability densities could be defined via ensemble average of level-deltas

$$\begin{aligned}\rho(t,\boldsymbol{r},q,\boldsymbol{p})&=\langle\Phi(t,\boldsymbol{r},q,\boldsymbol{p}|q_0)\rangle\\ &=\langle\delta(q(\boldsymbol{r},t)-q)\,\delta(\nabla q(\boldsymbol{r},t)-\boldsymbol{p})\rangle\ .\end{aligned} \tag{29}$$

One could show that function ρ is nothing but the one-point joint p.d.f. of q and $\boldsymbol{p}=\nabla q$, the same way as (25) gives a p.d.f. of stochastic process (23). In particular, all one-point Lagrangian and Eulerian statistics of (1-2) could be computed through ρ. For instance, mass centroid and moments

$$\begin{aligned}\langle\boldsymbol{r}\rangle&=\textstyle\iiint \boldsymbol{r}\rho(\ldots)\,d\boldsymbol{r}\,dq\,d\boldsymbol{p}\\ \langle\boldsymbol{r}_i\boldsymbol{r}_j\rangle&=\textstyle\iiint \boldsymbol{r}_i\boldsymbol{r}_j\rho(\ldots)\,d\boldsymbol{r}\,dq\,d\boldsymbol{p}\\ &\ldots\end{aligned}$$

density moments

$$\begin{aligned}\langle q\rangle&=\textstyle\iiint q\rho(\ldots)\,d\boldsymbol{r}\,dq\,d\boldsymbol{p}\\ \langle q^2\rangle&=\textstyle\iiint q^2\rho(\ldots)\,d\boldsymbol{r}\,dq\,d\boldsymbol{p}\\ &\ldots\end{aligned}$$

and similarly computed gradient moments and mixed moments.

A variety of other geometric data pertinent to the level-contour structure could also be expressed through Φ as well, hence (after statistical averaging) through ρ (see [15]). We bring a few examples:

- isolevel arclength and its "higher moments"

$$L(t,q) = \int d\boldsymbol{r} \int d\boldsymbol{p}\, |\boldsymbol{p}(\boldsymbol{r},t)|\, \delta(q(\boldsymbol{r},t) - q)$$
$$L_m(t,q) = \int d\boldsymbol{r} \int d\boldsymbol{p}\, |\boldsymbol{p}(\boldsymbol{r},t)|^m\, \delta(q(\boldsymbol{r},t) - q)$$

- Iso-area enclosed by the q-contour

$$S(t,q) = \tfrac{1}{2} \int d\boldsymbol{r} \int d\boldsymbol{p}\, \boldsymbol{r} \cdot \boldsymbol{p}(\boldsymbol{r},t)\, \delta(q(\boldsymbol{r},t) - q)$$

  -conserved for incompressible flows

- The mean number of q-contours viewed in a typical direction, a "degree of striation"

$$N(t,q) = \int_0^\infty d\boldsymbol{r}\, |\boldsymbol{p}(\boldsymbol{r},t)|\, \delta(q(\boldsymbol{r},t) - q) \ .$$

The reduced p.d.f. $\rho(t,\boldsymbol{r},q,\boldsymbol{p})$ can be shown [15] to obey the following *Fokker-Planck equation*, a consequence of the Furutsu-Novikov formula (20),

$$\partial_t \rho = L_r[\rho] + M_p[\rho] \ . \tag{30}$$

Here differential operators L (in variable $\boldsymbol{r}$) and M (in variable $\boldsymbol{p}$) depend on the mean flow U and the effective coefficients $D_1; D_2$ introduced in section III. Precisely,

$$\begin{aligned} L_r &= -U \cdot \partial_r + D_1 \Delta_r \\ M_p &= \partial_p (\boldsymbol{p} \cdot U)\, \partial_p + D_2 \left\{ (N+1)\, \Delta_p \boldsymbol{p}^2 - 2\partial_p \cdot \boldsymbol{p} - 2\,(\partial_p \cdot \boldsymbol{p})^2 \right\} \ . \end{aligned} \tag{31}$$

Operator L clearly implements the standard effective diffusion in the $\boldsymbol{r}$-space (4), while M could be viewed as higher order (gradient) correction to the process.

Remark 1 *The F-P equation (30) was derived for zero molecular diffusivity $\nu = 0$, essential in the very definition of contour-deltas (27) and their transport properties (28). Indeed, for $\nu > 0$, the q-isocontours are no longer conserved and basic equations (28), (30) would change dramatically. The approximate F-P (30) remains valid only in a limited time-range, until stretched contours and steep gradients switch on strong molecular fluxes and thus lead to next stage.*

Next we shall outline a few examples of (30) and draw some consequences.

B Zero-mean flow

Our first example is zero mean flow $U = \vec{0}$. Its Lagrangian moments yield

$$\partial_t \langle \boldsymbol{r} \rangle = \vec{0} \qquad (32)$$
$$\partial_t \langle \boldsymbol{r}_i \boldsymbol{r}_j \rangle = \int L^* [\boldsymbol{r}_i \boldsymbol{r}_j] \rho ... = 2D_1 \delta_{ij} .$$

Here L^* denotes the adjoint differential operation to L (31). The immediate consequences of (32) are zero centroid $\langle \boldsymbol{r} \rangle = \vec{0}$, and linear in t variances: $\langle \boldsymbol{r}_i \boldsymbol{r}_j \rangle = 2D_1 t \delta_{ij}$- a typical diffusive picture.

The Eulerian moments of q also bring little surprise

$$\partial_t \langle q \rangle = 0$$
$$\partial_t \langle q^2 \rangle = 0$$
$$...$$

-conservation of q by incompressible flow, as expected. The new features appear in the gradient moments

$$\begin{aligned} \partial_t \langle \mathbf{p} \rangle &= \textstyle\int M^* [\mathbf{p}] \rho ... = \vec{0} \\ \partial_t \langle p_i p_j \rangle &= \textstyle\int M^* [p_i p_j] \rho ... = 2D_2 [(N+1)\delta_{ij} - 2] \langle p_i p_j \rangle . \end{aligned} \qquad (33)$$

So the second moments $\langle p_i^2 \rangle = e^{\lambda t}$ increase, while $\langle p_i p_j \rangle = e^{-\mu t}$ $(i \neq j)$ fall off exponentially (isotropization). The real meaning of exponential growth of gradients, however is to set the time scale on the initial "explosive" phase and the onset of diffusivity

$$t_{\exp} = \mathcal{O}\left(\tfrac{\infty}{\lambda}\right) = \mathcal{O}\left(\frac{\infty}{\in \mathcal{D}_{\in} (\mathcal{N} - \infty)}\right)$$

where $N = 2, 3$-space dimension.

C Linear shear flow

Our next example is a randomly perturbed horizontal linear shear flow $U = (\alpha y, 0)$ in the plane, with δ-concentrated initial density

$$q_0 (x, y) = \delta (x - x_0) \delta (y - y_0)$$

The Lagrangian centroid

$$\left\langle \begin{matrix} x \\ y \end{matrix} \right\rangle = \begin{pmatrix} x_0 \\ y_0 \end{pmatrix} + \begin{pmatrix} \alpha y_0 t \\ 0 \end{pmatrix} = \begin{pmatrix} \bar{x} \\ \bar{y} \end{pmatrix} \qquad (34)$$

is drifted horizontally by the mean flow, as expected. But the second moments (variances)

$$\begin{array}{l} \left\langle (x-\bar{x})^2 \right\rangle = 2D_1 t \left(1+\frac{\alpha^2 t^2}{3}\right) \\ \left\langle (x-\bar{x})(y-\bar{y}) \right\rangle = \alpha D_1 t^2 \\ \left\langle (y-\bar{y})^2 \right\rangle = 2D_1 t \end{array}$$

exhibit anomalous (superdiffusive) stretching along the x-axis. This could be compared to the pipe flow ([25,1]), or the β-plane anisotropy ([3]).

The Eulerian moments of q are still conserved,

$$\begin{array}{l} \partial_t \langle q \rangle = 0 \\ \partial_t \langle q^2 \rangle = 0 \end{array}$$

and the first gradient moments are linearly stretched in time, similar to (34)

$$\begin{array}{ll} \langle p_1(t) \rangle & = p_1(0) \\ \langle p_2(t) \rangle & = p_2(0) - \alpha p_1(0)\, t \,. \end{array}$$

The second gradient moments are found to obey a linear coupled ODE system

$$\begin{array}{l} \partial_t \langle p_1^2 \rangle = D_2 \left\{ 2\langle p_1^2 \rangle + 6 \langle p_2^2 \rangle \right\} \\ \partial_t \langle p_1 p_2 \rangle = -4D_2 \langle p_1 p_2 \rangle - \alpha \langle p_1^2 \rangle \\ \partial_t \langle p_2^2 \rangle = D_2 \left\{ 6\langle p_1^2 \rangle + 2 \langle p_2^2 \rangle \right\} - 2\alpha \langle p_1 p_2 \rangle \end{array} \tag{35}$$

whose matrix has a positive eigen $\lambda > 0$ (with exponentially increasing solution), and a pair of negative eigens $\mu_{1,2} < 0$. The matrix and eigenvalues depend on the correlation coefficient D_2 (17) and the shear rate α. By analogy with the effective diffusivity and Péclet number (10-12) we could introduce a dimensionless parameter

$$P_2 = \frac{\alpha}{D_2} = \frac{\text{shear slope}}{\langle \text{enstrophy} \rangle\, \tau_0}$$

-"Péclet-two", and give an approximate expression of exponential rate λ in two extreme cases,

$$\lambda \approx \begin{cases} 8D_2 + \frac{\alpha^2}{12 D_2} = 8D_2\left(1+\frac{1}{96}P_2^2\right) & \text{small } P_2 \ll 1 \\ \\ \left(12\alpha^2 D_2\right)^{1/3} = C\, D_2 P_2^{1/3} & \text{large } P_2 \gg 1\,. \end{cases} \tag{36}$$

Such dependence of λ on P_2 resembles of the expected form of the effective diffusivity (13a) in terms of ν and the standard Péclet number.

Once again exponent λ (36) sets a bound for the initial "explosive" phase at $t_{\exp} = \mathcal{O}\left(\frac{\infty}{\lambda}\right)$. A few other qualitative observation could be drawn from our analysis:

- the initial transport phase is characterized by the broad and increasing fluctuations, as opposed to the asymptotic (terminal) phase, where fluctuation are expected to die out,

- shear statistically enhances the effect of random velocity noise and is statistically unstable,

- the instability exponent λ is proportional to the effective rate D_2 in both cases (zero and linear mean- shear), and exhibits some scaling dependence on the dimensionless parameter P_2,

- It would be instructive to compare the time scale t_{exp} with the mixing time t_{m} of section II D. We expect $t_{\mathrm{exp}} \approx t_{\mathrm{m}}$, but have no proof.

VI PERTURBATION ANALYSIS

The goal is to overcome the principal limitation of the previous section: zero molecular diffusivity and δ-correlation. We do not have yet a complete theory. Here we shall outline a few preliminary results and indicate possible directions.

A The effect of molecular diffusion

We start with molecular diffusivity. While negligible at the early stages it should come into play at time scales $\mathcal{O}(\sqcup_{\mathrm{exp}})$ or t_{m}, to arrest the exponential growth of gradients (sections V B and V C). The techniques of the "F-P reduction" however don't work in the presence of $\nu > 0$ (see remark in section V A).

One could set a perturbation scheme for mixed moments $\langle q^n \boldsymbol{p}^m \rangle$, starting with

$$\begin{cases} (\partial_t + U \cdot \partial_r) \langle q^n \rangle = (\nu + D_1) \Delta \langle q^n \rangle - \nu n (n-1) \langle q^{n-2} \boldsymbol{p}^2 \rangle \\ (\partial_t + U \cdot \partial_r) \langle q^{n-2} \boldsymbol{p}^2 \rangle = (D_1 \Delta + 2(N+2)(N-1) D_2) \langle q^{n-2} \boldsymbol{p}^2 \rangle \ . \end{cases} \tag{37}$$

Here ν enters the first equation, but is dropped in the second one. Such first-order perturbation scheme has secular behavior at times set up by the length scale l_0 of q_0 along with other parameters (see [15]). Namely, for the n-th moment

$$t_n \approx \tfrac{1}{2(N+2)(N-1)D_2} \ln \left(\frac{D_2 l_0^2}{\nu n^2} \right) \ .$$

To go beyond this range would require some form of renormalization technique (cf. [7]).

Turning to the long-range dynamics in the presence of molecular diffusivity we shall briefly mention one solvable case: random fluctuations $\tilde{q}$ of constant density gradient $G = \nabla q$. This example drew considerable attention (see [26,11,27,15]), as the transport evolution could be reduced to a stationary source-problem for perturbation $\tilde{q} = q - G \cdot \boldsymbol{r}$,

$$(\partial_t + \vec{u} \cdot \partial_r) \tilde{q} = -\vec{u} \cdot G + \nu \Delta \tilde{q} \,. \tag{38}$$

Two basic issues here are the equilibrium solution $\tilde{q}_e$ of (38) and the relaxation time T to equilibrium. Once again the F-P equation allows us to compute the moments of $\tilde{q}$ and its gradient $\tilde{p}$. In particular, we get

$$\langle \tilde{p}_e^2 \rangle_{\text{stationary}} = D_1 \tfrac{G^2}{\nu}$$

$$T \approx \tfrac{1}{2D_2(N+2)(N-1)} \ln \left(\tfrac{\nu + D_1}{\nu} \right) \,.$$

B Finite time-correlation via "diffusion approximation"

Another limitation of section V has to do with δ-correlated hypothesis on $\vec{u}$. Here we shall outline a way to relax it to finite correlation radius τ_0, following [15]. The following procedure, known as the diffusion approximation [13], bears some similarity to the EDQNM - method used in the turbulence and transport problems [18].

This time we shall work with two basic equations (19)-(21), for q and $\frac{\delta q}{\delta u'}$, and replace random velocity $\vec{u}$ in both by its mean value U. In doing so we assume that random velocity fluctuations have negligible effect on the essential statistics of q. Such first-order perturbation scheme can be justified within the time-range $0 < t < \tau_0$-velocity correlation radius. The resulting equations

$$\begin{aligned} &q_t + U \cdot \nabla q = \nu \Delta q \\ &q|_{t_0} = q_0 (\boldsymbol{r}) \end{aligned} \tag{39}$$

and

$$\begin{aligned} &(\partial_t + U \cdot \partial_r) \left(\tfrac{\delta q}{\delta u_i'} \right) = \nu \Delta_r \left(\tfrac{\delta q}{\delta u_i'} \right) \\ &\left. \tfrac{\delta q}{\delta u_i'} \right|_{t=t'} = (\partial_i q') \, \delta (\boldsymbol{r} - \boldsymbol{r}') \end{aligned} \tag{40}$$

become deterministic and randomness enters here only through the initial value.

The evolutions of both quantities (39)-(40) is given by differential operator $H = \nu \Delta - U \cdot \partial_r$. In particular, combining two evolutions we get an expression of the variational derivative through $q = q(\boldsymbol{r}, t)$

$$\frac{\delta q}{\delta u_i'} = e^{\tau H} \left\{ \delta (\boldsymbol{r} - \boldsymbol{r}') \, \partial_i' \right\} e^{-\tau H} [q] \, ; \tau = t - t'$$

that in essence allows to close the infinite chain of "variational means" (20).

The linear transform: $q \rightarrow \frac{\delta q}{\delta u_i'}$ is made of four operators: propagator $e^{-\tau H}$, followed by partial ∂_i' (in variable $\boldsymbol{r}_i'$), then multiplication with $\delta(\boldsymbol{r}-\boldsymbol{r}')$ and another propagator $e^{\tau H}$. We skip further details (see [15]) and bring the final result for the effective mean-field diffusivity tensor

$$
\begin{aligned}
D_{ij}^*(U) &= \int_0^\infty e^{\tau H}\left[b_{ij}(0,\tau)\right] d\tau \\
&= \int_0^\infty d\tau \int d\boldsymbol{r} \left\{ \exp\left[\tfrac{-\boldsymbol{r}^2}{2\nu\tau}\right] \Big/ (2\pi\nu\tau)^{N/2} \right\} b_{ij}(\boldsymbol{r}-U\tau,\tau) \ .
\end{aligned}
\tag{41}
$$

Here $b_{ij}(\boldsymbol{r},t)$ is the velocity correlation tensor (14). We recall that b_{ij} is expressed through the Fourier-transformed energy spectrum $E_{ij}(\boldsymbol{k},\tau)$ (15). So D_{ij}^* could be recast in the form

$$
D_{ij}^*(U) = \int_0^\infty d\tau \int d\boldsymbol{k}\, e^{\left(-\nu k^2 + ik\cdot U\right)\tau} E(\boldsymbol{k},\tau)\left(1-\tfrac{k_i k_j}{k^2}\right) \ . \tag{42}
$$

Formulae (41-42) give an approximate expression of the Taylor's diffusion coefficient (3), but they involve no "Lagrangian path" ambiguity, only mean-field U and the energy spectrum of fluctuation $\vec{u}$.

C Sedimentation problem

We conclude the paper with an application of (41-42) to a sedimentation problem. Here diffusing tracer acquires a constant downward (vertical) drift $U = (0,-U)$ - the balance of gravity and viscous friction. In addition the tracer is subjected to random velocity fluctuations (wind-shear) $\vec{u}$. For the sake of example we assume the energy spectrum of $\vec{u}$ to be $E(\boldsymbol{k},t) = E(\boldsymbol{k})\, e^{-t/\tau_0}$, τ_0-temporal correlation radius.

Then the horizontal and vertical diffusivities $D_{xx}^*; D_{yy}^*$ are computed in terms of the dimensionless parameter

$$
\sigma = \sigma(\boldsymbol{k},U) = \frac{(\boldsymbol{k}\cdot U)\,\tau_0}{1+\nu \boldsymbol{k}^2\tau_0}
$$

and a pair of special functions (see [15])

$$
\begin{aligned}
f_y(\sigma) &= \frac{\sigma\left(\tan^{-1}\sigma - 1\right) + \tan^{-1}\sigma}{\sigma^2} \\
f_x(\sigma) &= \tfrac{1}{2}\left\{\frac{\sigma\left(\tan^{-1}\sigma + 1\right) - \tan^{-1}\sigma}{\sigma^2}\right\} \ .
\end{aligned}
$$

Namely,

$$D_{xx}^{*} = \frac{4\pi}{U}\int_{0}^{\infty} dk\, kE(k)\, f_x(\sigma(\boldsymbol{k},U))$$

$$D_{yy}^{*} = \frac{4\pi}{U}\int_{0}^{\infty} dk\, kE(k)\, f_y(\sigma(\boldsymbol{k},U)) \ .$$

If σ is small ($U\tau_0 \ll l_0$, to the spatial correlation radius), both functions $f_x; f_y$ are close to $\frac{2\sigma}{3}$, which results in isotropic diffusion, independent of the drift velocity U. Another extreme -large σ ($U\tau_0 \gg l_0$), gives $f_y \approx 2f_x \approx \frac{\pi}{2}$, hence highly anisotropic diffusion with the vertical rate in excess of the horizontal one by factor 2.

More generally, one could extend the above analysis to anisotropic velocities (e.g. horizontal random shear) with energy tensor E_{ij}, and determine the combined effect of the vertical drift and horizontal shear. One could also use these methods to determine the spread and accumulation level of the deposited tracer in terms of its source, the mean flows and the effective diffusivities (41-42).

VII CONCLUSION

Turbulent transport by random Gaussian velocity fields $\vec{u}$ has many time scales and distinct phases, like temporal correlation radius τ_0 and mixing time scale τ_{m} .

The long-range asymptotics of mean-field $\langle q \rangle$ is always diffusive, and its effective rate D^* depends on the Péclet number, molecular diffusivity ν, and the statistics (mean, correlation) of $\vec{u}$. The onset of diffusive behavior, however could be delayed to infinite times, which could lead to anomalous (fractional) diffusive regime.

The early "transport phase" of the turbulent advection allows to neglect molecular flux. It can be characterized by exponential growth of statistical moments (fluctuations), fractalization of isocontours and steepening density gradients.

The effective parameters that describe such process include time-integrated velocity correlation tensor, as well as its derivatives at $r = 0$. The fluctuating component of random solution q obeys the Fokker-Planck equation (30) in the $(\boldsymbol{r}, q, \boldsymbol{p})$-phase space, derived via variational calculus and the Furutsu-Novikov formalism. A few first-order perturbation schemes are offered to incorporate molecular diffusivity and allow finite temporal correlation of $\vec{u}$ (rather than δ-correlation). Several applications of our method are given to the turbulent shear flow and randomly advected sedimentation.

The early "explosive" phase should eventually (beyond the mixing time-scale) give way to the "outer" diffusive regime where fluctuations die out, however the details of such transition remain beyond the scope of our method.

Finally we shall list a few other open problems and challenges

- higher order perturbation schemes
- long-range temporal correlation of $\vec{u}$
- compressible flows (the work is currently in progress)
- nonlinear transport problems, like reaction-diffusion models etc.

Acknowledgments

We are pleased to acknowledge the hospitality of organizers of the "Research Workshop on Two-dimensional Turbulence in Fluids and Plasma" at the Australian National University in Canberra.

This work was supported by the U.S. Civilian Research and Development Foundation under Award No. RM1-272. The second author was also supported by the Russian Foundation for Basic Research (Projects 95-05-14247, 96-05-65347 and 96-15-98527).

REFERENCES

1. Aris, R., *Proc. R. Soc. London*, Ser. A, **235**, 67 (1956).
2. Avellaneda, M., Majda, A., *Com. Math. Phys.*, **131**, 381, (1990).
3. Bartello, P., Holloway, G., *J. Fluid Mech.*, 223, 521, (1991).
4. Batchelor, G.K., *Proc. Camb. Phyl. Soc.*, **48**, 354 (1952).
5. Bouchard J-P., George A., *Phys. Rep.*, **195**, 127, (1990).
6. Constantin P., Procaccia I., *Nonlinearity*, **7**, 1045, (1994).
7. Fairhill A.L., Gat O., L'vov V., Procaccia I., *Phys. Rev. E.*, **53**, 4, 3518, (1996).
8. Isichenko, M.B., *Rev. Mod. Phys.*, 64 (4), 961, (1992).
9. Isichenko, M.B., Kalda J., *J. Nonlinear Sci.*, **1**, 375, (1991).
10. Isichenko, M.B., Kalda J., *Sov. Phys. JETP.*, **72**, 126 (1991a).
11. Holzer M., Purim A., *Phys. Rev. E.*, **47**, 1, 202, (1993).
12. Kimura, Y., Kraichnan R.H., *Phys. Fluids*, A, **5** (9), 2264, (1993).
13. Klyatskin, V. I., *Stochastic equations and waves in Random media*, Moskwa, Nauka, 1980.
14. Klyatskin, *Physics-Uspekhi*, **37**,5, 501, (1994).
15. Klyatskin, V. I.,Woyczynski, W., Gurarie, D., *J. Stat. Phys.*, **84**,3/4,797, (1996).
16. Kraichnan, R.H., *Phys. Fluids*, **13**, 22, (1970).
17. Kraichnan R.H., *Phys. Rev. Lett.*, **72**, 1016, (1994).
18. Lesieur, M., *Turbulence in fluids*, Kluwer, Boston, 1990.
19. McLaughlin, D., Papanicolau, G., Pironneau, O., *SIAM J. Appl. Math.*, **45**, 780, (1985).

20. Moffatt H.K., *Rep. Prog. Phys.*, **46**, 621, (1983).
21. Monin, S. A., Yaglom A.M., *Statistical fluid mechanics*, MIT Press, Cambridge, 1975.
22. Richardson, L. F., *Proc. R. Soc. London*, Ser. A, **110**, 709 (1926).
23. Solomon, T.H., Golub, J.P., *Phys. Fluids*, **31**, 1372, (1988).
24. Tatarinova, E.B., Kalugin, P.A., Sokol, A.V., *Eusophys. Let.*, **14**, 773, (1991).
25. Taylor, G.I., *Proc. R. Soc. London*, Ser. A, **225**, 471 (1953).
26. Purim A., Shraiman B.I., Siggia E.D., *Phys. Rev. Let.*, **66**, 23, 2984, (1991); *Phys. Rev. E*, **55**, 2, 1263, (1997).
27. Shraiman B.I., Siggia E.D., *Phys. Rev. E.*, **49**, 4, 2912, (1994); *Phys. Rev. Let.*, **77**, 12, 2463, (1996).
28. Taylor, G.I., *Proc. Lond. Math. Soc.*, Ser. 2, **20**, 196 (1921).
29. Zaslavsky, G.M., in *New approaches and concepts in turbulence*, 165, Birkhauser, 1993.
30. Zeldovich, Ya. B., *Dokl. Akad. Nauk USSR*, **7**, 1466, (1937).

Multiple Time Scales in Anisotropic Magnetohydrodynamics

Rodney M. Kinney and James C. McWilliams

Institute of Geophysics and Planetary Physics
UCLA, Los Angeles, CA 90095-1567
and
National Center for Atmospheric Research
Boulder, CO 80307-3000

Abstract. The consequences of a multiple time-scale resonance theory of magnetohydrodynamics with a strong (vertical) background field are examined and compared with direct numerical solutions at high Reynolds number. The background field defines a characteristic wavenumber k_L. Modes with vertical wavenumber above k_L are passively driven by the $k = 0$ mode, even when the majority of the energy is contained in the passive wavenumbers. The passive modes do not cascade to higher vertical wavenumbers, so the vertical wavenumber spectrum is not a power law and does not extend to dissipation scales. The $k = 0$ mode evolves with two-dimensional dynamics, forming coherent current structures which are mirrored by the passive modes.

I INTRODUCTION

Observations and laboratory measurements have shown that turbulent fluctuations against a strong background field tend to have longer length scales along the field direction and that the fluctuating component in the direction of the field is smaller than the components perpendicular to the field. The dynamical mechanism by which these anisotropies develop is still not well understood.

In magnetohydrodynamics (MHD), a background magnetic field gives rise to shear Alfvén waves. The interaction of Alfvén waves is unusual in that waves traveling in the same direction do not interact at all (barring compressible and dissipative effects). Thus, it has been argued that wave triad interactions are in fact empty unless one of the modes in the triad has zero component along the mean magnetic field [1,2]. This has the consequence that energy is therefore more efficiently transferred to perpendicular wavenumbers. It has also been argued [3] that the cascade is instead determined by the four-wave

CP414, *Two-Dimensional Turbulence in Plasmas and Fluids:* Research Workshop
edited by R. L. Dewar and R. W. Griffiths

resonant interactions [4]. Direct numerical tests of the interaction between pairs of wave packets, however, have shown a nonzero contribution from the triad interactions [5].

Here, we demonstrate that nontrivial turbulent evolution can occur based on triad interactions. In a multiple time-scale analysis, the slow evolution of Alfvén wave amplitudes is derived in a form that may be directly verified in fully turbulent solutions. In the case where the anisotropy is not strong, scaling arguments are used to predict the turbulent spectrum and its different dependence on vertical and horizontal wavenumber. The predictions are confirmed by comparison with numerical solutions.

II MULTIPLE TIME-SCALES IN REDUCED MHD

Reduced MHD (RMHD) [6,9] is a reduction of MHD in three spatial dimensions. The horizontal field components are self-consistently determined, with the vertical components obeying a passive relationship [7]. It is the simplest plasma description with all the necessary elements for studying nonlinear Alfvén wave turbulence, and may be written

$$\partial_t \boldsymbol{v} + \boldsymbol{v}\cdot\boldsymbol{\nabla}\boldsymbol{v} - \boldsymbol{B}\cdot\boldsymbol{\nabla}\boldsymbol{B} = B_0\partial_z\boldsymbol{B} - \boldsymbol{\nabla}\left(p + \frac{1}{2}B^2 + B_0\hat{\boldsymbol{z}}\cdot\boldsymbol{B}\right) \tag{1}$$

$$\partial_t \boldsymbol{B} + \boldsymbol{v}\cdot\boldsymbol{\nabla}\boldsymbol{B} - \boldsymbol{B}\cdot\boldsymbol{\nabla}\boldsymbol{v} = B_0\partial_z\boldsymbol{v} \tag{2}$$

$$\boldsymbol{\nabla}\cdot\boldsymbol{v} = \boldsymbol{\nabla}\cdot\boldsymbol{B} = 0 \tag{3}$$

$$\boldsymbol{\nabla} = \hat{\boldsymbol{x}}\partial_x + \hat{\boldsymbol{y}}\partial_y \tag{4}$$

The velocity field $\boldsymbol{v} = (v_x, v_y)$ is purely horizontal, while the magnetic has a constant vertical component B_0 and a spatially-varying horizontal component $\boldsymbol{B} = (B_x, B_y)$. Because of the incompressibility constraint, the system has no compressional modes, and the only linear waves are shear Alfvén waves which propagate along $\hat{\boldsymbol{z}}$ with phase speed B_0.

In addition to the linear waves, RMHD includes nonlinear interactions among those waves. The form of the nonlinearity can be most easily seen by introducing the Elsasser currents $\Omega^\pm = \hat{\boldsymbol{z}}\cdot\boldsymbol{\nabla}\times(\boldsymbol{v}\pm\boldsymbol{B})$ and the potentials defined such that $\Omega^\pm = \nabla^2\psi^\pm$. In terms of these fields, RMHD can be written as

$$\partial_t\Omega^\pm + \boldsymbol{\nabla}\cdot[\psi^\mp, \boldsymbol{\nabla}\psi^\pm] = B_0\partial_z\Omega^\pm. \tag{5}$$

in terms of the horizontal Jacobian $[f,g] = \partial_x f\partial_y g - \partial_y f\partial_x g$. From eq. (5) it is clear that Ω^+ represents an upward-traveling wave and Ω^- represents a downward-traveling wave, and that the only nonlinear interaction is between upward and downward waves.

The competition in anisotropic dynamics is between the linear and nonlinear terms in eq. (5). However, the magnitude of these terms depends not only on the strength of the fields but on their characteristic scale. We introduce the vertical Fourier transform of $\Omega^\pm$,

$$\Omega^\pm = \sum_k \Omega_k(\boldsymbol{x}, t)e^{ikz}, \tag{6}$$

and write eq. (5) as

$$\partial_t \Omega_k^\pm = -\boldsymbol{\nabla}\cdot\sum_{k'}[\psi_{k'}^\mp, \boldsymbol{\nabla}\psi_{k-k'}^\pm] \pm ikB_0\Omega_k^\pm. \tag{7}$$

If the typical horizontal scale and amplitude of all $\psi_k^\pm$ modes are comparable, then the question of whether the linear term dominates over the nonlinear term is dependent on the value of k. Let us define a cutoff wavenumber for linear behavior,

$$k_L \sim \frac{|\boldsymbol{\nabla v}|}{B_0}. \tag{8}$$

modes for which $k \ll k_L$ evolve mostly via the nonlinearity, while those for which $k \gg k_L$ evolve mostly linearly. The development of the system from isotropic initial conditions will depend strongly on the scale content relative to k_L.

A Resonant behavior

Let us first consider the case in which all modes in the initial excitation have $k > k_L$ with the exception of $k = 0$. This will be the case for any finite system with a minimum wavenumber $> k_L$. Then we may divide our time variable into a slow time scale τ and a fast time scale t' so that

$$\Omega_k^\pm = \Omega_k(x, y, \tau)e^{\pm ikB_0t'}. \tag{9}$$

A time-averaging analysis, in which one uses the above ansatz in eq. (5) and projects onto the k components, gives a result for the slow time-scale evolution of the individual mode amplitudes,

$$\partial_\tau \Omega_k^\pm + \boldsymbol{\nabla}\cdot[\psi_0^\mp, \boldsymbol{\nabla}\psi_k^\pm] = 0. \tag{10}$$

A similar result follows from an analysis of fully 3D incompressible MHD [8].

Eq. (10) describes the slow WKB evolution of Alfvén wave amplitudes as functions of x and y. Inspection of eq. (10) shows that the $k = 0$ mode evolves independently of all the finite k modes. Furthermore, the dynamics of the $k = 0$ mode (which we refer to as the "mean mode") are just those of 2D MHD. That the mean mode can be associated with the slow time-scale dynamics of anisotropic MHD has been known for some time [10]. An important point

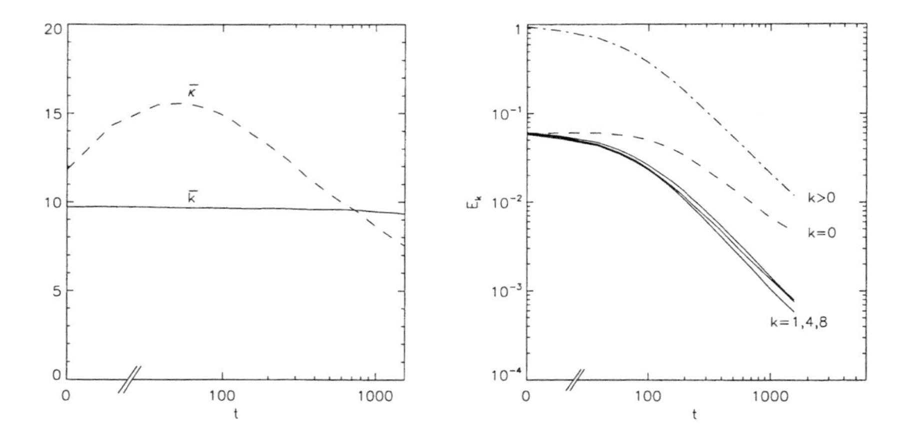

FIGURE 1. (a) Vertical wavenumber moment $\bar{k}$ (solid) and horizontal moment $\bar{\kappa}$ (dashed) from decaying RMHD solution with $B_0 = 5$. The vertical wavenumber shows almost no evolution, while the horizontal wavenumber shows a normal cascade. (b) Evolution of energy contained in various modes. The $k = 0$ mode decays more slowly than the finite k modes and will eventually dominate the solution.

is that the plasma need not be weakly z-dependent; the independence of the mean mode arises from the short interaction time of oppositely-propagating waves. In effect, the only field seen by a propagating wave is that of the mean mode, whose presence is felt continuously.

Each of the finite k components (the "wave modes") has a separate subdynamics which is driven by the mean mode without influencing it. An additional consequence of (10) is that

$$\partial_\tau \int |\nabla \psi_k^\pm|^2 d\boldsymbol{x} = 0, \tag{11}$$

i.e., the energy in each wave mode is conserved individually. Thus, there is no cascade in vertical wavenumber on the slow time-scale either. In the horizontal directions, however, there may is a cascade on the slow time-scale which is entirely controlled by the mean mode. Because the mean mode will undergo the forward energy cascade characteristic of 2D MHD, the wave modes will also be driven to small perpendicular wavelengths on the slow time scale. The significance of eq. (10) is that the evolution of the entire spectral range is determined solely by the 2D dynamics of the mean mode. We seek to verify the predictions of eq. (10) by calculating numerical solutions of eq. (4). A finite-difference code is used to solve RMHD with periodic boundary conditions. Freely decaying initial-value problems are solved in which the initial conditions are a broad-band spectrum with random phases, and the

Reynolds number is chosen sufficiently large that turbulent behavior develops as the solution decays. We adopt a normalization in which the total initial energy is 1, and the value of B_0 may be varied as a parameter.

Figure 1(a) shows the mean vertical and horizontal wavenumbers vs. time in a $B_0 = 5$ solution calculated on a 128^3 grid. No vertical cascade is evident, but a cascade proceeds normally in the horizontal wavenumbers. While $\bar{k}$ remains constant to within 2% during the entire solution, $\bar{\kappa}$ increases as the spectrum broadens. After reaching its maximum, $\bar{\kappa}$ decays due to dissipation and the absence of forcing. Since the mean mode is the only one with nonlinear dynamics, its behavior is fundamentally different from the other modes. Figure 1(b) shows the energy contained in a single vertical mode, E_k, as a function of time for various k. A clear difference between the mean and wave modes is visible. The energy of the mean mode E_0, decays more slowly than the others. The sum of all modes with $k > 0$ is also shown. In this solution, the mean mode begins with 6% of the total energy and ends with 40%. The $t \rightarrow \infty$ state is clearly one in which the mean mode will dominate all other modes.

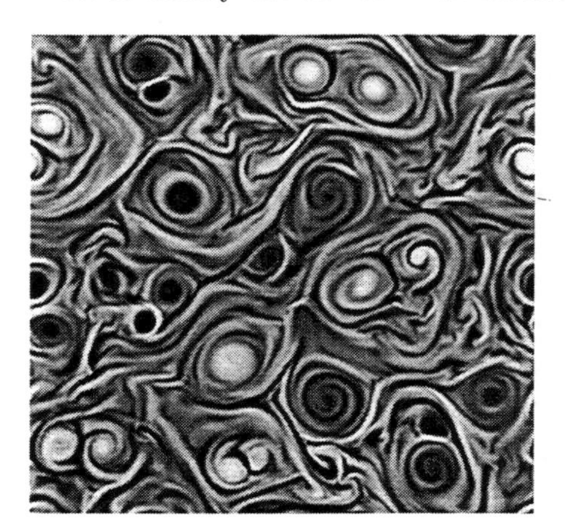

FIGURE 2. (a) Current of the $k = 0$ mode from a 3D RMHD solution. The structuring is characteristic of 2D MHD turbulence. (b) The magnitude $|\Omega^{+}_{k=2}|^2$ of the $k = 2$ mode. The passive dynamics of the finite k modes gives them structure inherited from the $k = 0$ mode.

The dynamics of the mean mode are recognizable as those of 2D MHD by features such as conservation of mean-squared potential, dominance of magnetic over kinetic energy, and coherent current vortex structures [11]. Visualizations of the current in the mean mode amplitudes is shown in Figure 2(a) from a $B_0 = 5$ solution on a $512^2 \times 32$ grid. An independent 2D solution, initialized with the mean mode at the time shown, remains closely correlated with the mean mode in this 3D solution over many eddy turnover times, to the extent that the two cannot be distinguished in visual plots.

A consequence of the passive dynamics of eq. (10) is that the wave mode amplitudes $\Omega^{\pm}_{k>0}$ acquire horizontal structure from the mean mode. The form of passive interaction in (10) is different from a simple passive scalar. Once the mean mode begins its direct cascade of energy and evolves coherent structures, the wave modes develop similar horizontal spectra and develop ghost structures. In Figure 2(b) is the mode amplitude $|\Omega^{+}_{k=2}|^2$ at the same time as in Figure 2(a). It is clear that the $\Omega^{+}_{k=2}$ cross-section mirrors certain vortex features apparent in j_0. The detailed time-dependent evolution of this mode can be checked against independent 2D solutions of (10) just as for the mean mode, and it is again found that the solutions are well-correlated over long times.

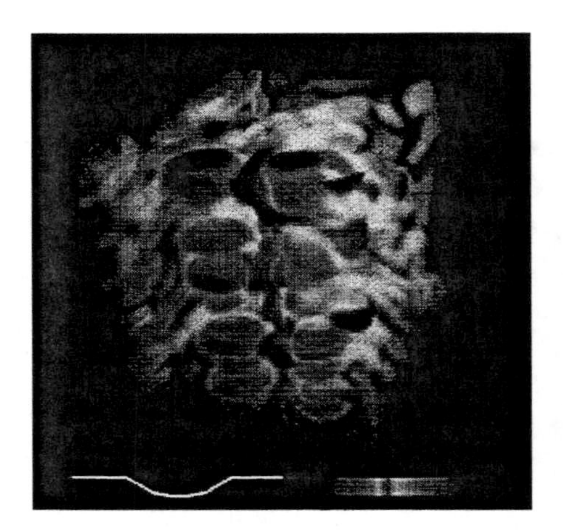
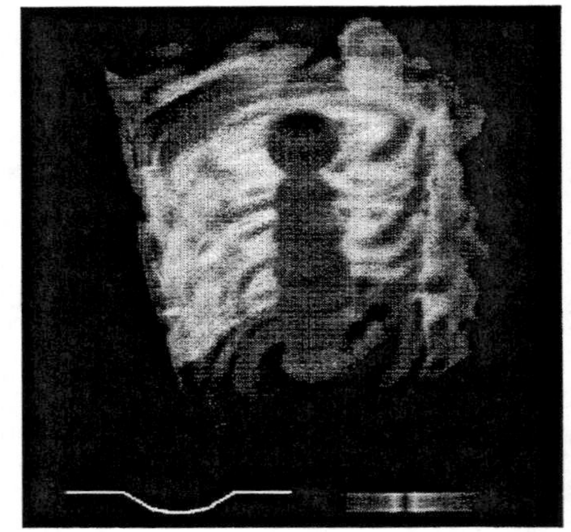

FIGURE 3. 3D visualizations of the total current in real space. Incoherent summation of the finite k modes gives a modulation of the 2D structure. Structures occur whether the $k = 0$ mode has comparable energy with the finite k modes (a) or is dominant (b).

The dynamics may be dominated by the mean mode even when it is not energetically dominant, as in Figure 1(b). When constructing the total field as a function of space, the wave modes will beat in such a way that they may cancel or reverse the sign of the current in the $k = 0$ mode. Examples of the total current field in the vicinity of a vortex structure are shown in Figure 3. The mean mode is not dominant in the case on the left. While it is possible to see that the vertically averaged current is structured, the total current can even reverse sign at particular levels, a result of beating by the wave modes. The right hand case is one in which the mean mode is dominant and the structures appear more vertically uniform.

B Non-Resonant Behavior

If the initial excitation is at wavenumbers less than the linear cutoff k_L, then the primary dynamics are nonlinear. In this case, the interaction time between wave packets is long enough that they may interact without being resonant and we expect a turbulent cascade to higher wavenumbers by the usual doubling in wavenumber space, i.e.

$$\partial_t \Omega_{2k}^{\pm} \sim \Omega_k^{\mp} \Omega_k^{\pm} + 2ikB_0 \Omega_{2k}^{\pm}. \tag{12}$$

Although initially unexcited modes grow in amplitude due to nonlinear coupling from lower-k modes, this is not a cascade in the traditional sense because of the characteristic amplitude B_0. We know from the previous subsections that if a mode Ω_{2k} grows to an amplitude such that the linear term in (12) is approximately as strong as the nonlinear term, the mode will begin to interact only in resonant triads (i.e. will be passively driven by the mean mode), and will no longer cascade to higher wavenumbers. Thus, the extent of the cascade in k is limited not by dissipation, but by the background field B_0. A mode will grow in amplitude until there is an approximate balance between the terms in (12). This implies a vertical spectrum with $\Omega_{2k} \sim \Omega_k^2$, which suggests an exponential form $\Omega_k \sim e^{ak}$. On dimensional grounds, the natural choice for the exponential scale factor is k_L, giving a vertical spectrum

$$\Omega_k^{\pm} \sim e^{-|k/k_L|}. \tag{13}$$

If k_L is small (B_0 large), then (13) prohibits the cascade from extending to dissipation scales.

We again confirm these arguments by direct numerical solution of RMHD, with B_0 moderate or small. In this case, none of the modes initially present are dominated by Alfvénic propagation, so that a nonlinear cascade begins to transfer energy to the higher wavenumbers. Figure 4(a) shows $\bar{k}$ and $\bar{\kappa}$ from three runs with $B_0 = 0.2,\ 0.5$, and 1.0 on a $128^2 \times 256$ grid. The horizontal cascade is indifferent to the strength of B_0, but the vertical cascade is not. As the initial spectrum expands into higher wavenumbers, the mean wavenumber $\bar{k}$ increases, with a maximum $\bar{k}$ that increases with decreasing B_0. After reaching its peak $\bar{k}$ does not decay like $\bar{\kappa}$, indicating that energy is not being dissipated by transfer to larger k.

Figure 4(b) show the E_k spectra for the same three runs as in Figure 4(a) at the times of maximum $\bar{k}$. Approximately exponential spectra are evident up to the largest wavenumbers, where the Fourier truncation prevents transfer to larger k, causing buildup at the resolution limit. Fits of the spectra to exponential curves for $20 \leq k \leq 100$ give scale factors in the ratio $2.8 : 1.5 : 1$, while the magnetic field strengths are in ratio $5 : 2 : 1$. Thus, the general arguments above give a correct order-of-magnitude estimate of the spectrum in eq. (13), but a more careful theory is required to predict the exponential scale factor more accurately.

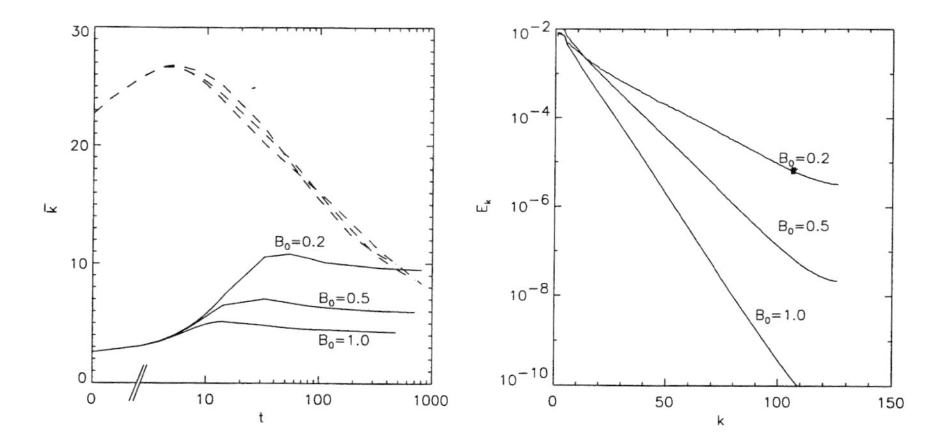

FIGURE 4. (a) Vertical (solid) and horizontal (dashed) wavenumber moments for three solutions with different $B_0 \leq 1$. The vertical cascade is limited by the magnetic field strength rather than by dissipation, while the horizontal cascade is unaffected by magnetic field strength. (b) Developed E_k spectra at time of maximum $\bar{k}$ from (a). The initial excitation is confined to $k \leq 4$, and the spectra become approximately exponential, with a scale factor that increases with B_0.

REFERENCES

1. J.V. Shebalin, W.H. Matthaeus, and D. Montgomery, J. Plasma Phys. **29**, 525 (1983).
2. D. Montgomery and W.H. Matthaeus, Astrophys. J. **447**, 706 (1995).
3. S. Sridhar and P Goldreich, Astrophys. J. **432**, 612 (1994).
4. P. Goldreich and S. Sridhar, Astrophys. J. **438**, 763 (1995).
5. C.S. Ng and A. Bhattacharjee, Astrophys. J. **465**, 845 (1996).
6. H. R. Strauss, Phys. Fluids **19**, 134 (1976).
7. R. Kinney and J.C. McWilliams, J. Plasma Phys, **57**, 93 (1997).
8. R. Kinney and J.C. McWilliams, "Turbulent Cascades in anisotropic magnetohydrodynamics", submitted to Phys. Plasmas.
9. D. Montgomery, Physica Scripta T**2/1**, 93 (1982).
10. D. Montgomery and L. Turner, Phys. Fluids **24**, 815 (1981).
11. R. Kinney, J.C. McWilliams, and T. Tajima, Phys. Plasmas **2**, 3623 (1995).

Structures in Driftwave Turbulence and some Comments on the related Particle Transport

V. NAULIN

Association EURATOM - Risø National Laboratory
Optics and Fluid Dynamics Department
Risø National Laboratory
PO Box 49
DK-4000 Roskilde

Abstract.
The structures of driven drift-wave turbulence in a doubly-periodic two-dimensional slab geometry are considered. The appearance and dynamics of such structures in saturated driven/damped systems is connected to the interpretation of the so-called anomalous transport observed in magnetically confined fusion-plasmas. Different techniques are applied to numerical data of drift-wave turbulence in the search for coherent structures. The combination of coherent sampling, tracking of maxima or minima, and usage of the Weiss field leads to a self-consistent picture of their transient appearance and the related particle transport caused by coherent structures.

I INTRODUCTION

The appearance of long lived coherent structures in turbulent drift-wave dynamics is a well known feature. Their existence is important for the anomalous plasma diffusion across magnetic field lines. For adiabatic electron density response stationary localized solutions are known and their self-organization was shown numerically [1]. Even though the properties of the turbulence were examined in detail (see e.g. [2] and the article in this volume) the dynamics of the structures in the flow is not yet completely examined nor understood. Considered is a model incorporating the dissipative instability leading to quasistationary self-sustained turbulence. It can be understood as a paradigm for 2D-turbulence excited by an internal instability at low k and energy dissipation at large k.

The paper is organized as follows: After a short description of the model and the numerics in section II, we will use tracking techniques to gain insight

CP414, *Two-Dimensional Turbulence in Plasmas and Fluids:* Research Workshop
edited by R. L. Dewar and R. W. Griffiths

into the statistical properties of the transient structures in the flow in section III. Section IV is devoted to the discussion of transport of trapped density by the structures. We finally use conditional sampling in section V to obtain knowledge of the average shape and the dynamics of the structures.

II MODEL AND NUMERICAL RESULTS

We consider a plasma with a density gradient in the x-direction in a strong, homogeneous magnetic field pointing in the z-direction. As the derivation of the model for different physical situations has been presented in detail before [3,4] we here just state the equation

$$\begin{aligned}&\partial_t(1-\nabla^2)\varphi+\partial_y\varphi+\mu\nabla^2\nabla^2\varphi+\{\nabla^2\varphi,\varphi\}\\&=\delta\partial_t\left(\partial_t+\partial_y\right)\varphi+\delta\left\{\varphi,\left(\partial_t+\partial_y\right)\varphi\right\}\ .\end{aligned} \tag{1}$$

The density fluctuations n are computed from the electrostatic field φ through

$$n=\varphi-\delta\left(\partial_t+\partial_y\right)\varphi\ . \tag{2}$$

$$\begin{aligned}&\frac{e\varphi}{T_e}\frac{L_n}{\rho_s}\to\varphi\quad,\qquad\frac{n}{n_{00}}\frac{L_n}{\rho_s}\to n\quad,\qquad\frac{tc_s}{L_n}\to t\quad,\\&\rho_s\,\nabla_\perp\to\nabla\quad,\qquad\frac{\mu_iL_n}{\rho_s^2c_s}\to\mu\quad,\end{aligned} \tag{3}$$

was used for normalization, where $\rho_s=c_s/\Omega_i$ is the ion gyroradius at electron temperature T_e and the sound speed $c_s=(T_e/m_i)^{1/2}$. An average density n_{00} is used for normalization, while within the so-called local approximation the background density n_0 is assumed to decay in the x-direction with a decay length L_n. Furthermore the Poisson bracket $\{\varphi,\psi\}\equiv\vec{z}\times\nabla\varphi\cdot\nabla\psi$ was used. An effective parallel length

$$L_{\|}=\left(\frac{L_n\,T_e}{m_e\,c_s\,\nu_{ei}}\right)^{1/2}\quad,\qquad L_{\|}\nabla_{\|}\to\nabla_{\|}\quad,\qquad\delta=\frac{1}{k_{\|}^2L_{\|}^2}\ . \tag{4}$$

is introduced to normalize $k_{\|}$ and to absorb the effects of parallel dissipation. Based on experiments showing a very long parallel correlation length, it is assumed that there is only one parallel mode exited, thus reducing the model to two dimensions and allowing the introduction of the coupling constant δ. The finite $k_{\|}$ makes the energy in the background density gradient accessible to the system by driving some modes at $k\approx1$ unstable, while the ion-viscosity μ dissipates energy in the large k region. With $\delta=\mu=0$ we get $n=\phi$ and the Hasegawa-Mima Equation

$$\partial_t(1-\nabla^2)\varphi + \partial_y\varphi + \{\nabla^2\varphi, \varphi\} = 0 \; . \tag{5}$$

This equation can be translated to the Charney-Obukhov equation for Rossby waves. The electrostatic potential then corresponds to the stream-function, while the background density gradient takes the same role as the gradient of the Coriolis force.

Equation (1) is solved numerically using a doubly-periodic spectral code. The time derivatives on the rhs of Eq. (1) are iterated with the HME (5).

Parameters are, if not otherwise stated, $\delta = 0.5$ and $\mu = 0.03$ while the domain of integration is 21×21 length units in size and initialized with low amplitude ($|\phi_k| \sim 10^{-7}$) random fluctuations, having a Gaussian shape spectrum centered at the zero mode in k-space. For about 25 modes around $k_x = 0$ and $k_y = \pm 1$ the linear dispersion relation gives positive growth rates with $\gamma_{\max} \approx 0.05$.

The overall picture is as follows: After the linear growth of the most unstable modes, leading to amplitudes of order unity, the nonlinearities become important and the system gets into a saturated turbulent state as depicted in Fig. 1. This turbulence is characterized by fluctuations up to 50% around the average values. For details see [5]. Visual inspection of the time development of the turbulence — looking at a movie — shows transient structures moving through the domain.

III TRACKING COHERENT STRUCTURES

We define coherent structures as entities trapping plasma-particles for a time much longer than the typical nonlinear time — corresponding to an internal eddy turnover time. This is given by $t_{nl} \approx 1/|\phi|_{\max} k_{\mathrm{av}}^2 \approx 1$, with the spectrum being peaked at an average wavenumber $k_{\mathrm{av}} \approx 1$. To have a confined area, nested closed isopotential-lines — which are the streamlines of the $E \times B$-velocity — are necessary, so that a coherent structure is localized by definition. An additional criterion to identify a coherent structure in the flow is found using the Weiss field

$$Q_{\mathrm{W}} = \frac{1}{4}\left(\sigma^2 - \omega^2\right) . \tag{6}$$

It balances the square of the strain $\sigma^2 := (\partial_x u - \partial_y v)^2 + (\partial_y u + \partial_x v)^2$ (u and v are the velocities in x and y respectively) and the square of the vorticity $\omega := \nabla^2\varphi$. This balance was introduced by Weiss [6] and separates the flow into elliptic regions defined by $Q_{\mathrm{W}} < 0$ and hyperbolic regions defined by $Q_{\mathrm{W}} > 0$. In the latter, fluid elements are separated from each other by the dominating strain, while in the elliptic regions the rotation dominates the dynamics and keeps fluid elements together. Hereby it is possible to distinguish between areas with and without particle confinement.

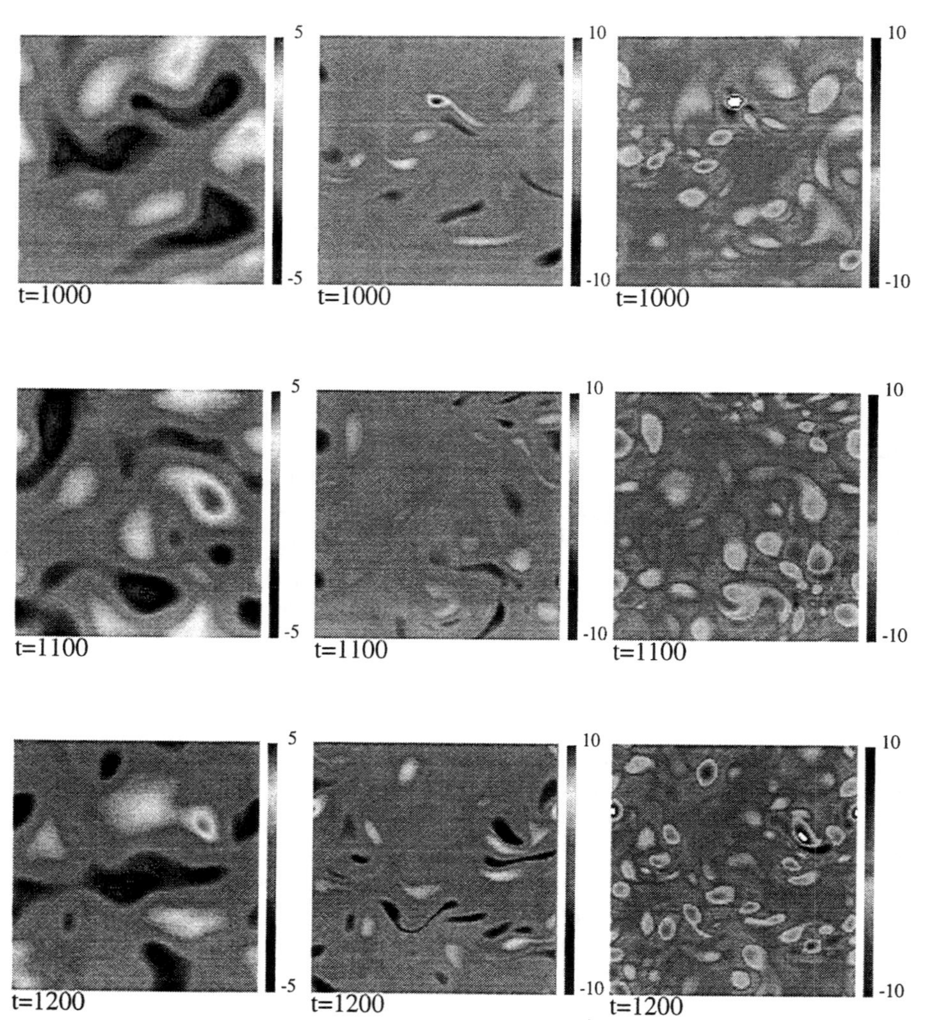

FIGURE 1. Potential ϕ (left), vorticity $\nabla^2\phi$ (center), and Weiss field (right) for times $t = 1000$ to $t = 1200$. (For reasons of clarity only values between 10 and -10 were colour-coded in the Weiss field, although it has values down to -25)

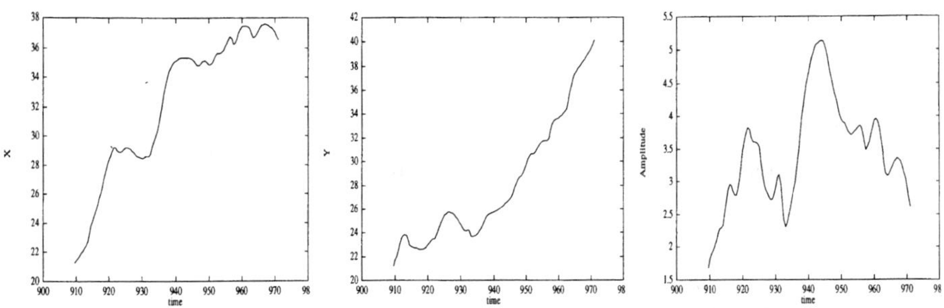

FIGURE 2. x-position (left), y-position (center) and amplitude (right) of a single structure over time.

However, as the Weiss field only analyses the frozen in flow, it is still necessary to take the time evolution of the system into account. This can be done by identifying and tracking structures using the Weiss field and setting a minimum lifetime before they qualify for the notation coherent structure.

Using the Weiss-field we can assign sizes $L_{x,y}$ to each structure over which $Q_{\mathrm{W}} < 0$, and — assuming a Boltzmann distribution $n \approx \phi$ to lowest order — a positive mass M is defined by

$$M = \mathrm{sgn}(n(\vec{r}_0)) \int_A n(\vec{r})\, H\left(-Q_{\mathrm{W}}(\vec{r})\right) d^2r \tag{7}$$

where $H(\zeta)$ is the Heaviside-function. The integration is carried out over an area A including only one event where it is assumed that the sign of the density fluctuation does not change over the extent of a structure. The exact position $\vec{R}$ of a structure is given by its center of mass

$$\vec{R} = \frac{1}{M} \int_A \vec{r}\, |n(\vec{r})|\; H\left(-Q_{\mathrm{W}}(\vec{r})\right) d^2r\ . \tag{8}$$

The track of a single longer lived event is shown in Figure 2. Average quantities taken from all events found can be taken from Table 1. Motivated by the interpretation of the Charney-Obukhov equation we call structures with positive potential (respectively density) ''anti-cyclones" and events with a negative potential ''cyclones". They appear symmetrical in the flow. The distribution of velocities is widespread (Figure 3), so that an individual structure path can be very different from the average picture we get. Further it is interesting to notice that structures of all masses (Figure 3) and lifetimes (Figure 4) can appear, and that there is only a weak correlation between the maximum mass $M_{\max}$ a structure achieves during its existence and its lifetime.

TABLE 1. Average lifetime t_l, velocity v, size L, Mass M and transport Γ of potential anti-cyclones and cyclones with standard deviation σ and number of observations N.

	Anti Cyclones			Cyclones		
	Value	σ	N	Value	σ	N
Lifetime t_l	14.5	12	1515	14.7	13	1531
v_x	0.15	0.47	43898	−0.16	0.47	45074
v_y	0.41	0.40	43898	0.41	0.39	45074
L_x	2.67	0.88	43898	2.66	0.88	45074
L_y	2.52	0.89	43898	2.48	0.86	45074
M	10.2	8.8	43898	9.9	8.6	45074
Γ	0.0051	4.0	43898	0.0047	3.8	45074

IV TRANSPORT

As the structures not only have a finite velocity component in the direction perpendicular to the background density gradient, but also parallel to the latter they account for particle transport. Regarding cyclones and anti-cyclones separately and averaging over all events the flux is evaluated:

$$\Gamma_{\mathrm{coh}} = \frac{1}{T} \int \sum_{\mathrm{C,AC}} M_{\mathrm{C,AC}} v_{x\,\mathrm{C,AC}}\, dt\ . \tag{9}$$

The total transport due to confined density in the structures is $\Gamma_{\mathrm{coh}} \approx 0.01$ compared to $\Gamma_{\mathrm{Fluc}} \approx 0.2$ for the fluctuation induced flux (taken from [4]), both at $\delta = 0.5$, so that $\Gamma_{\mathrm{coh}}/\Gamma_{\mathrm{Fluc}} = 5\%$. Even if this value is small, one has to note that structures provide a different kind of transport compared to the usually considered fluctuation-induced one. As the arrival of a structure is connected to a quite large transport event, the statistical properties of the transport are changed. This is seen by looking at the Probability Density Functions (PDF) of the various transport components. For the fluctuation-induced transport with an algebraic relation between density n and potential fluctuations ϕ (e.g. $n_k = [1 + i\ \epsilon(k)]\,\phi$) one evaluates

$$\Gamma_{\mathrm{Fluc}} = \int n v_x^{E\times B} dy \tag{10}$$

where v is given by the $E \times B$ velocity. One finds

$$p(\Gamma_{\mathrm{Fluc}}) = \frac{1}{\pi} \frac{\sqrt{1-\gamma^2}}{W} K_0 \left(\frac{|\Gamma_{Fluc}|}{W} \right) \exp\left(-\gamma \frac{\Gamma_{Fluc}}{W} \right)\ . \tag{11}$$

for the PDF. Here γ is a measure for the correlation between density and potential fluctuations and should approximately be of order minus one in the

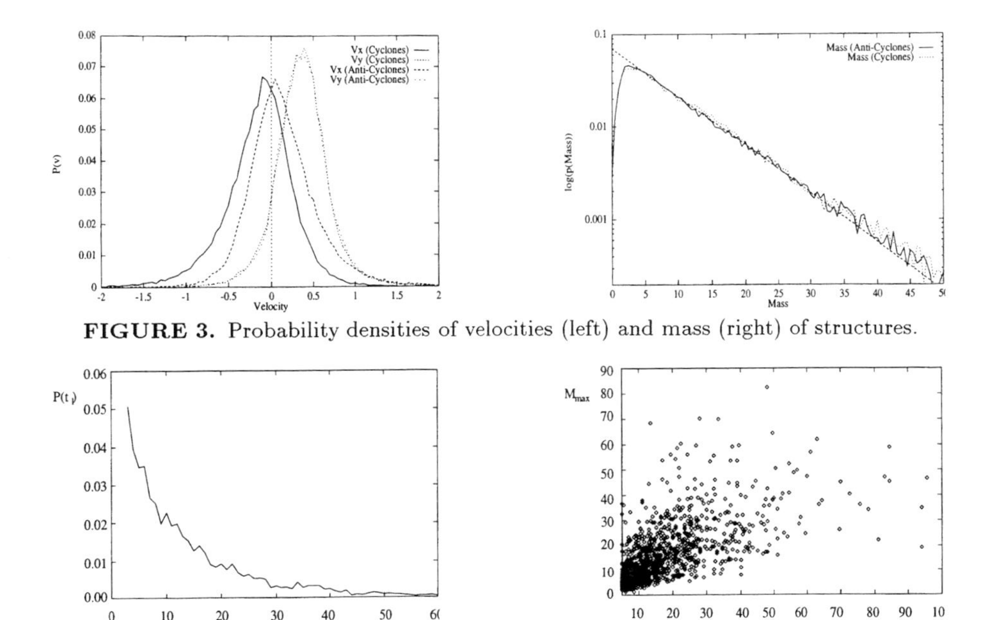

FIGURE 3. Probability densities of velocities (left) and mass (right) of structures.

FIGURE 4. Probability densities of life-time t_l (left) and maximum mass $M_{\max}$ (right) over life-time.

present case, as negative values of γ depict an outward flow. W is the product of the square variances of ϕ and n and also of order one. Details can be found in Carreras et al. [7]. In Figure 5 we show the transport-PDFs as calculated from (11) and numerical values, split into fluctuation-induced transport and the component caused by the structure movement.

The theory based on the density/potential relation is in good agreement with the data from numerical simulations for small transport events, but for large transport events deviations from the analytical expression become obvious. This is to be expected, as for large fluctuations an algebraic relation between density and potential is no longer given, while non-linear effects are important. Inclusion of the effect of trapping through structures in the PDF shows that this mechanism dominates for large transport events and arranges towards a power-law PDF, thus giving relatively high probabilities for very large transport events.

The cause of this transport due to the density trapped in structures is the polarization drift

$$\vec{v}^{\mathrm{pol}} = -\frac{cm_i}{eB^2}\frac{d}{dt}\nabla\phi. \tag{12}$$

It is well known that in the adiabatic limit $n = \phi$ — here $\delta = 0$ — the fluctuation-induced transport due to the $E \times B$ drift vanishes as can be easily

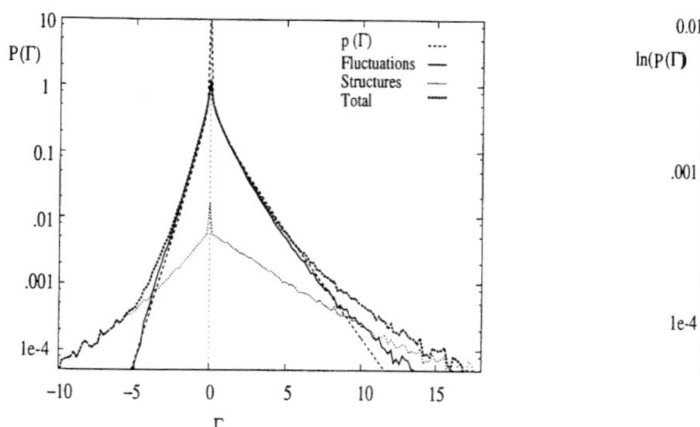

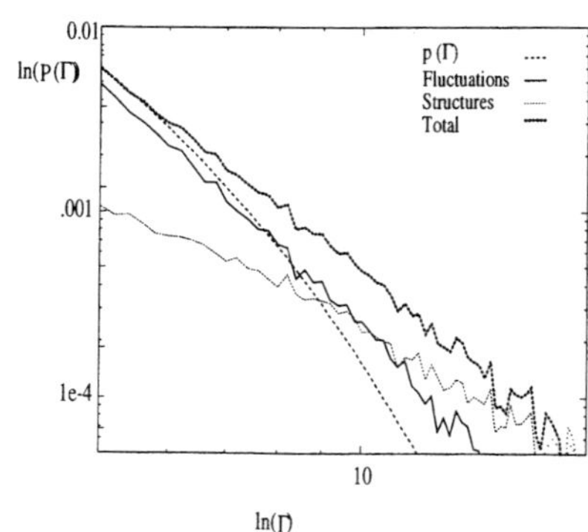

FIGURE 5. Log-normal (left) and log-log (right) plot of PDFs from Eq. (11) for $\gamma = -0.4$ and $W = 0.9$, fluctuations, structures and the sum of the latter two.

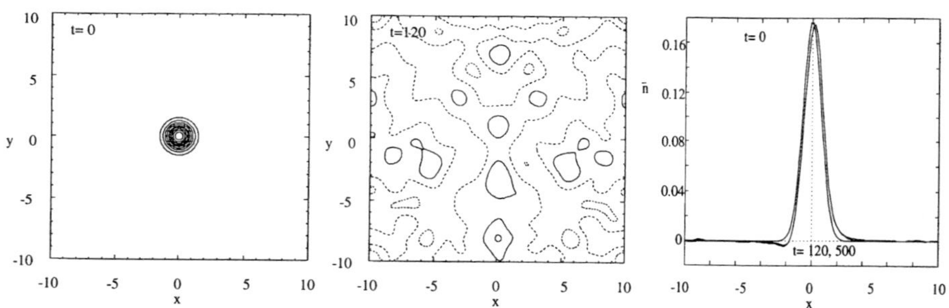

FIGURE 6. Isodensity contours for a small amplitude Gaussian (dz =0.02) at $t = 0$ (left) and $t = 120$ (middle) and corresponding fluctuation profile (right).

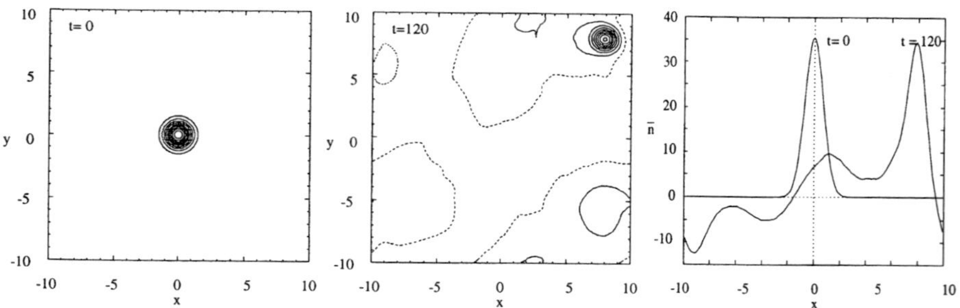

FIGURE 7. Isodensity contours for a large amplitude Gaussian (dz = 2) at $t = 0$ (left) and $t = 120$ (middle) and corresponding fluctuation profile (right).

seen from Eq. (10). So we define the fluctuation profile as

$$\bar{n}(x,t) = \int n(x,y,t)\,dy. \tag{13}$$

This quantity is constant in time when there is no transport. We therefore consider the time evolution of two Gaussian pulses according to the HME, one with a small amplitude, thus being in the linear regime, and the other with a large amplitude, being in the non-linear regime. For the linear Gaussian the profile stays constant (see Figure 6) over long times (after some initial adjustment). Even when the structure has long since dispersed into waves, the initial profile can still be retrieved from the flow, thus clearly indicating that there is no transport. In the nonlinear case however the fluctuation profile changes significantly as shown in Figure 7. This is due to the fact that now the polarization drift can no longer be neglected when considering the transport. The correponding flux is to lowest order given by

$$\Gamma_{\mathrm{coh}} = \int \frac{d}{dt}\frac{\partial}{\partial x}\phi\,dy. \tag{14}$$

For details see [8].

V CONDITIONAL SAMPLING AND SPECTRA

In this section we will focus on the spatial configuration of the structures and their spectral properties. To that extent, first an additional tool to analyze the flow is introduced: the conditional sampling technique (CST). CST has been often and successfully applied in plasma experiments and fluid dynamics [9,10], the basic ideas being discussed in [11]. Originally the CST is applied to derive, from independently taken time-series, a picture of the spatial organization of a flow.

Here we will use it in a different way: In each time-slice we again search for events E located at positions $\vec{r}_E$ and time t_E. Then a subsequence of time-slices is sampled around each event and normalized by the number of events N_E

$$\Phi_C(\vec{R},T) := \sum_E \phi(\vec{r}_E+\vec{R}, t_E+T)/N_E = < \phi(\vec{r},t)|E > \,. \tag{15}$$

If the dynamics in the vicinity of an event is reproducible, Φ_C should reflect the average dynamics. When used for the analysis of time-series, large scale structures, if they are reproducible in the experiment, can be found by this kind of analysis, while small scale erratic turbulence averages out. Here we know what kind of structures we are looking for, concerning sizes and velocities. As the spread in the velocities is high, we expect that the sampled coherent structures are smeared out when T is of the order $L/\sigma_v \approx 5$. While

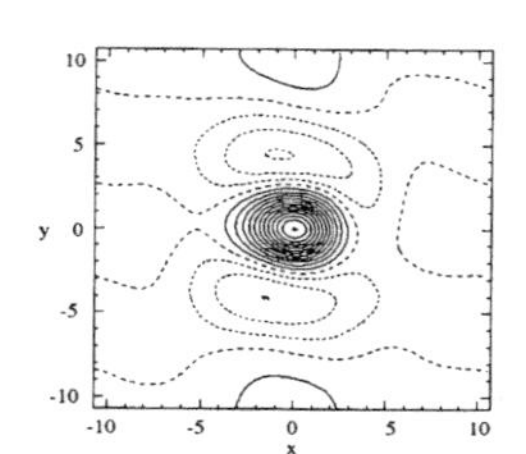

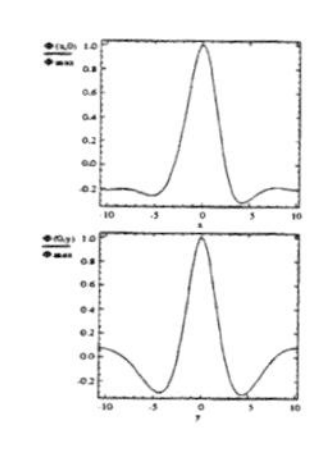
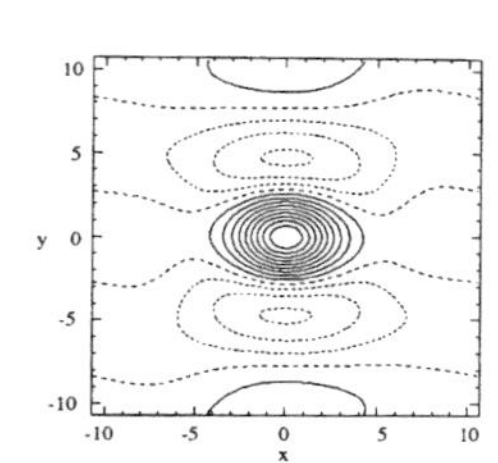

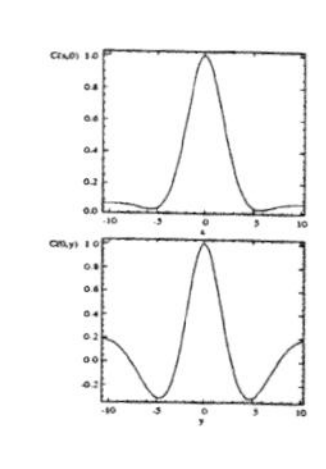

FIGURE 8. Normalized Conditional Sample (left) at $T = 0$ sampled using the maximum criterion as a trigger and Two Point Correlation (right). On each side: Contourplot (left) with 0.1 spacing between levels and sections in x (top, right) and y (bottom, right).

individual paths of the structures vary a lot, we can use sharper criteria for the sampling, thus trying to sample more similar situations. The cost for this is that with a smaller number of events to be sampled the statistics get worse. First we choose local maxima or minima as trigger for the sampling. The result is depicted — for the sampled maxima only, as the sampled minima look symmetric — in Figure 8 and compared for $T = 0$ to the two point correlation function

$$C(\vec{r}) = \frac{1}{< \phi^2 >} \int \exp(i\vec{k}\vec{r}) \left(\frac{1}{T} \int_0^T |\phi_k(t)|^2 dt \right) d\vec{k} \,. \tag{16}$$

The correlation function is — as can easily be seen — dominated by the structures in the flow as they account for most of the correlations and occupy a significant amount of the total area. The evolution of Φ_C with T reflects the movement of the maxima in the y and x direction as determined before. Due to the spread in velocity the area of positive correlation smears out as expected. Additionally one can see, that there is a large probability for other structures of the same size to be at the same y-position, whereas structures of the opposite sign appear with large probability below or over the reference point. This indicates the existence of a global zonal flow, with the localized structures riding on top of this background.

To investigate the behaviour of two monopolar structures which are aligned as as dipole, the maximum value of the electric field $\vec{E} = -\nabla\varphi$ in a direction anti-parallel to the x-axis was used to trigger the sampling, selecting dipolar structures where the positive ($v_x > 0$) and negative part ($v_x < 0$) are colliding. This criterion alone leads to insufficient results, as the dipole may be turning right or left, giving a very weak conditional sample for $T \neq 0$. Thus only events were considered for the sampling process, where the absolute value of the positive potential a length L_x away from the trigger point was more than 5% larger than that of the negative potential the same distance away in the other direction. The result of that sampling process is shown in Figure 9. It is seen, that in this combination the weaker monopole is torn around by the stronger one until both components are able to continue their way

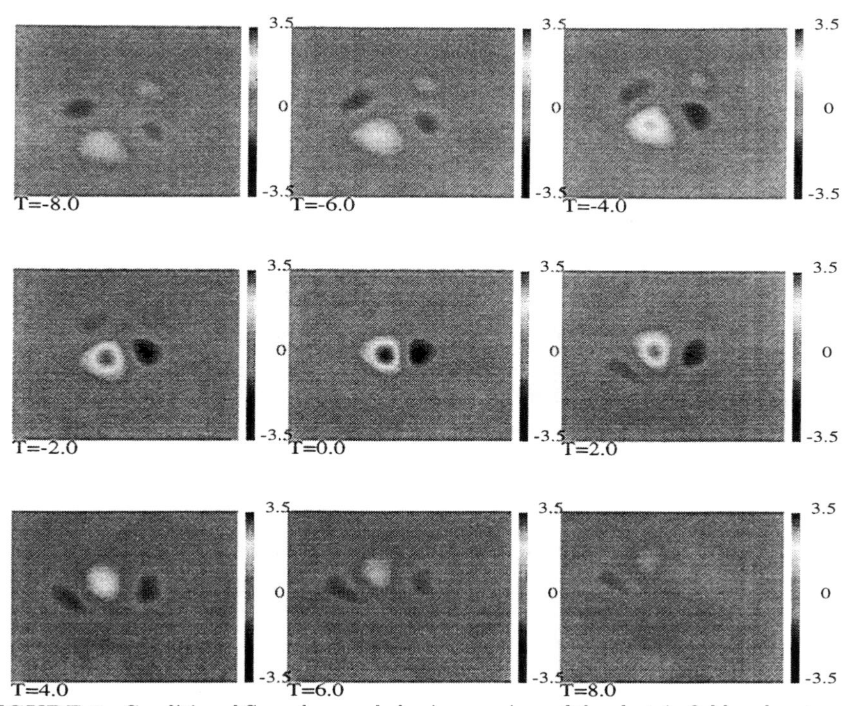

FIGURE 9. Conditional Sample sampled using maxima of the electric field and a stronger positive monopole for different time lags T.

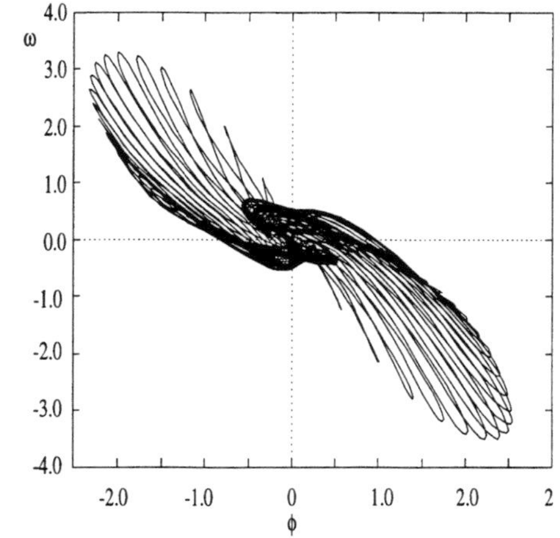

FIGURE 10. Scatter plot showing ω over ϕ for the conditional sample of Figure 9 at $T = 0$, values of same location x being connected.

in opposite x-directions. For $T = 0$ the sample reminds one strongly of the well-known dipole solution of the Hasegawa-Mima equation. In this context it is interesting to note that the conditional sampling at $T = 0$ shows a well organized scatter plot (Figure 10) with a nonlinear relation between ϕ and ω, demonstrating that the CST revealed a highly organized state.

VI SUMMARY

Using different techniques for the detection of structures we were able to clarify their behaviour in driven, quasistationary drift-wave turbulence. The following picture for the overall dynamics of structures emerges:
Transient monopolar structures move through the flow. Although the individual paths of cyclones and anti-cyclones are complicated, they move on the average in opposite directions parallel to the background density gradient and account for transport of the trapped particles. This transport changes the PDF of the transport to a more intermittent behaviour, as the structures are connected with large transport events. This transport is due to the polarization drift and has to be considered even when there is no phase shift between density and potential fluctuations.

Using the CST on the numerical data some averaged dynamics could be found. So in collisions the monopoles may form temporary dipoles, but these tilt and at latest after a rotation of ninety degrees the dipole breaks again into two monopoles. Thus no persistent dipoles could be found in the flow, but nevertheless, the temporary dipoles are highly organized.

Acknowledgement
Discussions with J.J. Rasmussen, K. Rypdal, A.H. Nielsen, and K.H. Spatschek are greatfully acknowledged. Part of the work was done under the European Union Marie Curie grant N° 5004-CT96-5023.

REFERENCES

1. M. Kono and E. Miyashita, Phys. Fluids **31**, 326 (1988).
2. W. Horton, Phys. Fluids **29**, 1491 (1986).
3. J. Crotinger and T.H. Dupree, Phys. Fluids B **4**, 2854 (1992).
4. V. Naulin and K.H. Spatschek, Phys. Rev. E **55**, 5883 (1997).
5. V. Naulin, Ph.D. thesis, Heinrich-Heine Universität, Düsseldorf, 1995.
6. J. Weiss, Physica D **48**, 273 (1991).
7. B.A. Carreras *et al.*, Phys. Plasmas **3**, 2664 (1996).
8. V. Naulin, to be published (1997).
9. A.H. Nielsen, H.L. Pécseli, and J.J. Rasmussen, Phys. Plasmas **3**, 1530 (1996).
10. H.L. Pécseli and J. Trulsen, Phys. Fluids B **1**, 1616 (1989).
11. R.J. Adrian, Phys. Fluids **22**, 2065 (1979).

Turbulent Countergradient Flow as a Problem in Kinetic Theory

R.E. Robson[a] and C.L. Mayocchi[b]

School of Computing, Mathematics and Physics
James Cook University, [a] Cairns 4870 and [b] Townsville 4811, Australia

Abstract. Dispersion of a passive additive in a turbulent medium is treated as a problem in the kinetic theory of gases. In particular, we solve the kinetic equation with an interaction term represented by a simple relaxation time or Krook model, and consider countergradient flow induced by large gradients in a layered source. Both analytic and numerical results are reported. A generalised flux-gradient relation, suitable for practical application, is derived.

I INTRODUCTION

Charged "test particles" or "swarms" in nonturbulent gases and plasmas [1], and "fluid particles" of "passive additive" in a turbulent fluid [2] have much in common from a theoretical point of view. In both cases, the particles do not interact with each other and do not disturb the (known) properties of the background fluid. In both cases a description through a phase space distribution function is possible, which may be obtained from the solution of an appropriate kinetic equation, eg, the Fokker-Planck equation [1,3] or the Krook "relaxation time" model equation [1,4–7]. All quantities of physical interest (eg, densities, fluxes, transport coefficients) then follow. Large-gradient effects [1,6,8–10] are of interest in both sets of literature, where they are variously labelled as "nonhydrodynamic" [1] and "near-field" [8] phenomena respectively. Certain well known approximations — "K-theory" in turbulent transport [2], the Chapman-Enskog scheme in the kinetic theory of gases [11] — are based upon the assumption that the length scale for variation of density is large compared with the dimension of the transfer mechanism (eddy size for turbulent transport, mean free path for collisional transport) and lead to simple flux-gradient relationships. Such theory is clearly inappropriate for dealing with large gradient effects. A rather striking breakdown of K-theory is evident, for example, in countergradient flow [6–9,12,13], first observed by Denmead and Bradley [12] in a forest canopy, and the theoretical understand-

CP414, *Two-Dimensional Turbulence in Plasmas and Fluids:* Research Workshop
edited by R. L. Dewar and R. W. Griffiths

ing of this peculiar phenomenon forms the thesis of this paper.

A The kinetic theory model

As a simplification we assume a steady-state, horizontally uniform, force free situation, in which the z-axis defines the vertical, and a simple relaxation time or Krook model [1,4,6] for the interaction term. Turbulence in the background fluid is assumed for simplicity to be stationary and homogeneous, and governed by Gaussian statistics. After integrating out the velocities in both the x and y directions we are left with the following kinetic equation in one spatial coordinate z and one velocity component c_z:

$$c_z \frac{\partial f}{\partial z} = -\frac{(f - nw(\alpha, c_z))}{\tau} + \bar{S}(z, c_z), \tag{1}$$

where τ is the characteristic (Lagrangian integral) time scale for turbulent action,

$$w(\alpha, c_z) = (\frac{\alpha^2}{2\pi})^{\frac{1}{2}} \exp(-\alpha^2 c_z^2/2), \tag{2}$$

is a Maxwellian characterising the equilibrium distribution function in which

$$\alpha = \frac{1}{u_*} \tag{3}$$

and u_* is the friction velocity.

The source distribution $\bar{S}$ represents the steady rate of production/loss of scalar per unit volume of phase space in the region under consideration. In general it depends upon velocity as well as position. The source term in configuration space is given by

$$S(z) = \int_{-\infty}^{\infty} \bar{S}(z, c_z) dc_z, \tag{4}$$

while the density is

$$n(z) = \int_{-\infty}^{\infty} f(z, c_z) dc_z. \tag{5}$$

The boundary condition is idealised as $\Gamma_z = 0$ at $z = 0$, corresponding to perfect reflection.

Initially we take a layered source distribution of thickness h at a height z_0 above a non-absorbing ground:

$$S(z) = \begin{cases} S_0, & z_0 < z < z_0 + h \\ 0, & \text{otherwise}, \end{cases} \tag{6}$$

where S_0 is the constant production of the scalar per unit time per unit thickness of the layer. Integrating (1) over velocity, c_z, we obtain

$$\frac{\partial}{\partial z}\Gamma_z = S(z), \tag{7}$$

where

$$\Gamma_z = \int_{-\infty}^{\infty} c_z f(z, c_z) dc_z \tag{8}$$

is the vertical flux of additive at height z. Integrating (7) further over z, we obtain expressions for the flux Γ_z in the three regions of interest:

$$\Gamma_z = \begin{cases} 0, & 0 \le z < z_0 \\ S_0(z - z_0), & z_0 \le z \le z_0 + h \\ S_0 h, & z_0 + h < z < \infty, \end{cases} \tag{9}$$

where the boundary condition has been employed. It is obvious that $\Gamma_z \ge 0$ in all regions, ie, flow is vertically upwards everywhere. Note also that in obtaining these results no assumptions or approximations have been made.

We now discuss modelling of the effect in two stages: Firstly, in Section 2, by direct numerical solution of a kinetic equation and secondly, in Section 3, by approximate analytic solution, completing the details outlined earlier in a brief report [6].

II K-THEORY AND NUMERICAL SOLUTIONS

A K-theory solution

If the gradient term on the left hand side of (1) is treated as a small perturbation, then the first Chapman-Enskog approximation (11) gives

$$\Gamma_z = -K\frac{\partial n}{\partial z}, \tag{10}$$

where

$$K(z) = u_*^2 \tau \tag{11}$$

is the classical diffusion coefficient. Note that in general, τ and therefore K, will be functions of z.

Using this in conjunction with (9) we obtain for the three regions

$$
n(z) = \begin{cases} n_0\,, & 0 \leq z < z_0 \\ n_0 - S_0 \displaystyle\int_{z_0}^{z} \frac{(z' - z_0)}{K(z')} dz'\,, & z_0 \leq z \leq z_0 + h \\ S_0 h \displaystyle\int_{z}^{\infty} \frac{dz'}{K(z')}\,, & z_0 + h < z < \infty. \end{cases} \tag{12}
$$

To determine n_0, the ground level density, we invoke continuity of $n(z)$ at $z = z_0 + h$, which yields

$$
n_0 = S_0 \left\{ \int_{z_0}^{z_0+h} \frac{(z' - z_0)}{K(z')} dz' + h \int_{z_0+h}^{\infty} \frac{dz'}{K(z')} \right\}. \tag{13}
$$

We now assume for simplicity that τ is a monotonically increasing function of height with the following power law dependence:

$$
\tau(z) = \tau_0 \left(\frac{z}{z_0} \right)^p, \tag{14}
$$

where p is a constant taken to be always greater than 1, corresponding to an unstable atmosphere. Other more sophisticated parameterization in terms of a Monin-Obukhov length is possible, but we do not investigate this here. The eddy dimension at height z,

$$
\ell(z) = u_* \tau(z) \tag{15}
$$

also has the same power law dependence on height as given by (14).

With these parameters, equations (12) and (13) can be integrated to obtain the K-theory profile, which is one where density decreases monotonically with height, the flux always being down the gradient (see Figure 1). We shall show that solution of the kinetic equation without the weak-gradient assumption inherent in K-theory indicates a region where density may actually *increase* with height and flow is actually *up* the gradient.

B Numerical solution

We now discuss numerical solution of (1). The spatial derivative is approximated using standard backward finite differencing techniques, while the scalar density is found from (5) using Hermite-Gauss quadrature. The boundary conditions used are $\Gamma_z = 0$ at the ground and $n = 0$ at infinity. It is convenient to use the following coordinate in place of z above the source, $z > z_0 + h$:

$$
X = \frac{z^{(1-p)}}{(p-1)}. \tag{16}
$$

Recalling that p is greater than 1, it is clear that $X \longrightarrow 0$ as $z \longrightarrow \infty$. We integrate the equation from infinity, where density is zero and $X = 0$, down to the top of the source region, and also simultaneously up from the ground to the top of the source Equation (1) is solved iteratively at each point: Starting with an initial estimate of $n(z)$, equation (1) is solved for $f(z, c_z)$, which is then substituted back in (5) to calculate $n(z)$, and so on. For these numerical calculations we have taken a source thickness of $h = 2z_0$, while in [14] we set $p = 1.2$. Under these conditions $\ell \sim h$ in the source region and large gradient effects are expected to come into play.

Thus we obtain the density profile given in Figure 1, where dimensionless quantities have been plotted for convenience. Above the source region ($z/z_0 > 3$), the profile closely approximates (to within 2%) the K-theory result (12), and below the source ($z/z_0 < 1$), both approaches lead to the same constant profile.

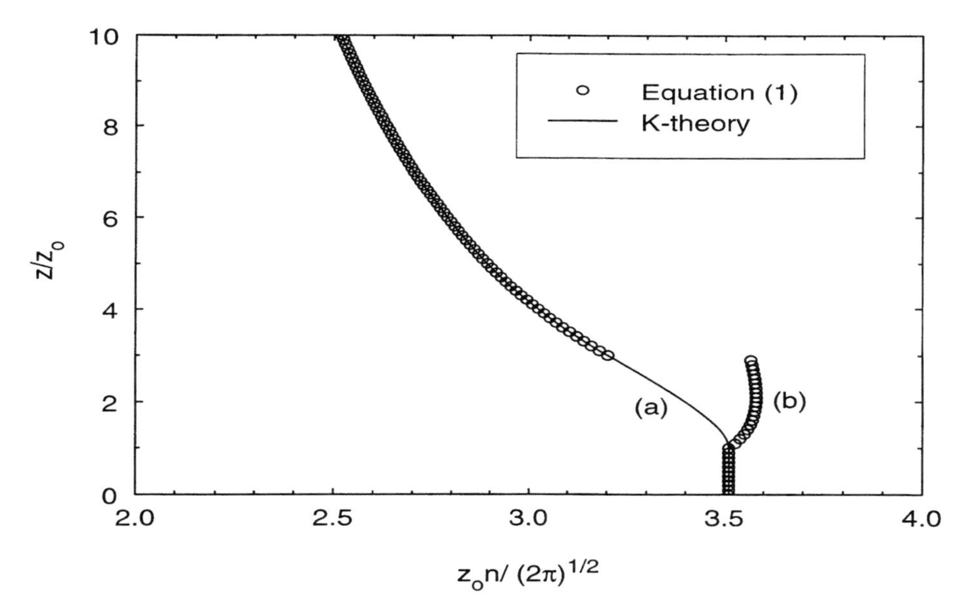

FIGURE 1. Dimensionless density as a function of dimensionless height z/z_0, calculated (a) analytically from K-theory, and (b) numerically from solution of the kinetic equation (1), produced by a uniform layered source of thickness 2, located between heights 1 and 3. Turbulence in the atmosphere is characterised by an eddy time scale that increases with height according to (14), with $p = 1.2$

We recall that the flux Γ_z is upward everywhere, including within the source region, and we observe that there ($1 < z/z_0 < 3$), density actually at first *increases* with height. That is, the prediction is that flux is *up* the gradient in part of the source region.

A comparison of the theoretical profile in Figure 1 with the experimental results of Denmead and Bradley [12] and Coppin *et al* [14] indicates good qualitative agreement. Both experimental and theoretical profiles show a maximum in the source region, with countergradient flow, and decay to zero above this. Below the source, densities vary only slowly.

A full quantitative comparison, which lies outside the scope of this paper, would require better modelling of the interaction term, eg, through use of a Fokker-Planck operator, rather than the simple relaxation time model of (1), more accurate representation of the background turbulence, especially in the source region, and more realistic representation of the surface boundary condition.

It is clear from Figure 1 that K-theory is hopelessly inadequate even from a qualitative perspective within the source region (unless one postulates a negative, unphysical value of K !), although it appears quite satisfactory outside that region. The fundamental difficulty stems from the fact that it effectively assumes the gradient term in the kinetic equation to be weak. Thus, a correct qualitative description of the countergradient flow phenomenon has more to do with the *left hand side* of the kinetic equation than the right hand side. This is why even such a simple kinetic model as (1) is capable of giving reasonable results.

The above calculations were for a uniform layer source. We now move on to consider another more realistic source distribution.

C Gaussian source distribution

A Gaussian source distribution with a mean height of 2 (dimensionless units) and standard deviation 0.1 was next considered (see Figure 2).

The resulting density profile is given in Figure 3.

As the width of the Gaussian profile diminishes, so too does the width of the density profile, while the maximum density attained increases, ie, the peak in n sharpens. In a similar manner, as the source is widened, the density profile maximum is reduced, as indicated by curves (a), (b) and (c) in Figures 2 and 3. These results confirm Raupach's observation [13] that the width of the source term is one of the most significant mechanisms in producing countergradient flow. Strong localized sources will give a more pronounced countergradient effect, while weaker, more diffuse sources may produce only small regions of increasing density with height. Note that all curves in Figure 3 approximate closely the K-theory profiles away from the source region.

If we move the source in Figure 2 to a higher mean height of 8, then we observe an increase in the maximum density attained (Figure 4). This is a manifestation of the monotonically increasing nature of eddy size with height:

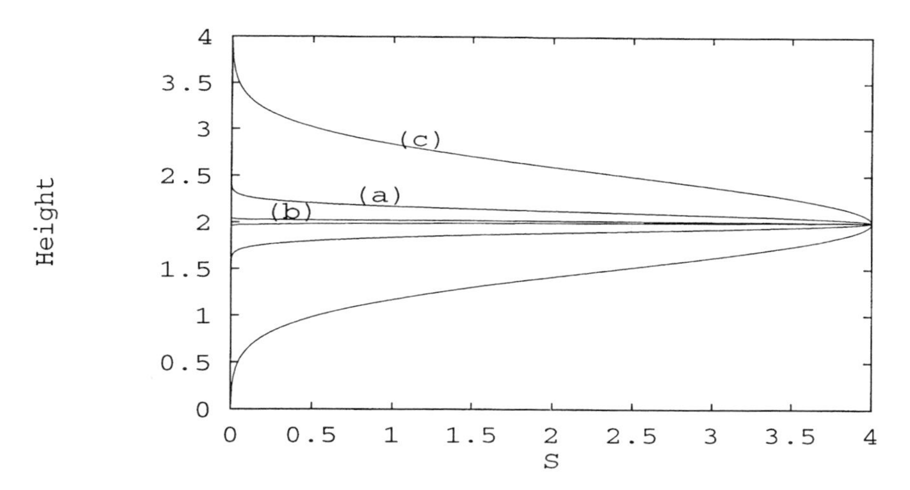

FIGURE 2. Gaussian source distributions S for which the mean height is 2 (dimensionless units) in all cases and standard deviations from the mean are (a) 0.1 (b) 0.01 and (c) 0.5 respectively.

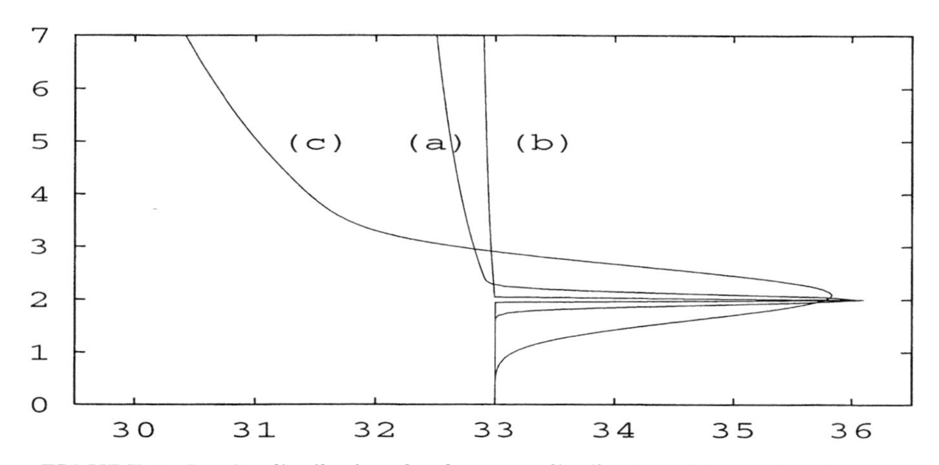

FIGURE 3. Density distributions for the source distributions of Figure 2, with the same turbulence parameters as prescribed in Figure 1.

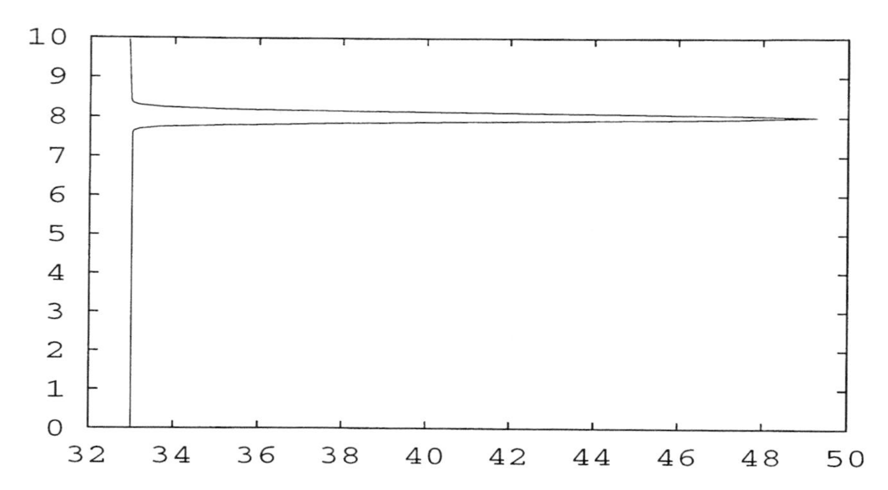

FIGURE 4. Density distribution for a Gaussian source distribution of mean height 8, standard deviation 0.1

At greater source heights, the eddy size is correspondingly larger, and countergradient flow effects are more pronounced.

III ANALYTIC SOLUTION: THE GENERALISED FLUX-GRADIENT RELATION

We introduce a vertical coordinate which is a generalisation of (16):

$$V(z) = -\int^{z} \frac{dz}{\tau(z)} \tag{17}$$

and leave $\tau(z)$ arbitrary for the present. Notice that $V \to \infty$, $\tau \to 0$ as $z \to 0$ and $V \to 0$, $\tau \to \infty$ as $z \to \infty$. Thus (1) becomes

$$c_z \frac{\partial f}{\partial V} - f = -n(V)w(\alpha, c_z) - \bar{S}(V, c_z), \tag{18}$$

where

$$\bar{S}(V, c_z) = \begin{cases} S_0 w(\alpha, c_z)\tau(V), & V_0 > V > V_1 \\ 0, & \text{otherwise} \end{cases} \tag{19}$$

and S_0 is the rate of generation of passive additive.

Taking the Laplace transform with respect to the new height variable V we have

$$(kc_z - 1)\bar{f} = c_z f(0, c_z) - \bar{n}w(\alpha, c_z) - \bar{\theta}(k)w(\alpha, c_z), \tag{20}$$

where

$$\begin{aligned}\bar{f}(k, c_z) &= \int_0^\infty e^{-kV} f(V, c_z)dV \\ \bar{n}(k) &= \int_0^\infty e^{-kV} n(V)dV = \int_{-\infty}^\infty \bar{f}(k, c)dc \\ \bar{\theta}(k) &= S_0 \int_{V_1}^{V_0} e^{-kV} \tau(V)dV.\end{aligned} \tag{21}$$

In order to continue, we must specify a value for $f(0, c_z)$ at $V = 0$, ie, $z = \infty$. To have zero density but non-zero flux, we assume

$$f(0, c_z) = c_z \alpha^2 \Gamma_\infty w(\alpha, c_z), \tag{22}$$

where Γ_∞ is the flux at $V = 0$ ($z = \infty$). We then have

$$\bar{n}(k) = \left(\int_{-\infty}^\infty \frac{[c_z^2 \Gamma_\infty \alpha^2 - \bar{\theta}(k)]w(\alpha, c_z)}{(kc_z - 1)} dc_z\right) / \left(1 + \int_{-\infty}^\infty \frac{w(\alpha, c_z)}{(kc_z - 1)} dc\right), \tag{23}$$

and using the following identities

$$\begin{aligned}\int_{-\infty}^\infty \frac{w(\alpha, c_z)dc_z}{kc_z - 1} &= \zeta Z(\zeta) \\ \int_{-\infty}^\infty \frac{\alpha^2 c_z^2 w(\alpha, c_z)dc}{kc_z - 1} &= 2\zeta^2(1 + \zeta Z(\zeta)),\end{aligned}$$

where $\zeta = \alpha/\sqrt{2}k$ and $Z(\zeta)$ is the plasma dispersion function of ref. [15], we obtain the following expression for the Laplace-transformed density:

$$\bar{n}(k) = \frac{\zeta[2\zeta\Gamma_\infty + (2\Gamma_\infty\zeta^2 - \bar{\theta}(k))Z(\zeta)]}{1 + \zeta Z(\zeta)}. \tag{24}$$

We now define the quantity

$$\bar{I}(k) = \frac{\alpha^2 \bar{\Gamma}(k)}{k\bar{n}(k)}, \tag{25}$$

which has the value unity if the flux-gradient relation (10) of K-theory is assumed. Deviation of $\bar{I}(k)$ from 1 therefore corresponds to departures from K-theory. If $I(V)$ represents the inverse Laplace transform of (25), ie,

$$\bar{I}(k) = \int_0^\infty e^{-kV} I(V)dV, \tag{26}$$

it then follows that

$$\Gamma(z) = -\int_z^\infty K(z,z')\frac{dn(z')}{dz'}dz', \tag{27}$$

where

$$K(z,z') = \frac{1}{\alpha^2}I(V(z) - V(z')). \tag{28}$$

Equation (27) is the generalized, non-local flux-gradient relation. In order to calculate the quantity $\bar{I}(k)$ explicitly from (25) we need to determine $\bar{\Gamma}(k)$. Integrating (18) over all c_z we have

$$\frac{\partial}{\partial V}\Gamma(V) - n = -n\int w(\alpha, c_z)dc_z + \int S_o w(\alpha, c_z)\tau(V)dc_z, \tag{29}$$

and since $\int w(\alpha, c_z)dc_z = 1$, we have

$$\frac{\partial}{\partial V}\Gamma(V) = \begin{cases} -S_0\tau(V), & V_0 > V > V_1 \\ 0, & \text{otherwise.} \end{cases} \tag{30}$$

After taking the Laplace transform we obtain

$$k\bar{\Gamma}(k) - \Gamma_\infty = -\bar{\theta}(k), \tag{31}$$

or equivalently

$$\bar{\Gamma}(k) = \frac{\Gamma_\infty - \bar{\theta}(k)}{k}. \tag{32}$$

Thus substituting (24) and (32) in the rhs of (25), we find

$$\bar{I}(k) = \frac{\alpha^2(\Gamma_\infty - \bar{\theta}(k))}{k^2\bar{n}(k)} \tag{33}$$

$$= \frac{\alpha^2(\Gamma_\infty - \bar{\theta}(k))(1 + \zeta Z(\zeta))}{k^2\zeta[2\zeta\Gamma_\infty + (2\Gamma_\infty\zeta^2 - \bar{\theta}(k))Z(\zeta)]}, \tag{34}$$

with

$$\Gamma_\infty = \bar{\theta}(0) = S_0\int_{V_1}^{V_0}\tau(V)dV. \tag{35}$$

These expressions are all exact to this point; however, to proceed any further, approximations are needed. The large V (small z) regime corresponds to small k, or large ζ. Therefore using the large ζ asymptotic expression for the plasma dispersion function (15)

$$Z(\zeta) \approx -\frac{1}{\zeta}\left(1 + \frac{1}{2\zeta^2} + \frac{3}{4\zeta^4} + \dots\right), \tag{36}$$

equation (34) becomes

$$\bar{I}(k) = \frac{-\alpha^2(\bar{\theta}(0) - \bar{\theta}(k))(\frac{1}{2\zeta^2} + \frac{3}{4\zeta^4} + \ldots)}{k^2(-2\bar{\theta}(0)\zeta^2(\frac{1}{2\zeta^2} + \frac{3}{4\zeta^4} + \ldots) + \bar{\theta}(k)(1 + \frac{1}{2\zeta^2} + \frac{3}{4\zeta^4} + \ldots)}$$

$$= \frac{\alpha^2(\bar{\theta}(0) - \bar{\theta}(k))(1 + \frac{3}{2\zeta^2} + \ldots)}{2\zeta^2 k^2 \left(\bar{\theta}(0) - \bar{\theta}(k) + \frac{1}{2\zeta^2}(3\bar{\theta}(0) - \bar{\theta}(k)) - \frac{3}{4\zeta^4} + \ldots\right)}. \quad (37)$$

We now expand $\bar{\theta}(k)$ as follows,

$$\bar{\theta}(k) = \bar{\theta}(0) - k\bar{\theta}'(0) + k^2\bar{\theta}''(0) - \ldots$$

and keep terms only to $O(k^2)$. Thus we obtain:

$$\bar{I}(k) \approx \left(1 + \frac{1}{2\zeta^2}\left[\frac{3\bar{\theta}(0) - \bar{\theta}(k)}{\bar{\theta}(0) - \bar{\theta}(k)}\right]\right)^{-1}$$

$$= \left(1 + \frac{2k\bar{\theta}(0)}{\alpha^2\bar{\theta}'(0)}\right)^{-1}, \quad (38)$$

or

$$\bar{I}(k) \approx \left(1 - \frac{\beta k}{\alpha}\right)^{-1}, \quad (39)$$

where

$$\beta = -\frac{2\bar{\theta}(0)}{\alpha\bar{\theta}'(0)}. \quad (40)$$

Taking the inverse Laplace transform we find,

$$I(V) \approx \frac{\alpha}{\beta} e^{\frac{-V\alpha}{\beta}}. \quad (41)$$

We may then write the transfer coefficient (28) as

$$K(z, z') = \alpha\beta \exp\left(\frac{-\alpha(v(z) - v(z'))}{\beta}\right), \quad (42)$$

and substitution back into (27) gives the approximate generalized flux-gradient relation

$$\Gamma_z - \beta\alpha K \partial_z \Gamma_z = -K\partial_z n, \quad (43)$$

where $K(z) = \tau(z)/\alpha^2$ is the familiar diffusion coefficient appearing in the conventional flux-gradient relation (10).

If we now denote the K-theory profile calculated in the previous section as $n_K(z)$, the solution of (43) can be written as

$$n(z) = \begin{cases} n_K(z) + \beta\alpha S_0(z - z_0) & , z_0 < z < z_0 + h \\ n_K(z) & , \text{outside source region.} \end{cases} \tag{44}$$

Thus within the source region the correction to the K-theory profile is linear, while outside this region it is zero. For the special model (14) equation (40) reduces to

$$\beta = \left(\frac{2(p-1)(2-p)}{\left(1 + \frac{h}{z_0}\right)^{2-p} - 1} \right) \left(\frac{h}{z_0} \right) \left(\frac{\ell_0}{z_0} \right), \tag{45}$$

where $\ell_0 = \tau(z_0)/\alpha$ is the eddy dimension at the base of the source. For $p = 1.2$ and $\ell_0 = z_0$ we find that $\beta \sim 0.4$ varies only slowly over a range of source dimensions.

Equation (43) produces a profile with all the features of the more accurate numerical solution discussed in the last section and shown in Figure 1, viz, $n(z) = n_K(z)$ outside the source, while countergradient flow exists within it. For this reason, the simplified generalised flux-gradient relation (43) is recommended for calculations in practical situations.

IV SUMMARY

We have discussed the solution of the relaxation-time model kinetic equation (1) under the assumption of steady state dispersion of a passive additive from a narrow source into an turbulently fluctuating atmosphere. Both accurate numerical and approximate analytic methods of solution were used. We showed that the large-eddy, large-gradient phenomenon of countergradient flow is fully explained by even this simple kinetic model, with theoretical results fully consistent with experiment. A generalised flux-gradient equation (43) has been proposed for practical calculation using experimental data.

REFERENCES

1. Kumar, K., Skullerud, H.R. and Robson, R.E., *Aust. J. Phys* **33**, 343 (1980).
2. Monin, A.S. and Yaglom, A.M., *Statistical Fluid Mechanics: Mechanics of Turbulence*, Cambridge, MA: M.I.T. Press, 1971, Vol. 1.
3. Monti, P. and Leuzzi, G., *Bound. Layer Met.* **80**, 311 (1996).
4. Lundgren, T.S., *Phys. Fluids* **12**, 485 (1969).
5. Uman, M. , *Introduction to Plasma Physics*, New York: McGraw-Hill, 1964.
6. Robson, R.E. and Mayocchi, C.L., *Phys. Fluids* **6**, 1952 (1994).

7. Mayocchi, C.L., *Ph.D. thesis*: James Cook University, 1994.
8. Kaimal, J.C. and Finnigan, J.J., *Atmospheric Boundary Layer Flows*, Oxford U.P., 1994, Ch.3.
9. Raupach, M.R., Finnigan, J.J. and Brunet, Y., *Bound. Layer Met.* **78**, 351 (1996).
10. Robson, R.E., *Aust. J. Phys.* **50**, 577 (1997).
11. Chapman, S. and Cowling, T.G., *The Mathematical Theory of Nonuniform Gases*, 3rd edition, Cambridge University Press, 1970.
12. Denmead, O.T. and Bradley, E.F., *Flux-Gradient Relationships in a Forest Canopy*, in *The Forest-Atmosphere Interaction*, edited by B.A. Hutchinson and B.B. Hicks, Dordrecht: Reidel, 1985, p. 421.
13. Raupach, M.R., *Quart. J. Roy. Met. Soc.* **113**, 107 (1987).
14. Coppin, P.A., Raupach, M.R. and Legg, B.J., *Bound. Layer Met.* **35**, 167 (1986).
15. Fried, B.D. and Conte, S.D.,*The plasma dispersion function*, New York: Academic Press, 1961.

Geophysical Localized Structures Induced by Non-uniform β and/or Zonal Shear Flow

Bhimsen K. Shivamoggi

University of Central Florida
Orlando, Florida 32816

Abstract. A generalized class of localized structures generated by a non-uniform β and/or a zonal shear flow is explored within the framework of the Charney-Hasegawa-Mima equation.

I INTRODUCTION

Quasi-geostrophic conservation of potential vorticity

$$q \equiv \nabla^2 \varphi - \frac{\hat{f}_0^2}{gH}\varphi + \beta y \tag{1}$$

where φ is the stream function, $\hat{f}_0$ is the local Coriolis parameter, β is the planetary vorticity gradient, H is the depth of the ocean (taken to be uniform) and q is the acceleration due to gravity, leads to the following equation (Charney [1]) for the stream function φ:

$$\frac{\partial}{\partial t}\left(\nabla^2 \varphi - \frac{\hat{f}_0^2}{gH}\varphi\right) + \beta \frac{\partial \varphi}{\partial x} + \left(\frac{\partial \varphi}{\partial x}\frac{\partial}{\partial y} - \frac{\partial \varphi}{\partial y}\frac{\partial}{\partial x}\right)\nabla^2 \varphi = 0. \tag{2}$$

Here,

- $\nabla^2 \varphi$ is the vertical component of the relative vorticity,
- $-\left(\hat{f}_0^2 / gH\right)\varphi$ is the vortex-stretching effect produced by a deformed free surface,
- βy is the planetary-vorticity variation with latitude.

CP414, *Two-Dimensional Turbulence in Plasmas and Fluids:* Research Workshop
edited by R. L. Dewar and R. W. Griffiths

Hasegawa and Mima [2] showed that this equation also describes nonlinear electro-static drift waves in a two-dimensional guiding-center plasma. Hence, equation (2) was dubbed the Charney-Hasegawa-Mima equation (Shivamoggi [3]).

When β is taken to be non-uniform or there is a zonal shear flow and one keeps only the quadratic nonlinearity, the Charney-Hasegawa-Mima equation admits monopole vortex solutions (Larsen [4], Clarke [5], Redekopp [6] and Lakhin et al. [7]). Here, we shall explore a generalized class of localized structures generated by a non-uniform β and/or a zonal shear flow.

II FLOW STRUCTURES DUE TO NON-UNIFORM β /ZONAL SHEAR FLOW

Let us first write equation (2) in the form

$$\frac{\partial}{\partial t}\left(\nabla^2\Phi - \Phi\right) + \left[\Phi, \nabla^2\Phi - \Phi + \hat{f}\right] = 0 \tag{3}$$

where,

$$\left[A, B\right] \equiv \frac{\partial A}{\partial x}\frac{\partial B}{\partial y} - \frac{\partial A}{\partial y}\frac{\partial B}{\partial x}.$$

If there is a zonal shear flow $U(y)$, i.e., one has

$$\Phi(x, y, t) = \phi_0(y) + \phi(x, y, t) \tag{4}$$

where,

$$\phi_0'(y) \equiv -U(y),$$

then equation (3) becomes

$$\frac{\partial}{\partial t}\left(\nabla^2\phi - \phi\right) + \left[\phi + \phi_0, \nabla^2\phi - \phi + \left(\hat{f} - \phi_0 + \phi_0''\right)\right] = 0 \tag{5}$$

Let us look for travelling-wave solutions of the form

$$\phi(x, y, t) = \phi(\xi, y) \tag{6}$$

where,

$$\xi \equiv x - ut.$$

Using (6), equation (5) becomes

$$\left[\phi+\phi_0+uy, \nabla^2\phi-\phi+\left(\hat{f}-\phi_0+\phi_0''\right)\right]=0. \tag{7}$$

III A COMPATIBILITY CONDITION

Multiply equation (7) through by ξ and integrate over the whole plane. This leads to the following compatibility condition -

$$\iint_{R^2}(u+\beta-U'')\phi\, d\xi\, dy=0 \tag{8}$$

In the constant-β, uniform zonal-flow limit, (8) leads to the Flierl-Stern-Whitehead Zero Angular Momentum Theorem (Flierl et al. [8]) -

$$\iint_{R^2}\phi\, d\xi\, dy=0. \tag{9}$$

(9) implies that the localized, travelling waves must have zero net angular momentum. The dipole-vortex is the simplest flow configuration compatible with this condition.

However, in the general non-uniform-β and/or zonal shear-flow case, (9) is inoperational. So, more general flow configurations are allowable.

IV LOCALIZED FLOW STRUCTURES

Equation (7) has the solution

$$\nabla^2\phi-\phi+\left(\hat{f}-\phi_0+\phi_0''\right)=F(\phi+\phi_0+uy) \tag{10}$$

F being an arbitrary function of its argument. F may be determined uniquely by the planetary vorticity and zonal flow profile on open streamlines extending to infinity.

For localized solutions, one has the boundary condition

$$|x|\Rightarrow\infty,\ y \text{ fixed}:\ \phi, |\nabla\phi|, |\nabla^2\phi|\Rightarrow 0 \tag{11}$$

This leads to

$$\nabla^2\phi-\phi+\tilde{f}\left(y+\frac{\phi_0}{u}\right)-\tilde{f}\left(y+\frac{\phi_0+\phi}{u}\right)=0 \tag{12}$$

where,

$$\tilde{f}\left(y+\frac{\varphi}{u}\right)\equiv\hat{f}(y)-\varphi(y)+\varphi''(y).$$

V. EQUILIBRIUM PROFILES

It should be noted that the equilibrium representation

$$\tilde{f}\left(y+\frac{\phi_0}{u}\right) \equiv \hat{f}(y) - \phi_0(y) + \phi_0''(y) \tag{13}$$

cannot lead to a determination of $\tilde{f}(\eta)$ for general profiles $\hat{f}(y)$ and $\phi_0(y)$, except for some monotonic profiles $\hat{f}(y)$ and $\phi_0(y)$.

EXAMPLE 1: Consider

$$\hat{f}(y) = \beta y, \phi_0(y) = \alpha e^{-\sqrt{\frac{\beta}{u}+1}\, y}. \tag{14}$$

Using (14), (13) leads to

$$\tilde{f}(\eta) = \beta\eta. \tag{15}$$

EXAMPLE 2: Consider

$$\hat{f}(y) = \alpha y^2 + \beta y, \phi_0(y) = ay^2 + by + c. \tag{16}$$

Using (16), (13) leads to

$$\tilde{f}(\eta) = \left(\frac{\alpha - a}{a}\right)u\eta + \left(2a - \frac{c\alpha}{a}\right) + \frac{(\beta+u) - \frac{\alpha(b+u)}{a}}{2a}\Big[-(b+u) + \pm\left\{(b+u)^2 - 4(c-u\eta)\right\}^{1/2}\Big]. \tag{17}$$

VI SAGDEEV POTENTIAL ANALYSIS

Let us write equation (12) in the form

$$\nabla^2\phi + \vartheta'(\phi) = 0 \tag{18}$$

where,

$$\vartheta'(\phi) \equiv -\phi + \tilde{f}\left(y+\frac{\phi_0}{u}\right) - \tilde{f}\left(y+\frac{\phi_0+\phi}{u}\right). \tag{19}$$

For localized waves to exist, one requires (Sagdeev (9))

$$\vartheta''(\phi) < 0, \text{ for small } \phi \tag{20a}$$

or

$$1 + \frac{1}{u}\tilde{f}'\left(y + \frac{\phi_0 + \phi}{u}\right) > 0 \text{ for small } \phi. \tag{20b}$$

VII DISCUSSION

In the present paper, we have explored a generalized class of localized structures generated by a non-uniform β and/or zonal shear flow within the framework of the Charney-Hasegawa-Mima equation. A generalized class of solitary waves becomes possible when both the quadratic and cubic nonlinearities associated with a non-uniform β and/or zonal shear flow are included. This will be the subject of a forthcoming paper.

ACKNOWLEDGMENTS

This work was started when the author was a Visiting Scientist at Risö National Laboratory, Denmark and was completed when the author was attending the Workshop on Two-dimensional Turbulence in Fluids and Plasmas at Canberra, Australia. The author is thankful Dr. Juul Rasmussen and Professor Robert Dewar for their hospitality and discussions.

REFERENCES

1. Charney, J.G., *Geophys. Publ. Kosjones Nor. Vidensk. Akad. Oslo* **17**, 3, (1948).
2. Hasegawa, A. and Mima, K., *Phys. Fluids* **21**, 87, (1978).
3. Shivamoggi, B.K., *Phys. Lett. A* **138**, 37, (1989).
4. Larsen, L.H., *J. Atmos. Sci.* **22**, 222, (1965).
5. Clarke, R.A., *Geophys. Fluid Dyn.* **2**, 343, (1971).
6. Redekopp, L.G., *J. Fluid Mech.* **82**, 725, (1977).
7. Lakhin, V.P., Mikhailovskii, A.B., and Onischenko, O.G., *Phys. Lett. A.* **119**, 348, (1987).
8. Flierl, G.R., Stern, M.E., and Whitehead, J.A., *Dyn. Atmos. Oceans* **7**, 233, (1983).
9. Sagdeev, R.Z., *Rev. Plasma Phys.* **4**, 23, (1966).

Effects of Sheared Rotation on Ion-Temperature-Gradient-Driven Instabilities

Hideo Sugama* and Wendell Horton†

*National Institute for Fusion Science, Toki 509-52, Japan
†Institute for Fusion Studies, The University of Texas at Austin, Austin, Texas 78712

Abstract. Effects of the sheared toroidal rotation on the ion-temperature-gradient-driven (ITG) modes are studied. The kinetic integral solution is derived for the ITG modes in the toroidally rotating plasma and is numerically solved to obtain the growth rate, the real frequency, and the mode eigenfunction. In the presence of the toroidal flow shear, the radial wavenumber in the ballooning representation depends on the time. Accordingly, the growth rate and the real frequency are also interpreted as functions of time. The mode eigenfunction moves poloidally and is stabilized in the good curvature region. In this way, the toroidal flow shear is expected to cause a significant reduction of the average growth rate.

INTRODUCTION

Particles and heat transport fluxes observed in fusion plasma devices greatly exceed the predictions of the collisional (classical and neoclassical) transport theories [1–4]. They are called anomalous transport [5] and considered to be due to turbulent fluctuations existing in plasmas. In magnetic confinement systems with electrostatic turbulence, turbulent ($\mathbf{E} \times \mathbf{B}$) velocity fields are given in the two-dimensional form by the magnetic field $\mathbf{B} = B\mathbf{b}$ and the potential fluctuation $\hat{\phi}$ as $\hat{\mathbf{v}}_{E\times B} = (c/B)\mathbf{b} \times \nabla\hat{\phi}$. These fluctuations result from various instabilities such as driven by density and temperature gradients. Suppression of the plasma turbulence and anomalous transport is the key to realize controlled fusion. Recently, as a means to the confinement improvement such as H-modes [6] and internal transport barriers [7] found in tokamaks, turbulence suppression by background sheared flows (or sheared radial electric fields) has been attracting considerable attention.

In this work, we consider ion-temperature-gradient (ITG) driven modes [8–11], which are micro-instabilities occurring in tokamak core plasmas, and

CP414, *Two-Dimensional Turbulence in Plasmas and Fluids:* Research Workshop
edited by R. L. Dewar and R. W. Griffiths

investigate effects of toroidal flow shear on them. Figure 1 shows schematically a toroidally rotating axisymmetric system, in which the magnetic field and the toroidal flow are written as

$$\mathbf{B} = \mathbf{B}_T + \mathbf{B}_P = I\nabla\zeta + \nabla\zeta \times \nabla\Psi$$
$$\mathbf{V}_0 = V_0\hat{\boldsymbol{\zeta}} = -Rc\frac{\partial\Phi_0}{\partial\Psi}\hat{\boldsymbol{\zeta}}. \tag{1}$$

Here, θ and ζ denote the poloidal and toroidal angles, respectively, and the poloidal flux Ψ is often used as a label of magnetic surfaces instead of the minor radius r. The background electrostatic potential is represented by Φ_0.

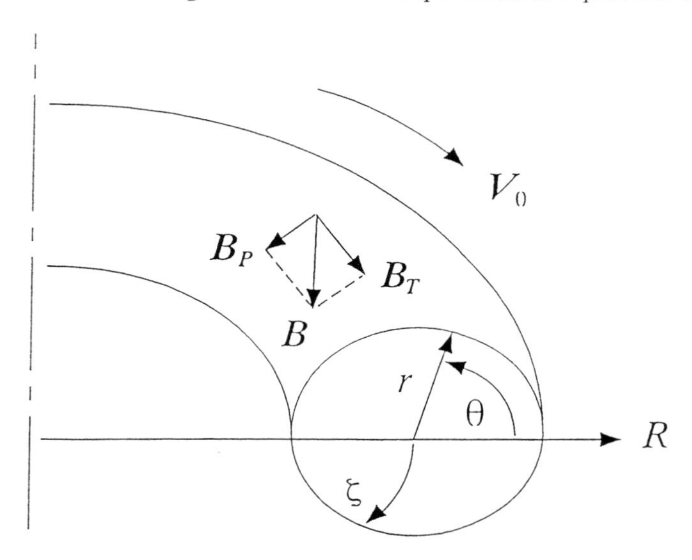

FIGURE 1. Toroidally rotating axisymmetric system. The magnetic field and the toroidal flow velocity are written as $\mathbf{B} = \mathbf{B}_P + \mathbf{B}_T$ and $\mathbf{V}_0$, respectively. The radial coordinate, poloidal and toroidal angles are denoted by r, θ, and ζ, respectively.

BALLOONING MODE REPRESENTATION FOR TOROIDALLY ROTATING SYSTEMS

In this section, following Dewar & Glasser [12] and Cooper [13], the ballooning mode representation for toroidally rotating systems is presented in preparation for the kinetic analysis of the toroidal ITG modes in the next section. Another description for the toroidal modes in the rotating system is given by Taylor & Wilson [14] while the Cooper's representation used here is equivalent to theirs at the leading order in a high mode number expansion.

Let us consider an arbitrary fluctuation field $\hat{F}$ with a small wavelength in the directions perpendicular to the magnetic field, which is written in the WKB (or eikonal) form as

$$\hat{F}(r,\theta,\zeta,t) = \hat{F}(r,\theta,t)\ \exp[iS(r,\theta,\zeta,t)]. \tag{2}$$

Here, the rapid variation in the perpendicular directions is described by the eikonal S which satisfies

$$\mathbf{B}\cdot\nabla S = 0. \tag{3}$$

For the toroidally rotating system with the toroidal velocity $\mathbf{V}_0$, another constraint is imposed on S as

$$\left(\frac{\partial}{\partial t} + \mathbf{V}_0\cdot\nabla\right) S = 0 \tag{4}$$

For a toroidal mode number n, the eikonal S satisfying Eqs. (3) and (4) is given by

$$S = -n\left[\zeta - q(r)\theta + \int \theta_{k0}(q)dq - V^{\zeta}t\right] \tag{5}$$

where $V^{\zeta} = V_0/R = -c\partial\Phi_0/\partial\Psi$ represents the toroidal angular velocity and $\int \theta_{k0}(q)dq$ appears as an integral constant. The safety factor $q \simeq rB_T/RB_P$ is also available for a radial coordinate instead of r. Then, the perpendicular wavenumber vector is written as

$$\mathbf{k}_\perp \equiv \nabla S = -n\left[\nabla\zeta - q\nabla\theta + \{\theta_k(t) - \theta\}\nabla q\right]. \tag{6}$$

Here the radial wavenumber component contains time dependence through the time-dependent ballooning angle defined by

$$\theta_k(t) \equiv \theta_{k0}(q) - \frac{\partial V^{\zeta}(q)}{\partial q}t. \tag{7}$$

Generally, the WKB (or eikonal) form in Eq. (2) with (5) is non-periodic function of the poloidal angle θ and is regarded as a quasimode defined in the ballooning (or covering) space $-\infty < \theta < +\infty$. The real physical mode, which is periodic in θ, is obtained from the quasimode through the ballooning representation:

$$\begin{aligned}
&\hat{\Phi}_n(r,\theta,\zeta,t)\\
&= \sum_{j=-\infty}^{\infty} \hat{F}_n(r,\theta+2\pi j,t)\exp\left(-in\left[\zeta - q(\theta+2\pi j) + \int \theta_k(t)dq\right]\right)\\
&= \mathrm{e}^{-in\zeta}\sum_{m=-\infty}^{\infty} \mathrm{e}^{im\theta}\int\frac{d\theta'}{2\pi}\hat{F}_n(r,\theta',t)\exp\left[i(nq-m)\theta' - in\int\theta_k(t)dq\right]
\end{aligned} \tag{8}$$

If we use the radial coordinate $x = r - r_0$ [r_0: the radial position of the rational surface defined by $q(r_0) = m/n$], the shear parameter $\hat{s} = (r/q)(dq/dr)$, the poloidal wavenumber $k_\theta = nq/r$, and the radial wavenumber variable $k' = \hat{s}k_\theta\theta'$, we find in Eq. (8) that

$$\int d\theta' \exp[i(nq - m)\theta'] \propto \int dk' \exp[ik'x] \tag{9}$$

which shows that the poloidal angle $\theta = k/(\hat{s}k_\theta)$ and the radial variable x are regarded as conjugate to each other. The radial structure of the ballooning mode is given by the Fourier transform of its poloidal structure.

KINETIC ITG MODE EQUATION FOR TOROIDALLY ROTATING SYSTEMS

Here, we assume that normalized fluctuating quantities and their characteristic wavenumbers and frequencies obey the gyrokinetic ordering:

$$\hat{f}_a/f_a \sim e_a\hat{\phi}/T_a \sim k_\parallel/k_\perp \sim (\partial/\partial t + \mathbf{V}_0 \cdot \nabla)/\Omega_a \sim \rho_a/L \tag{10}$$

where the subscript a denotes the particle species, and the ordering parameter ρ_a/L is given by the ratio of the thermal gyroradius $\rho_a \equiv v_{Ta}/\Omega_a$ [$v_{Ta} \equiv (2T_a/m_a)^{1/2}$: the thermal velocity, $\Omega_a \equiv e_aB/(m_ac)$: the gyrofrequency] to the equilibrium scale length L. The fluctuating distribution function is divided into adiabatic and non-adiabatic parts as

$$\hat{f}_a(\mathbf{k}_\perp) = \underbrace{-\frac{e_a\hat{\phi}(\mathbf{k}_\perp)}{T_a}f_{aM}}_{\text{adiabatic}} + \underbrace{\hat{h}_a(\mathbf{k}_\perp)e^{iL_a(\mathbf{k}_\perp)}}_{\text{non-adiabatic}} \tag{11}$$

where $L_a(\mathbf{k}_\perp) \equiv \mathbf{k}_\perp \cdot (\mathbf{v}' \times \mathbf{b})/\Omega_a$ ($\mathbf{v}' \equiv \mathbf{v} - \mathbf{V}_0$). The gyrokinetic equation for the toroidally rotating system [15,16] with the large aspect ratio $R/r \gg 1$ is written in the linear and collisionless form as

$$\left[\frac{(V_0 + v'_\parallel)}{Rq}\frac{\partial}{\partial\theta} + \frac{\partial}{\partial t} + i(\omega_E + \omega_{Da})\right]\hat{h}_a(\theta, t)$$
$$= f_{a0}J_0(\alpha)\left[\frac{V_0}{Rq}\frac{\partial}{\partial\theta} + \frac{\partial}{\partial t} + i(\omega_E + \omega_{*Ta})\right]\left(\frac{e_a\hat{\phi}(\theta, t)}{T_a}\right) \tag{12}$$

where $\alpha = k_\perp v'_\perp/\Omega_a$ represents the finite gyroradius effect, $\omega_E = \mathbf{k}_\perp \cdot \mathbf{V}_0 = -nV^\zeta$ is the Doppler shift due to the toroidal rotation, $\omega_{Da} = \epsilon_n\omega_{*a}(m_a/T_a)[\frac{1}{2}(v'_\perp)^2 + (V_0 + v'_\parallel)^2][\cos\theta + \hat{s}\{\theta - \theta_k(t)\}\sin\theta]$ is the magnetic (∇B and curvature) drift frequency, and $\omega_{*Ta} = \omega_{*a}[1 + \eta_a\{m_a(v')^2/2T_a -$

$\frac{3}{2}\} - (m_a v_{\|}'/T_a)(L_n dV_0/dr)]$ [$\omega_{*a} = k_\theta c T_a/(e_a B L_n)$: the diamagnetic drift frequency, $L_n = -(d\ln n/dr)^{-1}$: the density gradient scale length]. In Eq. (12), the differential operator $\partial/\partial\theta$ is taken along the magnetic field line. Note that the effect of the toroidal flow shear dV_0/dr is included in ω_{Da} through $\theta_k(t)$ [see Eq. (7)] and in ω_{*Ta}.

In order to obtain a closed system of linear equations, we use the gyrokinetic equation (12) for the ions ($a = i$), the Boltzmann relation for the electrons ($a = e$), and the charge neutrality: the latter two conditions are written as

$$\int d^3v\, \hat{f}_i = \hat{n}_i = \hat{n}_e = n_0(e\hat{\phi}/T_e) \tag{13}$$

Then, the linear behavior of the ITG mode is described by Eqs. (12) and (13). The main destabilizing sources of the ITG mode are given from the ion temperature gradient denoted by $\eta_i \equiv d\ln T_i/d\ln n$ in ω_{*Ti} and the magnetic curvature included in ω_{Di}.

In the case of no rotation ($\mathbf{V}_0 = 0$), we have $\theta_k(t) = \theta_{k0}$ from Eq. (7) and no explicit temporal dependence appears in the linear coefficients of Eqs. (12) and (13). Thus, we can do Fourier transform easily with respect to the time and replace $\partial/\partial t$ by $-i\omega$. Then, Eqs. (12) and (13) reduce to the eigenvalue problem with ω as a complex-valued eigenfrequency. Romanelli [9] and Dong *et al.* [10] solved this problem to obtain the kinetic dispersion relation of the ITG mode for $\mathbf{V}_0 = 0$ and $\theta_k = 0$. However, when the toroidal flow shear exists, Eqs. (12) and (13) should be solved as an initial value problem because of the explicit temporal dependence of the ballooning angle $\theta_k(t)$. In order to solve them more easily, we here assume that the temporal variation of $\theta_k(t)$ is much slower than the characteristic frequency of the mode observed in the rotating frame:

$$\frac{\partial V^\zeta}{\partial q} \sim \frac{V_0}{Rq} \ll \left(\frac{\partial}{\partial t} + i\omega_E\right) \sim \omega_{*e}. \tag{14}$$

We write the temporal dependence of the fluctuations as

$$\hat{\phi}(t) \propto \exp\left(-i\omega_E t - i\int \omega(t)dt\right)\,\hat{\varphi}(t) \tag{15}$$

where

$$\omega(t) \sim \omega_{*e} \gg \frac{1}{\omega(t)}\frac{d\omega(t)}{dt} \sim \frac{1}{\hat{\varphi}(t)}\frac{d\hat{\varphi}(t)}{dt} \sim \frac{V_0}{Rq}. \tag{16}$$

Then, $\omega(t) = \omega[\theta_k(t)]$, and $\hat{\varphi}(t) = \hat{\varphi}[\theta_k(t)]$ are regarded as a pair of an eigenvalue and an eigenfunction which depend on the time through $\theta_k(t)$. The stability of the mode should be judged from the average growth rate defined by

$$\gamma_{\text{ave}} = \lim_{T \to +\infty} \frac{1}{T} \int_0^T \gamma(t)\, dt = \frac{1}{2\pi} \int_0^{2\pi} \gamma(\theta_k)\, d\theta_k \tag{17}$$

where $\gamma(t) = \text{Im}\, \omega(t)$.

Using the approximation described in Eqs. (14)–(16), we obtained from Eqs. (12) and (13) the integral ITG mode equation which is given by

$$\left(1 + \frac{T_e}{T_i}\right) \hat{\phi}(k) = \int_{-\infty}^{+\infty} \frac{dk'}{\sqrt{2\pi}} K(k, k') \hat{\phi}(k') \tag{18}$$

with

$$K(k, k') = -i \int_{-\infty}^{0} \omega_{*e} d\tau \frac{\sqrt{2} e^{-i(\omega - \omega_E)\tau}}{\sqrt{a}(1+a)\sqrt{\lambda}} e^{-(k-k')^2/4\lambda}$$
$$\times \left[\frac{\omega}{\omega_{*e}} \tau_e + 1 - \frac{3}{2}\eta_i + \frac{\eta_i (k-k')^2}{4a\lambda} + \frac{2\eta_i}{(1+a)} \right.$$
$$\left. \times \left(1 - \frac{k_\perp^2 + k_\perp'^2}{2(1+a)\tau_e} + \frac{k_\perp k_\perp'}{(1+a)\tau_e} \frac{I_1}{I_0}\right) - \frac{\tau_e q(k-k')}{\epsilon_n \hat{s} \omega_{*e} \tau} \left(\frac{L_n}{c_s} \frac{dV_0}{dr}\right)\right] \Gamma_0(k_\perp, k_\perp') \tag{19}$$

where

$$\lambda = \frac{(\omega_{*e}\tau)^2}{\tau_e a} \left(\frac{\hat{s}}{q} \epsilon_n\right)^2, \qquad \theta - \theta_k = k/\hat{s}k_\theta, \qquad \theta' - \theta_k = k'/\hat{s}k_\theta$$
$$a = 1 + i \frac{2\epsilon_n}{\tau_e} \omega_{*e} \tau$$
$$\times \left(\frac{[(\hat{s}+1)(\sin\theta - \sin\theta') - \hat{s}\{(\theta - \theta_k)\cos\theta - (\theta' - \theta_k)\cos\theta'\}]}{(\theta - \theta')} \right)$$
$$\Gamma_0 = I_0 \left(\frac{k_\perp k_\perp'}{(1+a)\tau_e}\right), \qquad k_\perp^2 = k_\theta^2 + k^2, \qquad k_\perp'^2 = k_\theta^2 + k'^2. \tag{20}$$

Here, the wavenumber variables k_θ, k, and k' are normalized by ρ_s^{-1} ($\rho_s = \sqrt{2T_e/m_i}/\Omega_i$). Note that the kernel defined by Eq. (19) contains the flow shear effects through the explicit term ($\propto dV_0/dr$) and $\theta_k = \theta_{k0} - (\partial V^\zeta/\partial q)t$. If we put $V_0 = dV_0/dr = 0$, the integral ITG mode equation (18) with the kernel (19) reduces to the same one as given by Dong *et al.* [10]. In the next section, Eq. (18) is numerically solved to investigate the effects of the toroidal flow shear on the eigenfrequency $\omega = \omega[\theta_k(t)]$ and the eigenfunction $\hat{\phi}(k)$ of the ITG mode.

NUMERICAL RESULTS

The normalized growth rate γ/ω_{*e} and the normalized real frequency in the rotating frame $(\omega_r - \omega_E)/\omega_{*e}$ obtained by numerically solving Eq. (18)

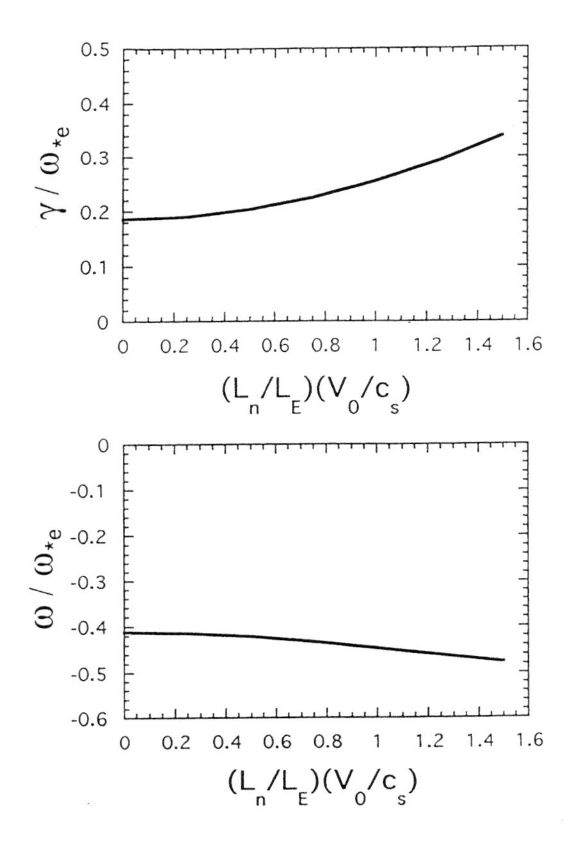

FIGURE 2. The normalized growth rate (top) and the normalized real frequency in the rotating frame (bottom) as functions of the normalized toroidal flow shear $(L_n/L_E)(V_0/c_s)$ for $\theta_k = 0$, $\eta_i = 2$, $k_\theta = 0.75$, $\epsilon_n = 0.2$, $q = 2$, $T_e/T_i = 1$, and $\hat{s} = r d\ln q/dr = 1$.

with Eq. (19) are shown in Fig. 2 as functions of the normalized toroidal flow shear $(L_n/L_E)(V_0/c_s)$ $[c_s \equiv (2T_e/m_i)^{1/2}$, $L_E \equiv -V_0/(dV_0/dr)]$ for $\theta_k = 0$, $\eta_i = 2$, $k_\theta = 0.75$, $\epsilon_n = 0.2$, $q = 2$, $T_e/T_i = 1$, and $\hat{s} = r d\ln q/dr = 1$. It is found that, when the ballooning angle is fixed as $\theta_k = 0$, the growth rate increases with increasing the toroidal flow shear. This behavior of the growth rate is similar to the destabilization of the quasitoroidal ITG mode due to the parallel flow shear [11]. In Eq. (12) where the large aspect ratio approximation is used, the perpendicular flow component of the toroidal flow is of $O(r/R)$ of its parallel component and is neglected. Thus, the toroidal flow shear dependence of the toroidal ITG mode for $\theta_k = 0$ shows the similar tendency to the parallel flow shear dependence of the quasitoroidal ITG mode.

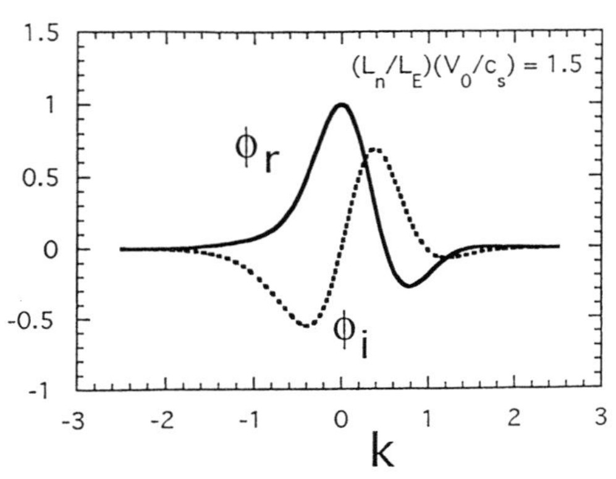

FIGURE 3. The ITG mode eigenfunction $\phi(k) = \phi_r + \phi_i$ for $(L_n/L_E)(V_0/c_s) = 1.5$ and $\theta_k = 0$. The other parameters are the same as in Fig. 2. The bad curvature region $|\theta| < \pi/2$ corresponds to $|k| < 3\pi/8 \simeq 1.18$.

If the perpendicular component of the toroidal flow is taken into account, the growth rate is expected to be reduced due to the perpendicular flow shear stabilization [11]. The real frequency has the negative sign (or the sign of the ion diamagnetic frequency) and its absolute value is an increasing function of the toroidal flow shear. Figure 3 shows the ITG mode eigenfunction $\phi(k) = \phi_r + i\phi_i$ for $(L_n/L_s)(V_0/c_s) = 1.5$ and $\theta_k = 0$. Here, the other parameters are the same as in Fig. 2 and the bad curvature region $|\theta| < \pi/2$ corresponds to $|k| < 3\pi/8 \simeq 1.18$. The eigenfunction is localized in the bad curvature region, which is characteristic of the ballooning mode structure. We see that the eigenfunction is asymmetric in k (or in θ), which is a contrast to the poloidally symmetric eigenfunction for the no flow case with $\theta_k = 0$ [10].

As mentioned in the previous section, when the flow shear exists, the growth rate and real frequency are functions of the time through $\theta_k(t) \equiv \theta_{k0}(q) - (\partial V^\zeta/\partial q)t$. Figure 4 shows the growth rate and real frequency as functions of $\theta_k(t)$ for $dV_0/dr = -V_0/L_E \to 0$ where the other parameters are the same as in Fig. 2. We find that the growth rate approaches to zero at $\theta_k(t) \simeq \pi/2$ and that the ITG mode is stabilized for $\theta_k(t) > \pi/2$. Here the growth rate and frequency for the stable case are not shown and we need to improve our numerical code to calculate the negative growth rate. The eigenfunction corresponding to the case of $\theta_k(t) = 0.3\pi$ in Fig. 4 is shown in Fig. 5. In this case, the bad curvature region is given by $-1.88 < k < 0.47$. The eigenfunction is centered at $\theta \simeq \theta_k = 0.3\pi$ $(k = 0)$ but it is partly contained in the good curvature region $k > 0.47$, which causes the reduction of the growth rate. Thus, with changing the ballooning angle $\theta_k(t)$, the eigenfunction moves poloidally along the magnetic field line and the average growth rate defined by Eq. (17) is expected to be significantly reduced.

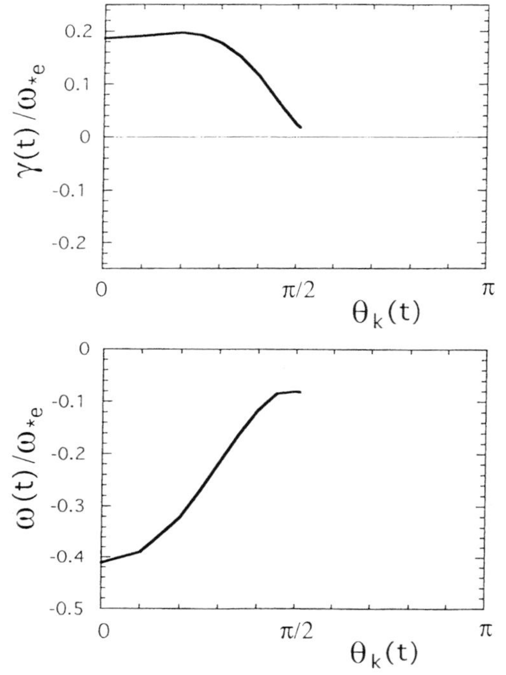

FIGURE 4. The normalized growth rate (top) and the normalized frequency in the rotating frame (bottom) as functions of $\theta_k(t)$ for $dV_0/dr = -V_0/L_E \to 0$. The other parameters are the same as in Fig. 2. The growth rate approaches to zero at $\theta_k(t) \simeq \pi/2$ and the ITG mode is stabilized for $\theta_k(t) > \pi/2$. The growth rate and frequency for the stable case are not shown.

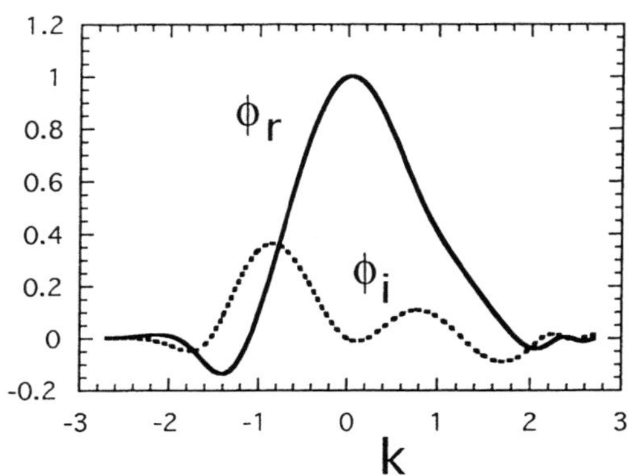

FIGURE 5. The eigenfunction $\phi(k) = \phi_r + \phi_i$ for the case of $\theta_k(t) = 0.3\pi$ in Fig. 4. The bad curvature region $|\theta| < \pi/2$ corresponds to $-1.88 < k < 0.47$.

SUMMARY

In this work, effects of the toroidal flow shear on the toroidal ITG modes are studied by numerically solving the kinetic integral solution. When the toroidal flow shear exists, the ballooning angle depends on the time as $\theta_k(t) = \theta_{k0} - (\partial V^{\zeta}/\partial q)t$. Then, the growth rate and real frequency of the modes also depend on the time through the ballooning angle $\theta_k(t)$. For $\theta_k(t) = 0$ fixed at the bad curvature region, the growth rate increases as the toroidal flow shear increases. The real frequency in the rotating frame has the sign of the ion diamagnetic frequency and its absolute value increases with increasing the flow shear. The poloidal symmetry of the mode structure is broken by the flow shear. As $\theta_k(t)$ increases, the mode structure moves from the bad to good curvature region. For $\theta_k(t) > \pi/2$, the mode is stabilized. The significant reduction of the average growth rate $\gamma_{\text{ave}} = \oint \gamma[\theta_k(t)]d\theta_k(t)/2\pi$ is expected especially for weak toroidal flow shear. Here, because of the large aspect ratio approximation, the perpendicular flow component of the toroidal flow is neglected and the toroidal flow shear for $\theta_k = 0$ shows similar effects to the parallel flow shear destabilization. The perpendicular component of the toroidal flow shear, if included, can reduce the growth rate. However, in this paper, the stabilizing effect by the toroidal flow shear through the rotating ballooning angle is more emphasized. Our results are still preliminary and combined effects of the (parallel and perpendicular) shear flows, the time-dependent ballooning angle, and the negative magnetic shear are to be studied as a future task. Also, the calculation of negative growth rates are necessary to obtain the average growth rate.

ACKNOWLEDGMENTS

The authors acknowledge useful discussions with Dr. J. Q. Dong on the numerical solution techniques for the integral equation. The author (HS) thanks Prof. M. Okamoto for his encouragement of this work. This work is supported in part by the Grant-in-Aid from the Japanese Ministry of Education, Science and Culture, and in part by the U.S. Department of Energy Grant DE-FG05-80ET-53088.

REFERENCES

1. Braginskii, S. I., *Reviews of Plasma Physics*, New York: Consultants Bureau, 1965, Vol. 1, p. 205.
2. Hinton, F. L., and Hazeltine, R. D., *Rev. Mod. Phys.* **42**, 239 (1976).

3. Hirshman, S. P., and Sigmar, D. J., *Nucl. Fusion* **21**, 1079 (1981).
4. Balescu, R., *Transport Processes in Plasmas*, Amsterdam: North-Holland, 1988, Vols. 1 and 2.
5. Connor, J. W., and Wilson, H. R., *Plasma Phys. Control. Fusion*, **36**, 719 (1994).
6. ASDEX Team, *Nucl. Fusion* **29**, 1959 (1989).
7. Koide, Y., Takizuka, T., Takeji, S., Ishida, S., Kikuchi, M., Kamada, Y., Ozeki, T., Neyatani, Y., Shirai, H., Mori, M., and Tsuji-Iio, S., *Plasma Phys. Control. Fusion* **38**, 1011 (1996).
8. Horton, W., Wakatani, M., and Wooton, A. J., *AIP Conference Proceedings 284, U.S.-Japan Workshop on Ion Temperature Gradient-Driven Turbulent Transport*, New York: AIP, 1994, ch. 1, p. 3.
9. Romanelli, F., *Phys. Fluids* B **1**, 1018 (1989).
10. Dong, J. Q., Horton, W., and Kim, J. Y., *Phys. Fluids* B **4**, 1867 (1992).
11. Dong, J. Q., and Horton, W., *Phys. Fluids* B **5**, 1581 (1993).
12. Dewar, R. L., and Glasser, A. H., *Phys. Fluids* **26**, 3038 (1983).
13. Cooper, W. A., *Plasma Phys. Control. Fusion* **30**, 1805 (1988).
14. Taylor, J. B., and Wilson, H. R., *Plasma Phys. Control. Fusion* **38**, 1999 (1996).
15. Artun, M., and Tang, W. M., *Phys. Plasmas* **1**, 2682 (1994).
16. Sugama, H., and Horton, W., *Phys. Plasmas* **4**, 405 (1997).

Drift waves and drift wave instabilities in dusty partially ionized plasmas

V. N. Tsytovich
General Physics Institute, Moscow 117942, Russia

S. V. Vladimirov[1]
School of Physics, The University of Sydney, NSW 2006, Australia

Abstract. The theory of drift waves in plasmas containing highly charged impurities (dust) is developed to include the specific features of dust, such as the attraction of dust grains as well as collisions between dust and neutral particles. Nonlinear equations taking into account variations of drift potential together with plasma electron, ion, and neutral densities, as well as dust charges, are obtained. The range of plasma parameters where the effects introduced by interactions of dust and neutrals can trigger a novel drift wave instability is found.

I INTRODUCTION

Plasmas containing highly charged impurities, or dust, are the subject of increasing interest [1] – [11]. It is well established that the presence of dust can considerably change collective processes in a plasma. For typical dusty plasmas, such as those in material processing devices [2] or in space [3], the dust grains are highly negatively charged, mainly by plasma currents [4]. The dust charging process introduces new physics by modifying plasma dielectric properties [5,6]. The most affected are plasma modes with frequencies of order or less than the dust charging frequency. The dust charging often appears as an important dissipative process leading to recombination of plasma electrons and ions on the dust grain surfaces. Thus the dust-plasma system is in general an open dissipative system, with sink and (if stationary) source of electrons and ions. This character of a dusty plasma affects development of dissipative instabilities (such as the drift wave instability), often leading to their considerable enhancement.

[1] Email: S.Vladimirov@physics.usyd.edu.au; http://www.physics.usyd.edu.au/~vladimi

CP414, *Two-Dimensional Turbulence in Plasmas and Fluids:* Research Workshop
edited by R. L. Dewar and R. W. Griffiths

The study of wave propagation in a dusty plasma is important for such applications as spread of heat flux to the walls in the scrape-off layers of tokamaks, as well as for understanding of processes in the Earth's ionosphere, space, and cometary plasmas [7] – [11]. For example, the presence of a relatively high-density dust component (with densities sometimes orders of magnitudes higher than was previously expected) in the lower ionosphere and the upper atmosphere demands a detailed investigation of the wave processes there. Accounting for the presence of dust in the near-wall regions of tokamaks is emphasized in connection with problems of high power loading in future machines. For example, dust can considerably influence the heat fluxes and power absorption in the edge tokamak plasmas [10].

There are also indications that the dust particles in a plasma can be self-contracted in dust clouds. Formation of such dust clouds can be supported by various mechanisms dust attraction [1]. The attraction forces can affect collective dissipative processes in dusty plasmas, e.g., development of instabilities. Thus one of the aims of the present article is to investigate the influence of the dust-dust attraction forces on development of drift wave instabilities. This is of special interest for the plasma physics of nuclear fusion devices as well as for ionosphere physics; we note here that in the Earth's atmosphere and lower ionosphere the presence of dust is due to man-made pollution as well as to other processes such as volcanic activity and meteoroid impacts. Other applications include plasma-assisted material processing, e.g., etching experiments, in the presence of an external magnetic field.

Another effect important for applications is dust-neutral collisions. These collisions can be either direct ion-neutral collisions influencing drift waves [12] or dust-neutral collisions affecting drift wave instabilities in the presence of dust. The second effect has not been investigated before. Thus here we aim to study drift wave instabilities in the presence of dust-neutral collisions. We note that in the case when the drift frequency is much less than the neutral-dust collision frequency, the dust grains can follow the neutral particles in their motion. Since the drift neutral vortices can have a significant potential, the dust can be trapped in these vortices. Theoretically, such a trapping is possible since the potential difference V created by plasma particles means a potential well $Z_d V$ for dust particles, where Z_d, which can exceed unity by orders of magnitude, is the dust charge in units of the elementary electron charge. Thus the solitary drift vortices can trap highly charged ($Z_d \sim 10^3 - 10^4$, as in many dusty plasmas) dust grains even when V is relatively small.

In the present paper, we derive the generalization of the Hasegawa–Wakatani [13] nonlinear equations for drift waves, taking into account the neutral-dust interactions and dust-dust attraction. Both effects can play an important role in the dust-cloud vortices. The most important point is that dust takes part in the motion of both vortices, namely: 1) in the neutral component (such as usual air) vortices, when the rate of dust-neutral collisions is sufficiently high, and the dust grains follow the vortex rotation; and 2) in the drift vortices (appearing due to the presence of inhomogeneity of plasma density), when the dust charge is sufficiently high, and the grains are trapped in the potential well of the drift wave. Such combined vortices, as was also discussed before [14], can in principle be responsible for the energy transfer from drift vortices towards air vortices in the lower ionosphere and the upper atmosphere. This effect competes with heating of the upper atmosphere by drift vortices, and, being a complementary process, needs dust grains to be present in the region of interaction (note that the local heating of the upper atmosphere can also be responsible for air convection, turbulence, and solitary vortices).

II FORCES BETWEEN DUST GRAINS

For the non-Debye screening at distances much larger than the Debye radius the electrons are practically Boltzmann-distributed (the relative role of their flux on dust particle is small), and their density is given by

$$n_e = n_{0e}\left(1 + \frac{e\phi}{T_e}\right) \tag{1}$$

where ϕ is the electrostatic potential around the dust grain and T_e is the electron temperature. For plasma ions, due to the shadowing, the ion density contains the additional contribution

$$n_i = n_{0i}\left(1 - \frac{e\phi_0}{T_i}\right) - n_{0i}\frac{a^2}{4r^2}\left(1 + \frac{2e^2 Z_d}{aT_i}\right), \tag{2}$$

where ϕ_0 is the potential on the surface of the grain, a is the grain radius, $-Z_d e$ is the grain charge, and T_i is the ion temperature. The quasineutrality condition allows us to write the dust repulsion potential which for distances much larger that the Debye length is given by (for simplicity, we assume ions to be single charged, $Z_i = 1$)

$$U_r = -Z_d e\phi_r = \eta_r \frac{Z_d^2 e^2 a}{2r^2}. \tag{3}$$

Here, the coefficient η_r is

$$\eta_r = \frac{T_e}{T_e + T_i}\left(1 + \frac{aT_i}{2Z_d e^2}\right) \approx 1. \tag{4}$$

The last approximation is written for $T_i \ll T_e$.

The attraction potential can be found by calculating the change in the ion momentum in the process of direct bombardment of the dust particle (subscript b) and the Coulomb scattering by the dust particle (subscript c) [1]

$$U_a = -(\eta_b + \eta_c)\frac{a^2}{\lambda_D^2}\frac{Z_d^2 e^2}{r}, \tag{5}$$

where λ_D is the plasma Debye length (note that $\lambda_D \approx \lambda_{Di} = (T_i/4\pi n_i e^2)^{1/2}$ for $T_i \ll T_e$), n_i is the ion density,

$$\eta_b = \frac{1}{2\sqrt{\pi}}\left(\frac{aT_i}{Z_d e^2}\right)^2 \int_{y_{\min}}^{\infty} \left(1 + \frac{Z_d e^2}{aT_i y^2}\right)^{5/2} y^4 \exp(-y^2) dy, \tag{6}$$

and

$$\eta_c = \frac{1}{4\sqrt{\pi}} \int_{y_{\min}}^{\infty} \left(1 + \frac{Z_d e^2}{aT_i y^2}\right) \ln L \exp(-y^2) dy. \tag{7}$$

In (6) and (7), we have

$$y_{\min} = \frac{a}{\lambda_D}\sqrt{\frac{Z_d e^2}{aT_i}}, \qquad L = \left(\frac{\lambda_D^2}{a^2} + \frac{Z_d^2 e^4}{4a^2 T_i^2 y^4}\right)\left[1 + \frac{Z_d e^2}{aT_i} + \frac{Z_d^2 e^4}{4a^2 T_i^2 y^4}\right]^{-1}. \tag{8}$$

The corresponding part of the linear dielectric permittivity is thus given by

$$\epsilon_{k\omega}^d = 1 - \frac{\omega_{pd}^2}{\omega^2 - (\pi/4)\eta_r \omega_{pd}^2 ka + (\eta_b + \eta_c)\omega_{pd}^2 a^2/\lambda_D^2}, \tag{9}$$

which takes into account the repulsion and attraction of dust particles.

III COLLISIONS OF DUST WITH PLASMA PARTICLES AND NEUTRALS

In general, the full electron(ion)-dust collision frequency must take into account the Coulomb elastic as well as charging collisions. While the former are typical for any plasma component, the latter are specific only for macro-size (comparing with sizes of electrons and ions) dust grains which can effectively absorb plasma particles on its surface. The presence of the absorbing charging collisions leads to the mentioned openness of the dust-plasma system, where a stationary state can only be maintained by an external ionization source compensating the dust charging loss of plasma particles. Another consequence of the charging process is that the effective collision frequencies entering the continuity and momentum equations for plasma species are different.

The standard [15] calculation of Coulomb elastic electron(ion)-dust collision rate (which enters Euler equation for the plasma component) gives us (see, e.g., [8,10])

$$\nu_{e(i)d}^{el} = \frac{4\sqrt{2\pi} Z_d^2 n_d e^4 \Lambda}{3m_{e(i)}^2 \bar{v}_{e(i)}^3}, \tag{10}$$

where n_d is the dust density, $m_{e(i)}$ is the electron(ion) mass, and $\bar{v}_{e(i)}$ is the average velocity of plasma particles defined as

$$\frac{1}{\bar{v}_{e(i)}^3} = -\left(\frac{\pi}{2}\right)^2 \frac{1}{n_{0e(i)}} \int \frac{1}{v^2} \frac{\partial}{\partial v} f_{0e(i)}(v) d^3 v. \tag{11}$$

Here, $v = |\mathbf{v}|$, $n_{0e(i)} \equiv \int d\mathbf{v} f_{0e(i)}(v)$ is the equilibrium density of the corresponding plasma component, and $f_{0e(i)}(v)$ is the equilibrium electron(ion) distribution function. When the equilibrium distribution is Maxwellian (which is not necessarily the case in the presence of dust impurities), we have $\bar{v}_{e(i)} = v_{Te(i)}$, where $v_{Te(i)} = (T_{e(i)}/m_{e(i)})^{1/2}$ is the electron(ion) thermal velocity. The Coulomb logarithm Λ in Eq. (10) is defined by the ratio of the maximum impact parameter, which is of order the plasma Debye length λ_D, to the minimum impact factor which we assume to be the radius $a \ll \lambda_D$ of the dust grains, $\Lambda = \ln(\lambda_D/a)$.

The effective frequency of collection of plasma particles by dust, which appears due to their bombardment of the grain surface, is determined by the cross section which in the simplest ballistic approximation [16] can be written as

$$\sigma_{e(i)}^{ch}(v) = \pi a^2 \left(1 + \frac{2 Z_d e q_{e(i)}}{a m_{e(i)} v^2}\right), \tag{12}$$

where $q_{e(i)} = \mp e$ (note that when considering magnetized plasmas, we also assume that the effective collision radius is much less than the Larmor radius of plasma particles). The averaged-over-angles capture frequency is then given by

$$\nu_{e(i)}^{ch} = \frac{n_d}{n_{0e(i)}} \int f_{0e(i)}(v) \left(1 + \frac{v}{3} \frac{\partial}{\partial v}\right) v \sigma_{e(i)}^{ch}(v) d^3 v. \tag{13}$$

For Maxwellian distributions, the calculation gives

$$\begin{aligned} \nu_e^{ch} &= \nu_i^{ch} 3(1+P)(\tau + \mathcal{Z}) \frac{4+\mathcal{Z}}{4\tau + 3\mathcal{Z}} \\ &= \nu_d^{ch} P \frac{\tau + \mathcal{Z}}{1 + \tau + \mathcal{Z}} \frac{4 + \mathcal{Z}}{\mathcal{Z}}. \end{aligned} \tag{14}$$

Here, we introduced the standard dimensionless parameters [5,6]

$$P = \frac{n_d Z_d}{n_{0e}}, \quad \tau = \frac{T_i}{T_e}, \quad \mathcal{Z} = \frac{Z_d e^2}{a T_e}, \tag{15}$$

which satisfy the balance equation of the electron and ion currents on the dust grain surface

$$\exp(-\mathcal{Z}) = \left(\frac{m_e}{m_i\tau}\right)^{1/2} (1+P)(\tau+\mathcal{Z}) \tag{16}$$

Furthermore, the dust charging frequency in Eq. (14) is given by

$$\nu_d^{ch} = \frac{\omega_{pi}^2 a}{\sqrt{2\pi} v_{Ti}}(1+\tau+\mathcal{Z}), \tag{17}$$

where $\omega_{pi}^2 = (4\pi n_{0i}e^2/m_i)^{1/2}$ is the ion plasma frequency.

Under the same assumption of the thermal particle distributions, the elastic collision frequencies (10) can be written as

$$\nu_{ed}^{el} = \frac{2\nu_{ch} P e^{\mathcal{Z}}}{3\mathcal{Z}(1+\tau+\mathcal{Z})}(\tau+\mathcal{Z})\Lambda \tag{18}$$

and

$$\nu_{id}^{el} = \frac{2\nu_{ch} P}{3z(1+\tau+\mathcal{Z})}\frac{\Lambda}{\tau(1+P)}. \tag{19}$$

The dust charging collisions lead to a sink term in the continuity equations for plasma electrons and ions. Assuming that in the zeroth approximation, the loss of the plasma particles on the dust grains is compensated by external sources, we write the continuity equation in the form

$$\partial_t n_{e(i)} + \nabla\cdot(n_{e(i)}\mathbf{v}_{e(i)}) = -\bar{\nu}_{e(i)d} n_{e(i)} + \bar{\nu}_{0e(i)d} n_{0e(i)}, \tag{20}$$

where

$$\bar{\nu}_{e(i)d} n_{e(i)} = \int n_d v \sigma_{e(i)}^{ch}(v) f_{e(i)}(v) d^3v. \tag{21}$$

For a thermal equilibrium distributions of plasma particles, we find [5,6]

$$\bar{\nu}_{ed} = \bar{\nu}_{id}(1+P) = \nu_d^{ch}\frac{P(\tau+\mathcal{Z})}{z(1+\tau+\mathcal{Z})}. \tag{22}$$

Note that in the presence of perturbations of dust density, we also have to take them into account in (20) and (21), see below.

IV EQUATIONS OF MOTION OF PLASMA PARTICLES, NEUTRALS, AND DUST

For plasma electrons, the starting equation is the Euler equation neglecting the electron inertia:

$$0 = -v_{Te}^2\nabla n_e + v_{Te}^2 n_e \nabla\phi - \Omega_e n_e \mathbf{v}_e \times \hat{\mathbf{z}} - n_e \nu_{ed}\mathbf{v}_e, \tag{23}$$

where the dimensionless electric field potential $\phi = e\varphi/T_e$, $v_{Te} = (T_e/m_e)^{1/2}$ is the electron thermal velocity, and $\Omega_e = eB_0/m_e c$ is the electron gyrofrequency (the magnetic field $\mathbf{B}_0$ is directed along the $\hat{z}$ axis). In the zeroth approximation, the electron diamagnetic drift velocity is given by

$$\mathbf{v}_{0e} = -\frac{v_{Te}^2}{\Omega_e n_{0e}} \hat{\mathbf{z}} \times \nabla n_{0e} = -v_s \frac{\rho_s}{L_n}, \tag{24}$$

where $\rho_s = v_s/\Omega_i$ is the ion Larmor radius at the electron temperature, $\Omega_i = eB_0/m_i c$ is the ion cyclotron frequency, and $v_s = (T_e/m_i)^{1/2}$ is the ion sound velocity. We assume that $\hat{x}$ is the direction of the density gradient, and the drift waves propagate in the direction $\hat{y}$; the characteristic length of the electron density inhomogeneity is given by

$$L_n^{-1} = \frac{1}{n_{0e}} \frac{\partial n_{0e}}{\partial x}. \tag{25}$$

Furthermore, we take into account linear and nonlinear perturbations up to quadratic (in the wave fields) nonlinearities. All zero-order equilibrium values can be inhomogeneous in space (in the present study, we consider only their dependences on x). Due to the quasineutrality in the equilibrium, the ion equilibrium density can be expressed via the electron equilibrium density as

$$n_{0i} = n_{0e}(1 + P). \tag{26}$$

Similar relation can be written for small perturbations $\delta n_i/n_{0i}$, assuming that the quasineutrality is maintained (which is the case when the wave length is much larger than the Debye length). We have

$$\delta n_i = \delta n_e + \delta Z_d n_{0d} + Z_{0d} \delta n_d \tag{27}$$

and therefore

$$\frac{\delta n_i}{n_{0i}} = \frac{\delta n_e}{n_{0e}} \frac{1}{1+P} + \left(\frac{\delta Z_d}{Z_{0d}} + \frac{\delta n_d}{n_{0d}} \right) \frac{P}{1+P}. \tag{28}$$

The further expansion procedure is standard although the new feature is that the new type of linear and nonlinear terms appear due to the variations of the dust charge δZ_d. Using the dust charging equation, we find

$$\frac{\partial}{\partial t} \frac{\delta Z_d}{Z_{0d}} = -\nu_d^{ch} \frac{\delta Z_d}{Z_{0d}} + \frac{1+P}{P} \bar{\nu}_i \left[\frac{\delta n_e}{n_{0e}} - \frac{\delta n_i}{n_{0i}} \right]. \tag{29}$$

In the next approximation, we find for the perpendicular electron velocity

$$\delta \mathbf{v}_{e,\perp} = \frac{v_{Te}^2}{\Omega_e} \left[\hat{\mathbf{z}} \times \nabla \phi - \hat{\mathbf{z}} \times \delta \left(\frac{1}{n_{0e}} \nabla n_{0e} \right) \right]. \tag{30}$$

In the same approximation, we have the parallel motion of plasma electrons given by

$$v_{1e,z} = \frac{v_{Te}^2}{\tilde{\nu}_{ed}} \partial_z \left(\phi - \frac{\delta n_e}{n_{0e}} \right). \tag{31}$$

The motion of ions is described by the momentum equation

$$\partial_t \mathbf{v}_i + \mathbf{v}_i \cdot \nabla \mathbf{v}_i = -\frac{v_{Ti}^2}{n_i} \nabla n_i - v_s^2 \nabla \phi + \Omega_i \mathbf{v}_i \times \hat{\mathbf{z}} - \nu_{id} \mathbf{v}_i. \tag{32}$$

For the unperturbed motion of ions, we have the ion diamagnetic drift velocity

$$\mathbf{v}_{0i,\perp} = \frac{v_s^2}{\Omega_i} \hat{\mathbf{z}} \times \nabla \phi. \tag{33}$$

Similarly to Eqs. (30) and (31), the next approximation gives us the perturbation of the perpendicular ion velocity

$$\delta \mathbf{v}_{i,\perp} = -\frac{v_s^2}{\Omega_i} (\partial_t + \nu_{id}) \hat{\mathbf{z}} \times \nabla \phi + \frac{v_s^4}{\Omega_i^2} \hat{\mathbf{z}} \times [(\hat{\mathbf{z}} \times \nabla \phi \cdot) \hat{\mathbf{z}} \times \nabla \phi] \tag{34}$$

as well as the velocity for the parallel motion of plasma ions

$$\delta v_{i,z} = \frac{v_{Te}^2}{\nu_{ed}} \partial_z \left(\phi - \frac{\delta n_e}{n_{0e}} \right). \tag{35}$$

In the linear approximation, we ignore the effects of the magnetic field on the motion of the dust component and invoke the linearized Euler equation

$$\partial_t \mathbf{v}_d = \frac{T_e Z_d}{m_d} \nabla \phi - \nu_{dn} (\mathbf{v}_d - \mathbf{v}_n) + \frac{\mathbf{F}_d}{m_d}, \tag{36}$$

where $\mathbf{F}_d$ includes the repulsion and attraction forces of the dust-dust interactions, and dust-neutral friction is taken into account. For the neutrals, we have in the same approximation

$$\partial_t \mathbf{v}_n = -\nu_{dn} (\mathbf{v}_n - \mathbf{v}_d), \tag{37}$$

where collision with plasma electrons and ions are neglected. Thus, assuming $\nu_{dn} = \nu_{nd}$, we find for the neutral velocity

$$\mathbf{v}_n = \frac{\nu_{dn}}{\partial_t + \nu_{dn}} \mathbf{v}_d. \tag{38}$$

Equation (36) can then be written as

$$\partial_t \left(1 + \frac{\mu \nu_{dn}}{\partial_t + \nu_{dn}} \right) \mathbf{v}_d = \frac{T_e Z_d}{m_d} \nabla \phi + \frac{\mathbf{F}_d}{m_d}, \tag{39}$$

where $\mu = m_n n_n / m_d n_d$. For the dust-dust interaction forces, we have from Eqs. (3)–(7) after Fourier transform

$$i\mathbf{k} \cdot \mathbf{F}_d = m_d \omega_{pd}^2 \frac{\delta n_d}{n_{0d}} \left(\frac{\pi}{4} \eta_r k a - \eta_a \frac{a^2}{\lambda_D^2} \right). \tag{40}$$

Thus we have derived a closed set of equations fully describing motion of plasma electron, ion, neutral, and dust components taking into account the dust charging effects and dust-dust interactions, as well as charging and elastic collisions.

V NONLINEAR EQUATION FOR DRIFT WAVES

Expanding the electron continuity equation in the parameter ϵ, we find

$$\partial_t \left(\frac{\delta n_e}{n_{0e}} \right) + \bar{\nu}_{ed} \left(\frac{\delta n_e}{n_{0e}} + \frac{\delta n_d}{n_{0d}} - \mathcal{Z} \frac{\delta Z_d}{Z_{0d}} \right) = -\frac{v_{Te}^2}{n_{0e}\Omega_e} \bar{\mathbf{z}} \times \nabla\phi \cdot \nabla_\perp n_{0e}$$
$$+ \frac{k_z^2 v_{Te}^2}{\nu_{ed}} \left(\phi - \frac{\delta n_e}{n_{0e}} \right) - \frac{v_{Te}^2}{\Omega_e} \nabla_\perp \left(\frac{\delta n_e}{n_{0e}} \hat{\mathbf{z}} \times \nabla\phi \right), \tag{41}$$

where $\nabla_\perp$ stands for the gradient perpendicular to the external magnetic field [such that $\nabla = (\nabla_\perp, \nabla_z)$], and we used the perturbation expansion of (21)

$$\frac{\delta \bar{\nu}_{ed}}{\bar{\nu}_{ed}} = -\mathcal{Z} \frac{\delta Z_d}{Z_{0d}} + \frac{\delta n_d}{n_{0d}}, \tag{42}$$

which can easily be found taking into account the quasineutrality condition (28) together with the perturbation of the charging equation (17).

For plasma ions, invoking Eqs. (28), (29), and (41), and using equation for the perturbations of the ion capture rate [similar to (42)]

$$\frac{\delta \bar{\nu}_{id}}{\bar{\nu}_{id}} = -\frac{\mathcal{Z}}{\tau + \mathcal{Z}} \frac{\delta Z_d}{Z_{0d}} + \frac{\delta n_d}{n_{0d}}, \tag{43}$$

the continuity equation can be written as

$$\frac{P}{1+P} \partial_t \left(\frac{\delta n_d}{n_{0d}} \right) - \partial_t \left(1 + \rho_s^2 \nabla_\perp^2 \right) \phi + \frac{v_s^4}{\Omega_i^3} \nabla \cdot \hat{\mathbf{z}} \times \left[\left(\hat{\mathbf{z}} \times \nabla\phi \cdot \nabla \right) \hat{\mathbf{z}} \times \nabla\phi \right]$$
$$= -\frac{v_s^2}{(1+P) n_{0e} \Omega_i} \bar{\mathbf{z}} \times \nabla\phi \cdot \nabla_\perp (n_{0d} Z_d) - \frac{v_s^2}{(1+P)\Omega_i} \nabla_\perp \cdot \left(P \frac{\delta Z_d}{Z_{d0}} + P \frac{\delta n_d}{n_{d0}} \right) \bar{\mathbf{z}} \times \nabla\phi$$
$$+ \frac{k_z^2 v_s^2}{\tilde{\nu}_{id}} \left[\phi + \frac{\tau}{1+P} \frac{\delta n_e}{n_{0e}} + \frac{\tau P}{1+P} \left(\frac{\delta Z_d}{Z_{0d}} + \frac{\delta n_d}{n_{0d}} \right) \right] - \frac{k_z^2 v_s^2}{(1+P)\nu_{ed}} \left(\phi - \frac{\delta n_e}{n_{0e}} \right), \tag{44}$$

Furthermore, we introduce the dimensionless variables

$$t \to \Omega_i t \frac{\rho_s}{L_n}, \quad \mathbf{r} \to \frac{\mathbf{r}}{\rho_s}, \quad \zeta \to \frac{\delta Z_d}{Z_{0d}} \frac{L_n}{\rho_s}, \quad n_{e(d)} \to \frac{\delta n_{e(d)}}{n_{0e(d)}} \frac{L_n}{\rho_s}, \quad \phi \to \phi \frac{L_n}{\rho_s}. \tag{45}$$

Thus we find from (44)

$$\partial_t (1 + \nabla_\perp^2)\phi - s\partial_y \phi - \frac{P}{1+P}\partial_t n_d + c_i \left[\phi + \frac{\tau}{1+P} n_e + \frac{\tau P}{1+P}(\zeta + n_d)\right] - \frac{c_e}{1+P}(\phi - n_e)$$
$$= \left\{\nabla_\perp^2 \phi, \phi\right\} - \frac{P}{1+P}\{\zeta + n_d, \phi\}, \tag{46}$$

where the Poisson bracket is defined by

$$\{A, B\} = \partial_x A \partial_y B - \partial_x B \partial_y A \tag{47}$$

and the following parameters were introduced:

$$c_e = \frac{k_z^2 v_{Te}^2 L_n}{\nu_{ed} \Omega_i \rho_s}, \quad c_i = \frac{k_z^2 v_s^2 L_n}{\nu_{id} \Omega_i \rho_s}, \quad s = \frac{P}{1+P} \frac{L_d^{-1}}{1 + P - P L_d^{-1}}, \quad L_d^{-1} = \frac{\partial \ln n_{0d} Z_{0d}}{\partial \ln n_{0i}}. \tag{48}$$

Electron continuity equation (41) can thus be rewritten as

$$\partial_t n_e + \alpha(n_e + n_d - \mathcal{Z}\zeta) + \partial_y \phi - c_e(\phi - n_e) = \{n_e, \phi\}; \tag{49}$$

Charging equation (17) is now given by

$$\partial_t \zeta + \beta \zeta = \alpha(1 + P)(n_e - n_d), \tag{50}$$

where

$$\beta = \frac{L_n(\bar{\nu}_{id} + \nu_d^{ch})}{\Omega_i \rho_s} = \alpha(1+P)\left(1 + \frac{\nu_d^{ch}}{\bar{\nu}_{id}}\right). \tag{51}$$

Note the term containing n_d on the right hand side of Eq. (50) appearing due to perturbation of dust motion. Finally, the combination of the dust continuity equation together with (39) gives us

$$\partial_t^2 \left(1 - \gamma^{-1}\partial_t\right) n_d = -\nabla \cdot \mathbf{f} - \kappa \nabla^2 \phi, \tag{52}$$

where κ can be expressed via the dust sound velocity $v_{ds} = (T_e/m_d)^{1/2}$:

$$\kappa = \frac{v_{sd}^2}{v_s^2} \frac{L_n^2}{\rho_s^2}, \tag{53}$$

and γ describes the rate of damping of the dust motions due to the dust-neutral collisions

$$\gamma = \frac{\nu_{dn}}{\Omega_i}\frac{L_n}{\rho_s}\frac{\mu+1}{\mu}. \tag{54}$$

Furthermore, in Eq. (52) the divergence of the force $\mathbf{f}$ can be written in the form

$$\nabla \cdot \mathbf{f} = \frac{\hat{\omega}_{pd}^2}{\Omega_i^2}\frac{L_n^2}{\rho_s^2}\hat{K} n_d, \tag{55}$$

where the operator $\hat{K}$ is given by

$$\hat{K} = \frac{\pi}{4}\frac{\eta_r ka}{\rho_s} - \eta_a \frac{a^2}{\lambda_D^2} \tag{56}$$

and we have

$$\hat{\omega}_{pd}^2 = \frac{4\pi n_{0d} Z_{0d}^2 e^2}{\hat{m}_d}, \qquad \hat{m}_d = m_d\left(1 + \frac{\mu\nu_{dn}}{\partial_t + \nu_{dn}}\right). \tag{57}$$

Note that the coefficient c_e describes the non-adiabaticity introduced by dust (namely, because of collisions of plasma electrons, moving along the magnetic field lines, with dust grains). The smaller c_e the larger the non-adiabaticity. The coefficient α is connected with the charging process and describes the influence of the charging collisions. The non-adiabaticity introduced by ions and described by c_i is usually small, and below we neglect it in our stability analysis.

The set of nonlinear equations for the drift wave potential (46), the electron density perturbation (49), the perturbation of the charging process (50), and dynamics of the dust density perturbations (52) [connected also with the perturbation in the motion of neutral particles via (38)] is the general result of the consideration. Below, we present the stability analysis of the linearized equations found on the basis of the general nonlinear set.

VI LINEAR INSTABILITIES OF DRIFT WAVES

In the linear approximation, neglecting the ion non-adiabaticity (i.e., putting $c_i = 0$), we obtain the following dispersion relation for the frequency and the instability rate of the drift waves in a weakly ionized dusty plasma

$$\omega(1+k_\perp^2) = k_y s - \frac{c_e}{1+P} - \frac{\kappa k^2}{\Omega_1^2}\frac{P\omega}{1+P} + \frac{1}{-i\Omega_2}\frac{1}{1+P}\left[c_e + P\tau\frac{\alpha(1+P)}{-i\omega+\beta}\right] \times \left[c_e - ik_y + \frac{\alpha\kappa k^2}{\Omega_1^2}\left(1 + \alpha\mathcal{Z}\frac{1+P}{-i\omega+\beta}\right)\right]. \tag{58}$$

Note that the third term on the right hand side of this equation (which contains Ω_1) is new comparing with the previous study [8]; it appears due to the motion of dust and also includes the dust-dust interaction. The presence of this term crucially affects the dissipative drift wave instability in a partially ionized dusty plasma.

Equation (58) is solved numerically taking into account dust charging and interaction, and some dependencies of the instability rate *vs.* the wave number of the drift wave are presented in Figs. 1–3. Note that in these figures the dimensionless wave frequency is measured in the units of $\omega = \Omega_i \rho_s / L_n$, and the dimensionless wave vector is measured in the units of $k \equiv k_y \rho_s$ (for simplicity, we assume $k_y \gg k_x$).

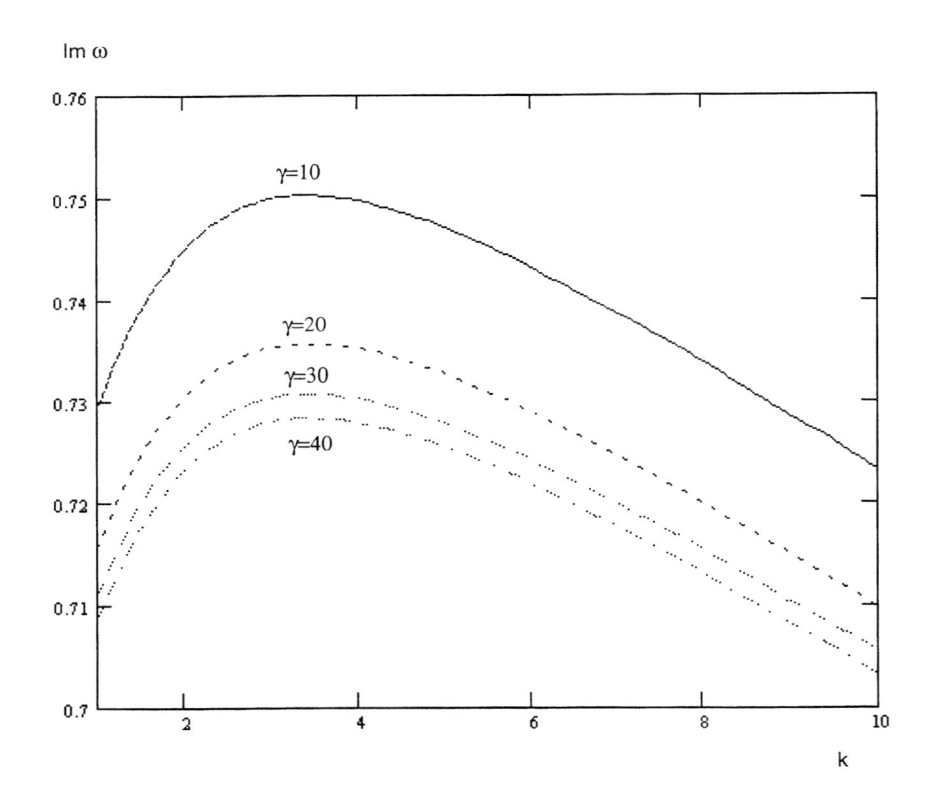

FIGURE 1. The growth rate of the dissipative drift instability for four different values of the damping of dust density perturbations γ. Other parameters are: $P = 10$, $\Omega \equiv \omega_{pi}^2/\hat{\omega}_{pd}^2 = 10$, $c_e = 10$, $\alpha = 0.1$, and $\tau = 0.01$.

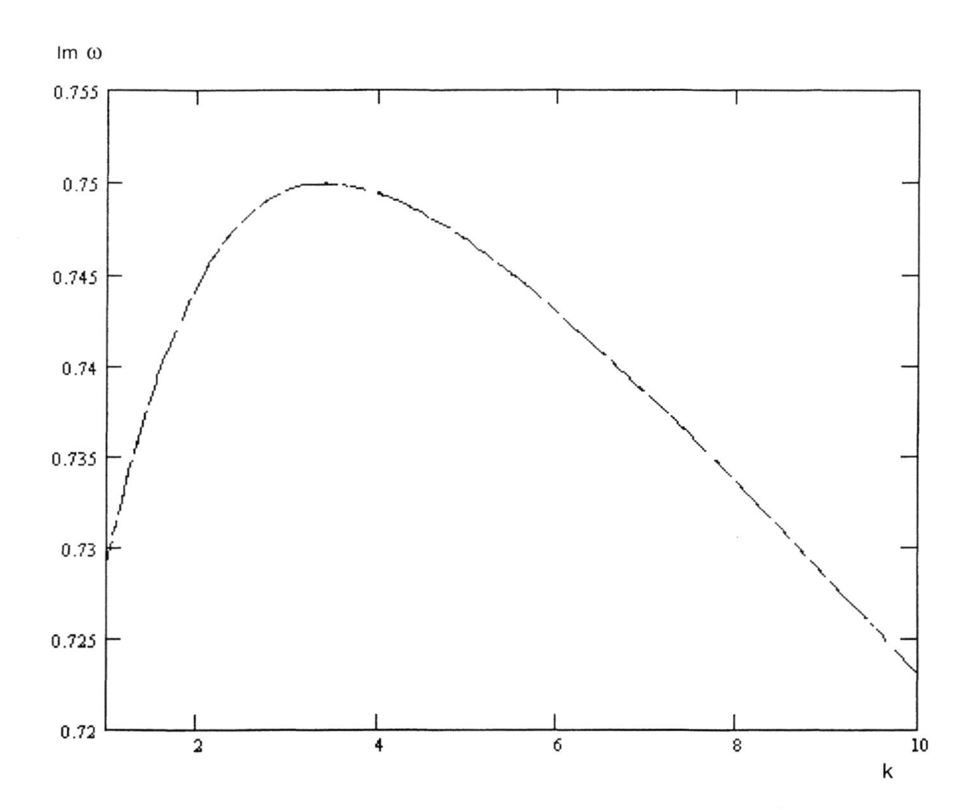

FIGURE 2. The growth rate of the dissipative drift instability for four different values of the electron adiabaticity parameter $c_e = 10, 100, 10^3, 10^4$ and $\gamma = 10$, $P = 10$, $\Omega = 10$, $\alpha = 0.1$, and $\tau = 0.01$. Note almost no dependence of the growth rate on c_e: the four curves are almost exactly superimposed on each other.

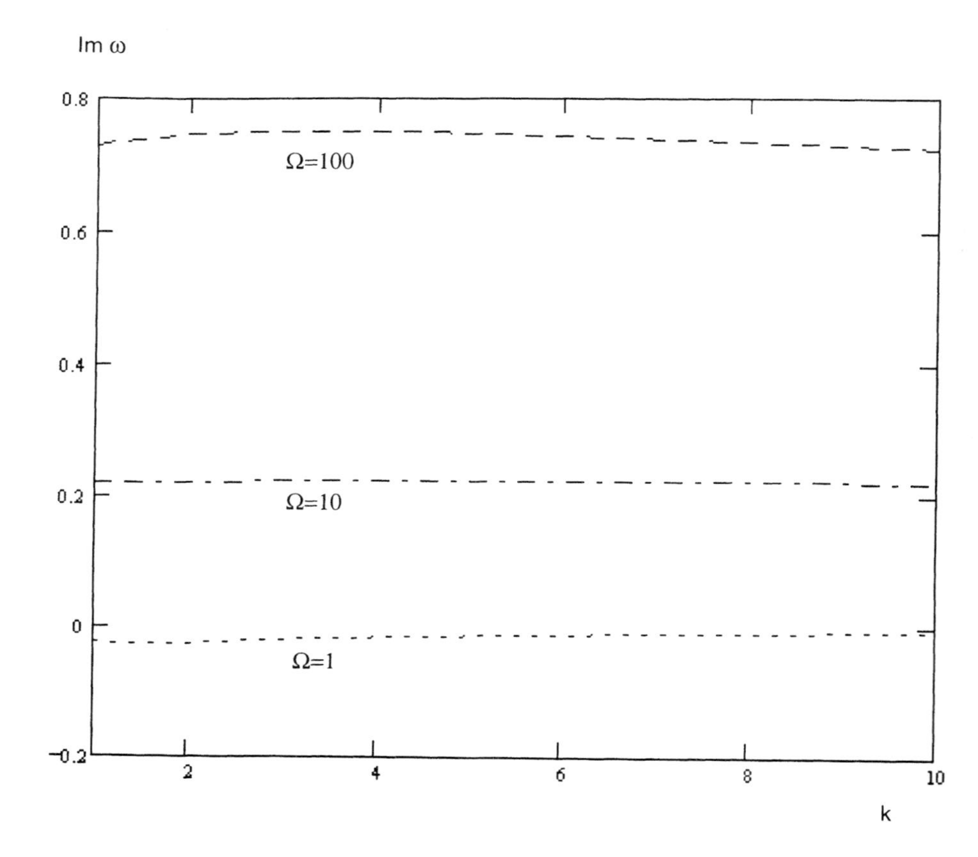

FIGURE 3. The growth rate of the dissipative drift instability for three different values of the parameter $\Omega \equiv \omega_{pi}^2/\hat{\omega}_{pd}^2 = 1, 10, 100$ (describing ratio of the ion plasma frequency to the dust plasma frequency) and $\gamma = 10$, $P = 10$, $c_e = 10$, $\alpha = 0.1$, and $\tau = 0.01$. Note strong dependence of the growth rate on Ω: the wave is stable for $\Omega = 1$ and becomes unstable with increasing Ω.

The analysis demonstrates that for the case considered there are two important parameters: 1) γ [given by (54)], which is the damping of the dust density perturbations due to their friction with neutrals; and 2) $\Omega = \omega_{pi}^2/\hat{\omega}_{pd}^2$ which describes the ratio of the ion plasma and dust plasma frequencies [note that the latter takes into account dust mass renormalization because of the interactions with neutrals, see Eq. (57)]. We also stress that the dissipative instability can develop for relatively large values of the parameter P [given by (15), which is the ratio of the total charge on the dust to the total charge of plasma electrons. This feature is completely new comparing with the results [8] in a fully ionized dusty plasma, when the drift wave instability was suppressed with increasing of P, such that there was no instability for $P > 1$.

Previously, it was shown in [8] that in a dusty plasma the drift-wave instability can be substantially enhanced due to the increase of the dissipation rate along the magnetic field lines as compared to the dust-free case. In this paper, we demonstrated that because of the dust-neutral friction and dust-dust interaction, the instability in a partially ionized dusty plasma can develop in new parameter range, i.e., with those values which correspond to stable drift waves in a fully ionized dusty plasma. The developed instability can lead to formation of nonlinear combined structures including the drift wave vortices coupled with the vortices of the neutral gas component via dust grains.

ACKNOWLEDGMENTS

A part of this work has been done during the Workshop on Turbulence and 2D Structures in Canberra in June-July 1997. The authors are grateful to R.L. Dewar for hospitality and discussions during the Workshop. This work was also partially supported by the University of Sydney and the Australian Research Council. One of the authors (VNT) is grateful to D. Melrose for hospitality during his stay in Sydney.

REFERENCES

1. V.N. Tsytovich, Sov. Phys. Uspekhi **40**, 53 (1997).
2. G.S. Selwyn, J.E. Heidenreich, and K.L. Haller, Appl. Phys. Lett. **57**, 1876 (1990).
3. C.K. Goertz, Rev. Geophys. **27**, 271 (1989).
4. J.E. Allen, Phys. Scripta **45**, 497 (1992).
5. V.N. Tsytovich and O. Havnes, Comm. Plasma Phys. Contr. Fusion **15**, 267 (1993).
6. S.V. Vladimirov, Phys. Plasmas **1**, 2762 (1994); Phys. Rev. E **50**, 1422 (1994).
7. S. Benkadda, V.N. Tsytovich, and A. Verga, Comm. Plasma Phys. Contr. Fusion **16**, 321 (1995).

8. S. Benkadda, P. Gabbai, V.N. Tsytovich, and A. Verga, Phys. Rev. E **50**, 2717 (1996).
9. N.F. Cramer and S.V. Vladimirov, Phys. Scripta **53**, 586 (1996).
10. V.S. Tsypin, S.V. Vladimirov, A.G. Elfimov, M. Tendler, A.S. de Assis, and C.A. de Azevedo, Phys. Plasmas **4**, *No*.9 (1997), in press.
11. N.F. Cramer and S.V. Vladimirov, Publ. Astron. Soc. Aust. **14**, 170 (1997).
12. R.N. Sudan, J. Geophys. Res. **88**, 4853 (1983).
13. A. Hasegawa and M. Wakatani, Phys. Fluids **27**, 611 (1984).
14. V.N. Tsytovich, S. Benkadda (private communication).
15. S.I. Braginskii, in *Review of Plasma Physics*, ed. M.A. Leontovich (Consultants Bureau, New York 1965) **1**, p.205.
16. Spitzer, L., Jr., *Physical Processes in the Interstellar Medium* (John Wiley, New York 1978), p. 168.

Author Index

AIP Conference Proceedings

	Title	L.C. Number	ISBN
No. 288	Laser Ablation: Mechanismas and Applications-II: Second International Conference (Knoxville, TN 1993)	93-73040	1-56396-226-8
No. 289	Radio Frequency Power in Plasmas: Tenth Topical Conference (Boston, MA 1993)	93-72964	1-56396-264-0
No. 290	Laser Spectroscopy: XIth International Conference (Hot Springs, VA 1993)	93-73050	1-56396-262-4
No. 291	Prairie View Summer Science Academy (Prairie View, TX 1992)	93-73081	1-56396-133-4
No. 292	Stability of Particle Motion in Storage Rings (Upton, NY 1992)	93-73534	1-56396-225-X
No. 293	Polarized Ion Sources and Polarized Gas Targets (Madison, WI 1993)	93-74102	1-56396-220-9
No. 294	High-Energy Solar Phenomena: A New Era of Spacecraft Measurements (Waterville Valley, NH 1993)	93-74147	1-56396-291-8
No. 295	The Physics of Electronic and Atomic Collisions: XVIII International Conference (Aarhus, Denmark, 1993)	93-74103	1-56396-290-X
No. 296	The Chaos Paradigm: Developments an Applications in Engineering and Science (Mystic, CT 1993)	93-74146	1-56396-254-3
No. 297	Computational Accelerator Physics (Los Alamos, NM 1993)	93-74205	1-56396-222-5
No. 298	Ultrafast Reaction Dynamics and Solvent Effects (Royaumont, France 1993)	93-074354	1-56396-280-2
No. 299	Dense Z-Pinches: Third International Conference (London, 1993)	93-074569	1-56396-297-7
No. 300	Discovery of Weak Neutral Currents: The Weak Interaction Before and After (Santa Monica, CA 1993)	94-70515	1-56396-306-X
No. 301	Eleventh Symposium Space Nuclear Power and Propulsion (3 Vols.) (Albuquerque, NM 1994)	92-75162	1-56396-305-1 (set) 156396-301-9 (pbk. set)
No. 302	Lepton and Photon Interactions/ XVI International Symposium (Ithaca, NY 1993)	94-70079	1-56396-106-7

	Title	L.C. Number	ISBN
No. 303	Slow Positron Beam Techniques for Solids and Surfaces Fifth International Workshop (Jackson Hole, WY 1992)	94-71036	1-56396-267-5
No. 304	The Second Compton Symposium (College Park, MD 1993)	94-70742	1-56396-261-6
No. 305	Stress-Induced Phenomena in Metallization Second International Workshop (Austin, TX 1993)	94-70650	1-56396-251-9
No. 306	12th NREL Photovoltaic Program Review (Denver, CO 1993)	94-70748	1-56396-315-9
No. 307	Gamma-Ray Bursts Second Workshop (Huntsville, AL 1993)	94-71317	1-56396-336-1
No. 308	The Evolution of X-Ray Binaries (College Park, MD 1993)	94-76853	1-56396-329-9
No. 309	High-Pressure Science and Technology—1993 (Colorado Springs, CO 1993)	93-72821	1-56396-219-5 (set)
No. 310	Analysis of Interplanetary Dust (Houston, TX 1993)	94-71292	1-56396-341-8
No. 311	Physics of High Energy Particles in Toroidal Systems (Irvine, CA 1993)	94-72098	1-56396-364-7
No. 312	Molecules and Grains in Space (Mont Sainte-Odile, France 1993)	94-72615	1-56396-355-8
No. 313	The Soft X-Ray Cosmos ROSAT Science Symposium (College Park, MD 1993)	94-72499	1-56396-327-2
No. 314	Advances in Plasma Physics Thomas H. Stix Symposium (Princeton, NJ 1992)	94-72721	1-56396-372-8
No. 315	Orbit Correction and Analysis in Circular Accelerators (Upton, NY 1993)	94-72257	1-56396-373-6
No. 316	Thirteenth International Conference on Thermoelectrics (Kansas City, Missouri 1994)	95-75634	1-56396-444-9
No. 317	Fifth Mexican School of Particles and Fields (Guanajuato, Mexico 1992)	94-72720	1-56396-378-7
No. 318	Laser Interaction and Related Plasma Phenomena 11th International Workshop (Monterey, CA 1993)	94-78097	1-56396-324-8
No. 319	Beam Instrumentation Workshop (Santa Fe, NM 1993)	94-78279	1-56396-389-2
No. 320	Basic Space Science (Lagos, Nigeria 1993)	94-79350	1-56396-328-0

	Title	L.C. Number	ISBN
No. 321	The First NREL Conference on Thermophotovoltaic Generation of Electricity (Copper Mountain, CO 1994)	94-72792	1-56396-353-1
No. 322	Atomic Processes in Plasmas Ninth APS Topical Conference (San Antonio, TX)	94-72923	1-56396-411-2
No. 323	Atomic Physics 14 Fourteenth International Conference on Atomic Physics (Boulder, CO 1994)	94-73219	1-56396-348-5
No. 324	Twelfth Symposium on Space Nuclear Power and Propulsion (Albuquerque, NM 1995)	94-73603	1-56396-427-9
No. 325	Conference on NASA Centers for Commercial Development of Space (Albuquerque, NM 1995)	94-73604	1-56396-431-7
No. 326	Accelerator Physics at the Superconducting Super Collider (Dallas, TX 1992-1993)	94-73609	1-56396-354-X
No. 327	Nuclei in the Cosmos III Third International Symposium on Nuclear Astrophysics (Assergi, Italy 1994)	95-75492	1-56396-436-8
No. 328	Spectral Line Shapes, Volume 8 12th ICSLS (Toronto, Canada 1994)	94-74309	1-56396-326-4
No. 329	Resonance Ionization Spectroscopy 1994 Seventh International Symposium (Bernkastel-Kues, Germany 1994)	95-75077	1-56396-437-6
No. 330	E.C.C.C. 1 Computational Chemistry F.E.C.S. Conference (Nancy, France 1994)	95-75843	1-56396-457-0
No. 331	Non-Neutral Plasma Physics II (Berkeley, CA 1994)	95-79630	1-56396-441-4
No. 332	X-Ray Lasers 1994 Fourth International Colloquium (Williamsburg, VA 1994)	95-76067	1-56396-375-2
No. 333	Beam Instrumentation Workshop (Vancouver, B. C., Canada 1994)	95-79635	1-56396-352-3
No. 334	Few-Body Problems in Physics (Williamsburg, VA 1994)	95-76481	1-56396-325-6
No. 335	Advanced Accelerator Concepts (Fontana, WI 1994)	95-78225	1-56396-476-7 (set) 1-56396-474-0 (Book) 1-56396-475-9 (CD-Rom)

	Title	L.C. Number	ISBN
No. 336	Dark Matter (College Park, MD 1994)	95-76538	1-56396-438-4
No. 337	Pulsed RF Sources for Linear Colliders (Montauk, NY 1994)	95-76814	1-56396-408-2
No. 338	Intersections Between Particle and Nuclear Physics 5th Conference (St. Petersburg, FL 1994)	95-77076	1-56396-335-3
No. 339	Polarization Phenomena in Nuclear Physics Eighth International Symposium (Bloomington, IN 1994)	95-77216	1-56396-482-1
No. 340	Strangeness in Hadronic Matter (Tucson, AZ 1995)	95-77477	1-56396-489-9
No. 341	Volatiles in the Earth and Solar System (Pasadena, CA 1994)	95-77911	1-56396-409-0
No. 342	CAM -94 Physics Meeting (Cacun, Mexico 1994)	95-77851	1-56396-491-0
No. 343	High Energy Spin Physics Eleventh International Symposium (Bloomington, IN 1994)	95-78431	1-56396-374-4
No. 344	Nonlinear Dynamics in Particle Accelerators: Theory and Experiments (Arcidosso, Italy 1994)	95-78135	1-56396-446-5
No. 345	International Conference on Plasma Physics ICPP 1994 (Foz do Iguaçu, Brazil 1994)	95-78438	1-56396-496-1
No. 346	International Conference on Accelerator-Driven Transmutation Technologies and Applications (Las Vegas, NV 1994)	95-78691	1-56396-505-4
No. 347	Atomic Collisions: A Symposium in Honor of Christopher Bottcher (1945-1993) (Oak Ridge, TN 1994)	95-78689	1-56396-322-1
No. 348	Unveiling the Cosmic Infrared Background (College Park, MD, 1995)	95-83477	1-56396-508-9
No. 349	Workshop on the Tau/Charm Factory (Argonne, IL, 1995)	95-81467	1-56396-523-2
No. 350	International Symposium on Vector Boson Self-Interactions (Los Angeles, CA 1995)	95-79865	1-56396-520-8
No. 351	The Physics of Beams Andrew Sessler Symposium (Los Angeles, CA 1993)	95-80479	1-56396-376-0
No. 352	Physics Potential and Development of $\mu^+ \mu^-$ Colliders: Second Workshop (Sausalito, CA 1994)	95-81413	1-56396-506-2

	Title	L.C. Number	ISBN
No. 353	13th NREL Photovoltaic Program Review (Lakewood, CO 1995)	95-80662	1-56396-510-0
No. 354	Organic Coatings (Paris, France, 1995)	96-83019	1-56396-535-6
No. 355	Eleventh Topical Conference on Radio Frequency Power in Plasmas (Palm Springs, CA 1995)	95-80867	1-56396-536-4
No. 356	The Future of Accelerator Physics (Austin, TX 1994)	96-83292	1-56396-541-0
No. 357	10th Topical Workshop on Proton-Antiproton Collider Physics (Batavia, IL 1995)	95-83078	1-56396-543-7
No. 358	The Second NREL Conference on Thermophotovoltaic Generation of Electricity	95-83335	1-56396-509-7
No. 359	Workshops and Particles and Fields and Phenomenology of Fundamental Interactions (Puebla, Mexico 1995)	96-85996	1-56396-548-8
No. 360	The Physics of Electronic and Atomic Collisions XIX International Conference (Whistler, Canada, 1995)	95-83671	1-56396-440-6
No. 361	Space Technology and Applications International Forum (Albuquerque, NM 1996)	95-83440	1-56396-568-2
No. 362	Two-Center Effects in Ion-Atom Collisions (Lincoln, NE 1994)	96-83379	1-56396-342-6
No. 363	Phenomena in Ionized Gases XXII ICPIG (Hoboken, NJ, 1995)	96-83294	1-56396-550-X
No. 364	Fast Elementary Processes in Chemical and Biological Systems (Villeneuve d'Ascq, France, 1995)	96-83624	1-56396-564-X
No. 365	Latin-American School of Physics XXX ELAF Group Theory and Its Applications (México City, México, 1995)	96-83489	1-56396-567-4
No. 366	High Velocity Neutron Stars and Gamma-Ray Bursts (La Jolla, CA 1995)	96-84067	1-56396-593-3
No. 367	Micro Bunches Workshop (Upton, NY, 1995)	96-83482	1-56396-555-0
No. 368	Acoustic Particle Velocity Sensors: Design, Performance and Applications (Mystic, CT, 1995)	96-83548	1-56396-549-6
No. 369	Laser Interaction and Related Plasma Phenomena (Osaka, Japan 1995)	96-85009	1-56396-445-7

	Title	L.C. Number	ISBN
No. 370	Shock Compression of Condensed Matter-1995 (Seattle, WA 1995)	96-84595	1-56396-566-6
No. 371	Sixth Quantum 1/f Noise and Other Low Frequency Fluctuations in Electronic Devices Symposium (St. Louis, MO, 1994)	96-84200	1-56396-410-4
No. 372	Beam Dynamics and Technology Issues for + - Colliders 9th Advanced ICFA Beam Dynamics Workshop (Montauk, NY, 1995)	96-84189	1-56396-554-2
No. 373	Stress-Induced Phenomena in Metallization (Palo Alto, CA 1995)	96-84949	1-56396-439-2
No. 374	High Energy Solar Physics (Greenbelt, MD 1995)	96-84513	1-56396-542-9
No. 375	Chaotic, Fractal, and Nonlinear Signal Processing (Mystic, CT 1995)	96-85356	1-56396-443-0
No. 376	Chaos and the Changing Nature of Science and Medicine: An Introduction (Mobile, AL 1995)	96-85220	1-56396-442-2
No. 377	Space Charge Dominated Beams and Applications of High Brightness Beams (Bloomington, IN 1995)	96-85165	1-56396-625-7
No. 378	Surfaces, Vacuum, and Their Applications (Cancun, Mexico 1994)	96-85594	1-56396-418-X
No. 379	Physical Origin of Homochirality in Life (Santa Monica, CA 1995)	96-86631	1-56396-507-0
No. 380	Production and Neutralization of Negative Ions and Beams / Production and Application of Light Negative Ions (Upton, NY 1995)	96-86435	1-56396-565-8
No. 381	Atomic Processes in Plasmas (San Francisco, CA 1996)	96-86304	1-56396-552-6
No. 382	Solar Wind Eight (Dana Point, CA 1995)	96-86447	1-56396-551-8
No. 383	Workshop on the Earth's Trapped Particle Environment (Taos, NM 1994)	96-86619	1-56396-540-2
No. 384	Gamma-Ray Bursts (Huntsville, AL 1995)	96-79458	1-56396-685-9
No. 385	Robotic Exploration Close to the Sun: Scientific Basis (Marlboro, MA 1996)	96-79560	1-56396-618-2
No. 386	Spectral Line Shapes, Volume 9 13th ICSLS (Firenze, Italy 1996)		1-56396-656-5

	Title	L.C. Number	ISBN
No. 387	Space Technology and Applications International Forum (Albuquerque, NM 1997)	96-80254	1-56396-679-4 (Case set) 1-56396-691-3 (Paper set)
No. 388	Resonance Ionization Spectroscopy 1996 Eighth International Symposium (State College, PA 1996)	96-80324	1-56396-611-5
No. 389	X-Ray and Inner-Shell Processes 17th International Conference (Hamburg, Germany 1996)	96-80388	1-56396-563-1
No. 390	Beam Instrumentation Proceedings of the Seventh Workshop (Argonne, IL 1996)	97-70568	1-56396-612-3
No. 391	Computational Accelerator Physics (Williamsburg, VA 1996)	97-70181	1-56396-671-9
No. 392	Applications of Accelerators in Research and Industry: Proceedings of the Fourteenth International Conference (Denton, TX 1996)	97-71846	1-56396-652-2
No. 393	Star Formation Near and Far Seventh Astrophysics Conference (College Park, MD 1996)	97-71978	1-56396-678-6
No. 394	NREL/SNL Photovoltaics Program Review Proceedings of the 14th Conference— A Joint Meeting (Lakewood, CO 1996)	97-72645	1-56396-687-5
No. 395	Nonlinear and Collective Phenomena in Beam Physics (Arcidosso, Italy 1996)	97-72970	1-56396-668-9
No. 396	New Modes of Particle Acceleration— Techniques and Sources (Santa Barbara, CA 1996)	97-72977	1-56396-728-6
No. 397	Future High Energy Colliders (Santa Barbara, CA 1997)	97-73333	1-56396-729-4
No. 398	Advanced Accelerator Colliders Seventh Workshop (Lake Tahoe, CA 1996)	97-72788	1-56396-697-2 (set) 1-56396-727-8 (cloth) 1-56396-726-X (CD-Rom)
No. 399	The Changing Role of Physics Departments (College Park, MD 1996)	97-74866	1-56396-698-0
No. 400	High Energy Physics First Latin Symposium (Yucatan, México 1996)	97-73971	1-56396-686-7

	Title	L.C. Number	ISBN
No. 401	Thermophotovoltaic Generation of Electricity Third NREL Conference (Colorado Springs, CO 1997)	97-74374	1-56396-734-0
No. 402	Astrophysical Implications of the Laboratory Study of Presolar Materials (St. Louis, MO 1996)	97-74679	1-56396-664-6
No. 403	Radio Frequency Power in Plasmas 12th Topical Conference (Savannah, GA 1997)	97-74472	1-56396-709-X
No. 404	Future Generations Photovoltaic Technologies First NREL Conference (Denver, CO 1997)	97-74386	1-56396-704-9
No. 405	Beam Stability and Nonlinear Dynamics (Santa Barbara, CA 1996)	97-74676	1-56396-731-6
No. 406	Laser Interaction and Related Plasma Phenomena 13th International Conference (Monterey, CA 1997)	97-76763	1-56396-696-4
No. 407	Deep Inelastic Scattering and QCD 5th International Workshop (Chicago, IL 1997)	97-74677	1-56396-716-2
No. 408	The Ultraviolet Universe at Low and High Redshift (College Park, MD 1997)	97-76762	1-56396-708-1
No. 409	Dense 2-Pinches 4th International Conference (Vancouver, Canada 1997)	97-76959	1-56396-610-7
No. 410	Proceedings of the 4th Compton Symposium (Williamsburg, VA 1997)	97-77179	1-56396-659-X (set)
No. 411	Applied Non-Linear Dynamics Near the Millenium (San Diego, CA 1997)	97-77035	1-56396-736-7
No. 412	Intersections Between Particle and Nuclear Physics: 6th Conference (Big Sky, MT 1997)	97-0564	1-56396-712-X
No. 413	Towards X-Ray Free Electron Lasers (Garda Lake, Italy 1997)	97-06161	1-56396-744-8
No. 414	Two-Dimensional Turbulence in Plasmas and Fluids (Canberra, Australia 1997)	97-06162	1-56396-764-2